A. Delyannis   E.-E. Delyannis

Seawater and Desalting

Volume 1

Springer-Verlag
Berlin Heidelberg New York 1980

Prof. Dr. Anthony Delyannis
Dr. Euridike-Emmy Delyannis
Plastira-St. 3

GR – Amarousion-Pefki

Library of Congress Cataloging in Publication Data.

Delyannis, A. Seawater and desalting. Bibliography: p. Includes index. 1. Saline water conversion. 2. Sea-water. I. Delyannis E., joint author. II. Title. TD479.D44 628.1'67 80-19090

ISBN-13: 978-3-642-67739-7   e-ISBN-13: 978-3-642-67737-3
DOI: 10.1007/978-3-642-67737-3

© Springer-Verlag Berlin Heidelberg 1980.

Softcover reprint of the hardcover 1st edition 1980

2152/3140-5 4 3 2 1 0

Foreword

The literature on water and desalination, dealing with basic research as well as engineering aspects, is spread in a large number of journals. In addition the vaste patent literature augments the citations to an extent, which makes information of researchers and engineers quite time consuming.

In order to produce a systematic literature survey, the authors have compiled the relevant citations in a more or less complete form.

The present volume contains about 2170 citations, including 652 patents, of the most recent publications up to 1979. Condensed abstracts are given of the main publications, as well as titles or highlights of any other article or patent, which might be of interest to the reader. The book is intended to serve as a handy guide to scientists and engineers.

More detailed information is given on novel developments of the various desalination processes and some details are included on desalting plants recently erected in various parts of the world. As the literature survey extends also to fields related to desalination, the book might also be useful to scientists of other disciplines, as Geoscientists, Oceanologists, Analysts, Ecologists and Corrosion Engineers, as data on the composition, the physical properties, the analytical chemistry of seawater and the corrosion of metals in ocean water are also included.

In view of many new ideas particular attention was given to the patent literature. Patents are cited separately in each section of the book and are summarized in the Patent Index.

We intend to cover the further development of the field in additional volumes. About every second year such a volume will be published.

Prof. Dr. Anthony Delyannis   Dr. Euridike-Emmy Delyannis

Table of Contents

1A. The Water Problem   11
    Water quality   11
    Water management   11
    Water reuse   11
    Literature to 1A.   12

1B. Desalting Processes   13
    International meetings   13
    International activities   13
      Egypt, France, Germany W.,
      Israel, Japan, Libya, Nether-
      lands, Saudi Arabia, Spain,
      U.S.A.
    Education and training   15
    Literature to 1B.   16

1C. Raw Material Seawater   17
    1. Mineral content of seawater   17
      Aluminum, Ammonium, Arsenic,
      Cerium, Chromium, Cobalt,
      Copper, Halogens, Hardness,
      Helium, Iron, Lead, Manganese,
      Mercury, Nitrates, Organics,
      Oxygen, Phosphoric acid, Plu-
      tonium, Radionuclides, Rheni-
      um, Strontium, Sulfates, Trace
      metals, Tritium, Uranium, Va-
      nadium, Zinc
    Trace metals patents   20
    Literature to 1C.1   21

    2. Physical Properties of Sea-
       water   23
      Activity and osmotic coeffi-
      cients, Conductance, Density,
      Dielectrics, Electrical con-
      ductivity, Evaporation, Heats
      of solution, Ion association,
      Molal volumes, Solubilities,
      Sound speed, Specific heats,
      Structure of water, Thermody-
      namic properties, Vapor press-
      ure, Viscosity
    Literature to 1C.2   26

    3. Analytical Chemistry   27
      General, Acidity, Alcalinity
      and pH, Aluminum, Ammonium,
      Arsenic, Barium, Beryllium
      and Bismuth, Boron, Cadmium,
      Caesium, Calcium, Carbon, Ce-
      rium, Chromium, Cobalt, Cop-
      per, Dissolved gases, Gold,

      Halogens, Iron, Lead, Li-
      thium, Magnesium, Manganese,
      Mercury, Molybdenum, Nitra-
      tes, Organics, Oxygen, Po-
      tassium, Radionuclides, Sa-
      linity, Sodium, Strontium,
      Sulfate, Sulfur, Trace ele-
      ments (Neutron activation,
      Electrochemical methods,
      Spectroscopy, Various meth-
      ods), Uranium, Zinc
    Patents   27, 31, 32, 34, 35
    Literature to 1C.3   35

2A. Distillation   40
      Patents   40
    Heat transfer   41
    Condensation   41
    Dropwise condensation   41
    Corrugated and finned tubes   42
      Patents   42
    Fluidized beds   42
    Direct contact heat transfer   43
      Patents   43
    Boiling   43
    Polluting materials   44
      Patents   44
    Biomedical aspects   44
    Stabilization of distillate   44
      Patents   44
    Disposal of effluents   44
    Literature to 2A.   45

2B. Components of Distillation
    Plants   47
    Intake Systems   47
      Patents   47
    Containment vessels   47
    Deaerators   47
      Patents   47
    Demisters   48
    Heat exchangers   48
      Patents   48
    Pumps   48
    Literature to 2B.   48

2C. Vertical Tube Evaporators   49
    Falling liquid film   50
    Rising liquid   50
      Patents   50
    Operating experience   51
    Literature to 2C.   51

8

2D. Horizontal Tube Evaporators 52
    Patents 52
    Literature to 2D. 53

2E. Flash Evaporation 53
    Flashing flow 53
    Multi-stage flash distillation 53
    Patents 54
    Operating experience 55
    Germany W., Japan, Kuwait, Libya, Mexico, Netherlands, Oman, Saudi Arabia, U.S.S.R.
    Literature to 2E. 55

2F. Other Distillation processes 56
    Vapor compression 56
    Patents 57
    Vertical multi-stage evaporators. Patents 57
    Vapor reheat 57
    Vertical MSF evaporator
    Patents 57
    Multi-stage flash fluidized bed evaporator 58
    Rotating evaporator. Patents 58
    Literature to 2F. 58

2G. Combined Distillation Plants 59
    VTE-MSF evaporation 59
    Patents 59
    VC-VTE-MSF combined plant 59
    Patents 59
    Literature to 2G. 60

2H. Dual Purpose Plants 60
    Patents 60
    Operating experience 61
    Libya, Saudi Arabia
    Literature to 2H. 61

2I. Waste Heat as Energy Source 61
    Patents 62
    Dehumidification 62
    Patents 62
    Literature to 2I. 62

2K. Nuclear Energy as Heat Source 63
    Nuclear vs. conventional power 64
    Safety and environment 64
    Literature to 2K. 64

2L. Geothermal Energy as Heat Source 65
    Literature to 2L. 66

2M. Solar Energy as Heat Source 67
    Patents 68
    Literature to 2M. 69

2N. Scale Formation and Prevention 70

Alkaline scale 71
    Patents 72
Sulfate scale 72
Removal of scale 73
    Patents 73
    Literature to 2N. 73

2O. Treatment of Feed Water 74
    Chlorination 74
    Acid injection 75
    Patents 75
    Treatment by phosphates 75
    Patents 75
    Additives 75
    Patents 76
    Seeding. Patents 77
    Treatment by ion exchange 77
    Patents 77
    Magnetic treatment 77
    Patents 78
    Literature to 2O. 78

2P. Corrosion 80
    Corrosion inhibition 81
    Patents 81
    Literature to 2P. 81

2Q. Materials of Construction 83
    C-steel and iron-base alloys 83
    Patents 83
    Stainless steels 84
    Patents 84
    Copper-base alloys 84
    Patents 85
    Aluminum 85
    Titanium 86
    Concrete 86
    Patents 87
    Non metallics 87
    Protective coatings 87
    Cathodic protection 87
    Patents 87
    Literature to 2Q. 88

3A. Ion Exchange 91
    Ion exchange technology 91
    Patents 92
    Sirotherm process 92
    Patents 92
    Literature to 3A. 93

3B. Membrane Processes 94
    Literature to 3B. 94

3C. Ion Exchange Membranes 95
    Preparation and properties 95
    Patents 96
    Ion selectivity 97
    Patents 98
    Ion and water transport 98

Electrical properties   99
    Patents   100
Concentration polarization   100
Electroosmosis   101
    Patents   101
Scaling and fouling   101
    Patents   102
Inorganic ion-exchange
membranes   102
Literature to 3C.   102

3D. Electrodialysis   108
    Patents   109
Electrodialysis and ion-
exchange   110
    Patents   110
Electrodialysis reversal   111
    Patents   111
High temperature electro-
dialysis   111
Hygienic evaluation   111
Operating experience   112
    Greece, Japan, Libya, U.S.A.
Treatment of waste waters   112
Separation of inorganics   112
    Patents   113
Separation of organics   113
    Patents   113
Other electrochemical
processes   113
Electrosorption   113
Donnan dialysis   114
Electrochemical pumping   114
Literature to 3D.   114

3E. Reverse Osmosis and Ultra-
filtration   117
Mechanism of reverse osmosis   118
Osmosis   118
    Patents   118
Thermoosmosis   119
Osmotic energy production   119
Pressure retarded osmosis   119
Literature to 3E.   120

3F. Reverse Osmosis Membranes   121
    Patents   121
Water transport, salt
rejection   122
Concentration polarization   124
Degradation and fouling   124
    Patents   125
Diffusion   126
Cellulose acetate membranes   126
Preparation and properties   126
Structure   128
    Patents   129
Modified cellulose acetate
membranes   129
    Patents   130
Polymer film membranes   130

Acrylic acid derivatives   131
Aromatic polyamid polymers   132
Ethylene derivatives   133
Poly(methyl-methacrylate)   133
Poly(vinyl alcohol) polymers   133
Polyvinylpyrrolidone polymers   133
Other polymeric materials   134
    Patents   134
Ultrathin, composite
membranes   136
    Patents   136
Other types of membranes   136
Dynamically formed membranes   136
    Patents   137
Charge membranes   138
Porous glass membranes   138
    Patents   138
Literature to 3F.   138

3G. Reverse Osmosis Process   145
    Patents   147
Pretreatment of feed water   148
    Patents   148
Energy considerations   149
    Patents   149
Plate and frame system   149
    Patents   150
Tubular module   150
    Patents   151
Spiral wound module   151
    Patents   152
Hollow fiber module   152
    Patents   153
Hygienic considerations   154
Operating experience   154
    France, Germany W., India,
    Italy, Japan, Netherlands,
    Saudi Arabia, Sweden, Uni-
    ted Kingdom, U.S.A., Vene-
    zuela
Literature to 3G.   156

3H. Other Applications of Reverse
    Osmosis and Ultrafiltration   160
Domestic and municipal wastes   160
Spacecraft wastes   161
Recycling of industrial wastes   161
Recycling in the food industry   161
    Patents   161
Hemodialysis and artifical
kidney   161
    Patents   162
Separation of inorganics   162
    Patents   163
Separation of organics   163
    Patents   164
Literature to 3H.   164

4. Other Desalting Processes   169
    Ice formation and melting   169

	Freezing processes	169		Magnesium	175
	Patents	170		Patents	175
	Hydrates	170		Potassium	175
	Bioconversion	170		Strontium	175
	Chelate treatment	171		Patents	175
	Literature to 4.	171		Uranium	175
				Patents	177
5.	Economic Considerations	172		Literature to·5.	177
	Recovery of byproducts	173			
	Patents	173	6.	Iceberg Utilization	180
	Common salt	174		Literature to 6.	181
	Patents	174			
	Halogens	174		Patent Index	182
	Patents	174			
	Lithium	174		Abbreviations	189
	Patents	174			

1A. The Water Problem

New editions of Wasser Kalender 1979 [1] and the Manual on Water [2] have been published. A bibliography on water resources in arid and semiarid regions contains 140 abstracts, 31 of which are new entries. The bibliography covers reports on water quality, resource management, potable water, consumption and irrigation, particularly those in the western regions of the U.S. [3]. Causes of the shortage of clean water were examined with heavy emphasis placed on the problems of the U.S. Among the solutions that the aurhor suggests are desalination and weather modification [4]. The trend of the future and solution to the water crisis will come primarily from increasing emphasis placed on tapping underground reservoirs and desalination of seawater. The solution will also involve judicions water·management [5].

WATER QUALITY

The 1978 Annual Book of ASTM Standards, Pt. 31: Water, has been published [6]. Reviews include the evaluation of drinking water and water quality standards [7], water qualities of water supplies [8], taste and chemical composition of drinking water [9], water quality control systems [10], criteria for the production of high purity water [11], water quality criteria standards [12], chemical investigation of the expression of water quality [13], operational achievement of drinking water quality [14], and continuous on-stream monitoring of water quality [15].

Two mathematical models were developed for computing water quality. The eutrophication model considers inputs-outputs, migration of the thermocline, nutrient fixation and mineralization and sediment-water interactions. The limnological model considers the dynamics of heat and mass transport, hydromechanisms and chemical and biological transformations [16]. An improved computational procedure for solving water quality management models was developed with respect to the specific problem of basin-wide thermal and organic pollution control to meet water quality standards [17]. The effects of nonmetallic materials on the water quality, such as concrete, rubbers, coatings, plastics, lubricants, sealants, bitumens etc. used in storage and transport of water are discussed [18]. Water quality criteria are issued for 27 of the 65 pollutants listed as toxic under the Clean Water Act [19].

WATER MANAGEMENT

Recommendations and resolutions of the United Nations Water Conference in Mar del Plata, Argentina, concerning many aspects of the development and use of water resources, particularly from the environmental and communal points of view, are summarized [20]. A discussion is given on the use of seawater resources, including water desalination, recovery of minerals and particularly U, and use of seawater for pollution control [21]. Other papers refer to the role of the Department of the Interior in total water resources management [22], implications of conservation, reclamation and desalting [23], joint venture for technology transfer in water resources development [24]. The primary energy requirements for three sectors of water management: municipal water supply, municipal sewage treatment and water for irrigation, were evaluated in six major cities of the United States [25].

In a report of the U.S. General Accounting Office to the Congress, desalting of water is considered as a possible solution to the lack of fresh water in many locations in the Nation. The Office of Water Research and Technology should develop and implement a comprehensive, well-defined saline water conversion program plan aimed at achieving a practical, low-cost desalting method [26].

WATER REUSE

A bibliography on water reuse contains 478 abstracts. Volume 6 covers the period from May 1976 to December 1977 [27].

Selected papers dealing with water reuse include wastewater characterization and process reliability for potable waste water reclamation [28], problems associ-

ated with the reuse of purified sewage effluents for power station cooling purposes [29], recycling of wastewater to make drinking water [30], modeling and simulation of wastewater reuse systems [31], improved methods of wastewater treatment [32], wastewater treatment for reuse and its contribution to water supplies
[33], treatment of purified wastewater to drinking water in Windhoek, Namibia [34],
contaminants associated with direct and indirect reuse of municipal wastewater[35],
water reuse and the membrane processes [36], reuse of wastewater in water supply
systems [37], health considerations for future water reuse[38], the role of desalination in water reuse [39], desalination and the reuse of agricultural drainage
waters in Saudi Arabia [40], new improvements in ultraviolet sterilization of desalted and reuse water [41].

LITERATURE TO 1A.

1. H. Huebner, editor (E. Schmidt Verlag [1978] 500 pp.)
2. C. E. Hamilton, editor (ASTM Special Technical Publication No 422, 4th edition
   [1978] 472 pp.)
3. R. J. Brown (NTIS Rept. PS-77/0947/0GA [1977])
4. D. E. Carr, Death of the Sweet Waters (W. W. Norton & Co, New York [1977] 257 pp.)
5. A. Van Dam (The Futurist 11, No 3 [1977] 163/167)
6. ASTM Standards, Pt. 31:Water (ASTM, Philadelphia, Pa. [1978] 1232 pp.)
7. K. Okazawa (Kenchiku Setsubi to Haikun Koji 14, No 5 [1976] 43/48)
8. M. Yamaguchi, K. Honda, K. Ueda (Nagasaki-ken Eisei Kogai Kenkyusho Ho 16 [1976]
   84/88.-C. A. 89 [1978] 203821)
9. B. C. J. Zoeteman, G. J. Piet, C. F. H. Morra, F. E. De Grunt (Proc. 6th Int.
   Symp. Olfaction Taste [1977] 421/427.-C. A. 89 [1978] 203838)
10. S. Yabe (Suishitsu Odaku no Jido Bunseki [1976] 231/246.-C. A. 89 [1978] 203822)
11. J. J. McCarthy (NTIS Rept. AD-A049032.- G. R. A. 78, No 8 [1978] 167)
12. R. J. Wells (Water Pollut. Control 77, No 1 [1978] 25/30.-C. A. 89 [1978] 94833)
13. I. Iwasaki (Kogyo Yosui 232 [1978] 13/32.-C. A. 89 [1978] 64763)
14. L. R. J. Van Vuuren, J. R. H. Hoffman (Water Pollut. Control 77, No 1 [1978]
    39/44.-C. A. 89 [1978] 94834)
15. H. C. Brinkhoff (Prog. Water Technol. 9 [1978] 173/182.-C. A. 89 [1978] 65015)
16. R. G. Baca, A. F. Gasperino, A. Brandstetter, A. S. Myhres (NTIS Rept. PB-275913.-
    G. R. A. 78, No 7 [1978] 180)
17. L. A. Rossman, F. T. Vanecek (Water Resour. Bull. 14 [1978] 842/855.-C. A. 89
    [1978] 135569)
18. N. P. Burman, J. S. Colbourne (J. Inst. Water Eng. Sci. 33, No 1 [1979] 11/18)
19. Environmental Protection Agency (Fed. Regist. 44, No 52 [1979] 15926/15981.-
    C. A. 91 [1979] 26975)
20. United Nations (Publication E/CONF. 70/29, New York [1977] 186 pp.)
21. H. Arida (Kagaku Keizai 25, No 8 [1978] 2/9)
22. J. S. Burton (Proc. 5th Ann. Conf. NWSIA, San Diego, Calif. [1977] 20 pp.)
23. R. B. Robie (Proc. 5th Ann. Conf. NWSIA, San Diego, Calif. [1977] 13 pp.)
24. Z. Bushnak (Proc. 5th Ann. Conf. NWSIA, San Diego, Calif. [1977])
25. M. Lounsbury, S. Hebenstreit, R. S. Berry (NTIS Rept. PB-288046/6GA [1978]
    210 pp.- G. R. A. 79, No 5 [1979] 138)
26. U. S. General Accounting Office, Report to the Congress, 1 May 1979 (NTIS Rept.
    PB-295318/0GA [1979])
27. Office of Water Research and Technology (NTIS Rept. PB-282607/1GA [1978] 420 pp.)
28. A. C. Petrasek (NTIS Rept. PB-274874 [1977].- C. A. 89 [1978] 64573)
29. R. A. Flook (Prog. Water Technol. 10, No 1-2 [1978] 105/111.-C. A. 89 [1978]
    152504)
30. A. Van Haute (Ingenieursblad 47, No 7 [1978] 176/187)
31. C. L. Smith (NTIS Rept. AD-A062710/9GA [1978])
32. E. Catier (Trait. Surf. 19, No 162 [1978] 9/27)
33. H. P. Warner, J. N. English (NTIS Rept. PB-280145 [1978] 51 pp.)
34. B. Jost (Gas, Wasser, Abwasser 58, No 3 [1978] 165/166.-C. A. 89 [1978] 135436)
35. SCS Engineers (NTIS Rept. PB-280482 [1978] 359 pp.)
36. R. Quinn (Proc. 6th Ann. Conf. NWSIA, Sarasota, Florida [1978] 5 pp.)
37. G. H. Boone (Proc. 6th Ann. Conf. NWSIA, Sarasota, Florida [1978] 18 pp.)

38. H. R. Pahren (Proc. 6th Ann. Conf. NWSIA, Sarasota, Florida [1978] 5 pp.)
39. O. K. Buros (Desalination 32 [1980] 305/308)
40. A. M. Al Taweel, A. Bushnak, W. Bawarith (Desalination 32 [1980] 341/352)
41. J. E. Cruver, M. Jhawar (Desalination 32 [1980] 365/372)

1B. Desalting Processes

An index of literature on desalination of brines covers the period 1936 to 1975 [1]. General reviews on desalination processes include present state and development trends in producing fresh water from the sea [2], recent progress in desalination [3], water desalination, a stringent problem [4], feasibility study of desalting seawater [5], economic and technological aspects of desalination methods [6], state of conversion of saline waters [7], assessment of the prospects of various methods of water desalination for municipal drinking water supply [8], desalting technology assessment under total water management [9], total water management the way of the future [10], state and development trends in seawater desalination [11], future of desalting in community water supplies [12] and potential of demineralization in water reuse [13].

An evaluation of technical information was conducted, based on eight criteria: interest and commitment, general site considerations, consideration of desalting needs, demographic and socio-economic considerations, water supply considerations and environmental considerations. Thirty-seven sites were evaluated during the study. Sites with critical water needs were identified and each of them represents a genuine potential site for a desalting demonstration plant [14].

Methods were presented for estimating the costs for conveying product water from desalination plants from 19 to 380 thousand m^3/day (5 to 100 Mgd) to distances up to 80 km. Costs include pipeline construction, pumping stations, power lines and electrical switchgear [15]. A procedure was described for techno-economical analysis of methods for desalting natural waters [16]. A review was presented on the various methods of seawater desalination [17].

The annual progress report on saline water conversion research at the University of California, covering the period 1 July 1977 to 30 June 1978 was published [18].

INTERNATIONAL MEETINGS

The 5th annual conference of the National Water Supply Improvement Association was held in San Diego California, 17 to 21 July 1977 [19] and the 6th annual conference in Sarasota, Florida, 16 to 20 July 1978 [20]. Technical Proceedings were published for both conferences, however, they do not contain all papers announced for presentation.

Two workshops were held in Düsseldorf, 17 November 1977, and Cologne, 12 December 1977, with subject seawater desalination and the required power supply. The proceedings contain 24 papers and were edited by the V.D.I. [21].

An International Congress on Desalination and Water Reuse was held in Nice, France, 21 to 27 October 1979, jointly sponsored by the International Desalination and Environmental Association (IDEA) and the Association pour le Developpement Industriel [22].

INTERNATIONAL ACTIVITIES

Egypt. Proposals have been made for step-by-step planning and development of the groundwater resources of the Western Desert. Experimentation with well spacing patterns and pumping procedures are under consideration. Programs of total water management, including desalination of wastewaters and their reuse, being coupled with exploitation of the Nubian sandstone aquifer are considered [23]. Available water resources in Egypt and possible solutions for the supply of fresh water required in various areas were outlined. The utilization of electric supply from the grid, at periods other than the peak loads, to operate electrodialysis plants running for 10 h/day would provide 2.6 million m^3/day at an operating cost of 10 c/m^3 in addition to capital costs [24].

France. An important program of desalination research and implementation was undertaken by the Commissariat ã l'Energie Atomique. French industry is supplying various types of desalination equipment, such as multiple-effect distillation, vapor compression, electrodialysis, etc. Research is performed on large scale vapor compression, use of low cost construction materials (plastics, aluminum, concrete), nuclear desalting with small reactors etc. [25].

Germany W. A Center aiming to improve and optimize thermal desalination plants (MSF, VC, VTE), as well as to test their components is operated by Thyssen Engineering G.m.b.H. [26].

Israel. The joint United States - Israel desalination project aims to construct a 4.6 Mgd intermediate test module, which will subsequently be incorporated into a 10 Mgd desalting plant coupled to a 50 MW turbogenerator. The project will use the horizontal aluminum tube multi-effect distillation process, which operates below 170° F [27]. A review on research and development in desalination includes Israel's available natural water supplies, the approaches to desalting which have been tried and the current state of development, including the large multiple effect plant being built at Aschod and work on membrane processes [28]. About 75% of the water in Israel goes to agriculture, 8 to 10% to industry and the balance to domestic consumers. Intensive R & D efforts leading to full-scale installations aim to provide water from unconventional sources. Flood cathment and aquifer recharge, sewage renovation are undertaken together with a desalination effort, including operation of multi-stage flash, multiple effect distillation, reverse osmosis, electrodialysis and improved membrane development [29].

Japan. Technology of producing fresh water from seawater by the reverse osmosis process has been developed since 1974. The two-stage process using reverse osmosis modules developed in Japan has recorded highly satisfactory and stable performances. Durability of the membranes is expected to be more than three years. The quality of the product water can be kept below 200 ppm TDS. Construction of a demonstration plant of 500 m^3/day is planned [30]. A review was presented on the status of the application of desalination techniques in Japan [31].

Libya. In 1977 there were seven desalination plants with total output of about 140000 m^3/day. It is intended to increase the capacity to about 225000 m^3/day by 1981 [32]. Multi-stage flash and reverse osmosis processes are the favored types of equipment [33]. The peak day demand for municipality water of the city of Tripoli is expected to increase from presently 0.26 to 0.6 Mm^3/day in the year 2000. Schemes to meet future demands are surveyed [34]. The two most important solutions to meet present and future libyan water demand, water reuse and desalination, are reviewed [35]. Difficulties encountered in erecting and operating waste water treatment plants in developing countries were discussed [36].

Netherlands. The expected consumption of water in the Netherlands in the year 2000 is estimated to be 3 billion m^3/yr, of which 1.9 billion m^3 can be supplied by groundwater and the rest by surface water, mainly from the Rhine and Meuse rivers [37].

Saudi Arabia. Saline Water Conversion Corporation (SWCC) is developing long-range plans to meet water and power demands. The general principles underlying the possible applications of these plans on a regional basis are described [38]. SWCC entered into an agreement in 1977 with the U.S. Department of the Interior, Office of Water Research and Technology, for technical cooperation in desalination. A joint team of desalting specialists was formed to work on the establishment of a Research, Development and Training Center to train Saudis in the operation and maintenance of desalting plants, and to carry out R & D directed to lowering the operating and maintenance costs, as well as increasing the reliability of the plants [39]. Considering MSF and electrodialysis as alternate technologies, multi-objective decision analysis was used to select the most preferred technology for Saudi Arabia. The methodology is based upon assessment of the attributes, such as reliability, operability, maintenability, manpower requirements and utilization of national resources [40]. Population of 2000 to less than 60 persons were found, in a survey, of necessity to be drinking brackish and semi-brackish water [41]. Two systems are used in the King Faisal Medical City in Riyadh for the collection of waste water, one of brine waste and one for water to be recycled. The latter after treatment is used for irrigation [42]. A desalination plant with a capacity of 380 m^3/h is scheduled for commercial operation in early 1981. In a completely arid area the plant will supply water during construction of power/desalination and petrochemical plants at the Yanbu Industrial City, on the east bank of the Red Sea. Subsequent units will follow to satisfy the ultimate requirement estimated to be as much as 4800 m^3/h [43].

Spain. A review is presented of various desalination processes used in Spain including distillation (VTE, MSF, VC, solar), crystallization (vacuum freezing, vacuum compression), membranes (electrodialysis, reverse osmosis) and ion exchange [44].

U. S. A. The emphasis in California's water supply planning is on reclamation of used water and on the conjunctive use of ground and surface water. Legal problems are involved [45]. The application of the total water management concept in the case of the Ventura County project has led to potential plans, which would not have been possible otherwise [46]. In future planning to assure adequate water supplies for South Florida a range of alternative systems has been evaluated. Lake Okeechobee, the second largest freshwater lake in the U.S., is impacted by numerous inflows which carry large amounts of nutritients, such as nitrogen and phosphorus. The placement of a large scale desalination plant at a strategic point prior to the water's input to the lake, could provide a reduction potential, which would allow the various standards to be met [47].

A massive market for desalting, 29.6 billion gpd, is predicted by the year 2000. Specific research and development needs are discussed and it is concluded that building of five demonstration plants in the U.S. was a good start, but much more needs to be done [48].

The saline waters of the Colorado river pose major management problems and are also important in relations with Mexico. Total water withdrawals exceed annual replenishments and total salinity in the lower reaches of the river are near or beyond threshold uses [49]. State and local management actions, which may be taken to reduce salinity in the Colorado river were proposed. Of over thirty possible state and local management actions considered, twelve have been proposed for implementation in the Colorado river basin [50]. Construction work on the parts of the Colorado river basin salinity control project downstream the Imperial Dam began in 1975. Construction of the upstream parts of the project were due to start in 1977 [51]. The Yuma plant will treat up to 129000 acre-feet of 3200 mg/l irrigation return flows and will deliver 283 mg/l product water to maintain a 115 mg/l differential between water delivered to U.S. users and to Mexico. Partial lime softening followed by multi-media filtration was chosen as pretreatment system. Two spiral wound reverse osmosis systems were chosen as most efficient from physical and economic viewpoint. Construction of the main plant and installation of desalting equipment is scheduled for award summer of 1980. Total cost of the desalting complex is approximately 190 million dollars [52].

To supplement dwindling groundwater and imported supplies in the Orange County Water District and to prevent seawater intrusion of the coastal aquifers, Water Factory 21 was built. The facility has a total capacity of 15 Mgd and features a 5 Mgd reverse osmosis demineralizer [53].

EDUCATION AND TRAINING

Difficulties may appear in developing countries for the period of time from taking over a desalination plant and the first two years of operation. It is recommended to use during this period proper consultants to supervise plant operation and train the operating, maintenance and management personnel [54].

A Desalination Technology Transfer Center was established at the West Indies

Laboratory of the Fairleigh Dickinson University [55]. A series of courses in various aspects of desalination are held. Accumulated experience from eight years offering of extension education courses in the desalination field were reported [56]. Courses on desalination technology are also given at the University of Glasgow, Scotland. The Master of Engineering course in Desalination Technology has been running for twelve years [57]. A recommended content of material, suggested for qualifying desalination plant operators, was presented [58].

LITERATURE TO 1B.

1. S. Seiitkurbanov, Desalination of brines (Ylum, Ashkhabad, Turkm. SSR [1977] 169 pp.)
2. R. Rautenbach, H. Hoeck, K. Rauch (NTIS Rept. PC-A03/MF-A01 [1976].-E.R.A. 3, No 17 [1978] 41233)
3. W. M. Walker (AMDEL Bull. No 22 [1977] 33/51.- E.I. 16, No 6 [1978] 40671)
4. G. Zuliani (Elettrificazione No 12 [1977] 561/564)
5. M. S. Wang (Yen Yeh Yu Yen Kung 28 [1977] 14/15)
6. V. P. Alekseev, L. F. Smirnov (Visn. Akad. Nauk Ukr. RSR No 2 [1978] 41/54)
7. A. Clerfayt (Acad. R. Sci. Outre-Mer, Cl. Sci. Tech., Brussels, 18, No 3 [1978] 1/86)
8. Yu. A. Rakhmanin, M. V. Sanin (Gig. Sanit. No 10 [1978] 18/23)
9. M. B. Bessler (5th Ann. Conf. Nat. Water Supply Improv. Assoc., San Diego, Calif., 1977)
10. P. C. Culler (6th Ann. Conf. Nat. Water Supply Improv. Assoc., Sarasota, Florida [1978] 8 pp.)
11. J. Jansky, G. Marotz, G. Tusel (Chem. Ztg. 102 [1978] 256/258)
12. I. Nusbaum (6th Ann. Conf. Nat. Water Supply Improv. Assoc., Sarasota, Florida [1978] 11 pp.)
13. R. D. Heaton (6th Ann. Conf. Nat. Water Supply Improv. Assoc., Sarasota, Florida [1978] 4 pp.)
14. D. H. Furukawa, K. F. Frank, R. E. Mannion, E. Papadopoulos, T. C. Ryan (NTIS Rept. PB-290388/3GA [1978].- E.R.A. 79, No 8 [1979] 66)
15. S. A. Reed, M. L. Marsh (NTIS Rept. PC-A02/MF-A01, ORNL/TM-6290 [1978] 19 pp.)
16. G. M. Davydov (Uch. Zap. Azerb. Inst. Nefti i Khimii No 5 [1977]79/85.- Ref. Zh., Teploenerg. [1978] 5 R 143)
17. M. A. Molleda Carbonell (Dyna, Bilbao, 53, No 11 [1978] 309/313)
18. University of California, Los Angeles (Water Res. Center Desal. Rept. No 67 [1978] 77 pp.)
19. National Water Supply Improvement Association, Technical Proceedings 5th Annual Conference, San Diego, California (NWSIA, Ipswich, Mass. 1977)
20. National Water Supply Improvement Association, Technical Proceedings 6th Annual Conference, Sarasota, Florida (NWSIA, Ipswich, Mass. 1978)
21. Verein Deutscher Ingenieure, Meerwasserentsalzung und ihre Energieversorgung (Fortschr. Ber. VDI Z 3, No 47 [1978] 249 pp.)
22. Proceedings in Desalination, Volumes 30, 31 [1979] and 32 [1980])
23. W. E. Warne (6th Ann. Conf. Nat. Water Supply Improv. Assoc., Sarasota, Florida [1978] 6 pp.)
24. M. F. El Fouly, E. E. Khalil (Desalination, 30 [1979] 205/212)
25. J. J. Libert, J. C. Vich (Desalination 30 [1979] 95/99)
26. J. Hapke, K. Osterholt (Desalination 30 [1979] 101/126)
27. L. Awerbuch, W. L. Barnes, R. H. Horowitz, A. Z. Barak (5th Ann. Conf. Nat. Water Supply Improv. Assoc., San Diego, Calif. [1977] 11 pp.)
28. E. Tal (Proc. 13th Ann. Desal. Conf., Nat. Coun. for R & D, Publ. No 8-78 [1978] XV/XXV)
29. D. Ashboren, M. Waldman (Desalination 30 [1979] 213/221)
30. Y. Kunisada, Y. Murayama (Japan Water Reuse Promotion Center [1978] 17 pp.)
31. T. Shirozu (Proc. Int. Water Supply Assoc. Congr. 12 [1978] S13/S18)
32. S. Kushad (Pure Water 6, No 3 [1977] 14/17)
33. A. H. Nasrat, T. V. Balakrishnan (Desalination 30 [1979] 187/202)
34. M. Aswed, A. Sunni (Desalination 30 [1979] 155/162)
35. H. El Hares, M. Aswed (Desalination 30 [1979] 163/173)

36. M. Hares, M. Abdel-Dayem (Desalination 30 [1979] 175/185)
37. P. L. Knoppert (Natuurkd. Voordr. 1975-1977 [1978] 54/55, 159/172)
38. I. M. R. Jamjoom (5th Ann. Conf. Nat. Water Supply Improv. Assoc., San Diego, Calif. [1977] 16 pp.)
39. I. M. R. Jamjoom, Y. H. Nassif, R. T. Heizer, A. Raschid Khan (Desalination 30 [1979] 247/257)
40. A. A. Bushnak, A. F. Abdul-Fattah, A. A. Husseiny, A. M. El-Nashar (Desalination 30 [1979] 483/498)
41. T. G. Temperley (Desalination 30 [1979] 319.-Abstract only)
42. F. X. Floyd (Desalination 30 [1979] 321.-Abstract only)
43. J. R. Stange, W. S. Hsieh (Desalination 31 [1979] 69.-Abstract only)
44. F. Troyano, M. Torres, A. Cajigas (Rev. Obras Publicas 124 [1977] 345/354)
45. J. E. Bryson (5th Ann. Conf. Nat. Water Supply Improv. Assoc., San Diego, Calif. [1977] 8 pp.)
46. V. E. Martin (5th Ann. Conf. Nat. Water Supply Improv. Assoc., San Diego, Cailf. [1977] 21 pp.)
47. J. R. Maloy (Desalination 30 [1979] 39/47)
48. P. J. Schroeder (Desalination 30 [1979] 5/13)
49. C. V. Moore, J. H. Snyder (NTIS Rept. PB-290784/8GA, [1979] 26 pp.)
50. J. G. Milliken, L. C. Lohman, S. A. Lyon, G. W. Sherk (NTIS Rept. PB-283134/5GA [1977] 371 pp.)
51. D. L. Krull (5th Ann. Conf. Nat. Water Supply Improv. Assoc., San Diego, Calif. [1977] 29 pp.)
52. M. Lopez (Desalination 30 [1979] 15/21)
53. P. K. Allen, G. L. Elser (Desalination 30 [1979] 23/38)
54. A. De Maio, F. Fioravanti, G. Odone (Desalination 30 [1979] 439/442)
55. R. Bakish (NTIS Rept. PB-291478/6GA [1979].-G.R.A. 79, No 9 [1979] 83)
56. R. Bakish, E. Catanzaro, W. Arthur (Desalination 30 [1979] 443/447)
57. W. T. Hanbury, W. S. McCartney, J. Cunningham, T. Hodgkiess (Desalination 30 [1979] 449/459)
58. R. Bakish, E. Catanzaro (Desalination 30 [1979] 461/466)

1C. Raw Material Seawater

1. MINERAL CONTENT OF SEAWATER

Volume 7 of Chemical Oceanography, edited by Riley and Chester, has been published in 2nd edition [1]. A representative group of papers give an insight into the subject of cycling of major elements in seawater, edited by Drever, as Volume 45 of Benchmark Papers in Geology [2]. An Introduction to Marine Science with reference to the chemistry of seawater was published [3].

The preparation of standard seawater was described, as well as the high precision method used for determining the chlorinity of the standard [4]. The chemistry of water and aqueous solutions was briefly reviewed, pointing out how in numerous cases water stands apart from other solvent systems. A discussion is also presented of the properties of ice, water and aqueous solutions [5]. Progressive heating of seawater to 350°C causes it to become increasingly acid and depleted in Ca, Mg and SO_4^{2-}, because anhydride and a previously undescribed magnesium oxysulfate are precipitated. Comparison of solution composition with theoretical solubilities indicates that carbonate minerals were supersaturated with increasing temperature [6].

Other work on the properties of seawater includes the sea as a chemical system[7], state of mineral components in surface waters [8], seawater chemical characteristics [9], sampling of seawater [10], chemical compounds in seawater [11] and evaluation of parameters of seawater [12].

A review of the content and properties of the most important mineral constituents in seawater is given in the following.

Aluminum. Dissolved aluminum in the Mediterranean Sea, offshore of Corsica, has been found 1.5 mg/l Al, with extreme values ranging from 0.5 to 4 mg/l Al [13]. Other work includes the biological control of dissolved aluminum in seawater [14].

18

Ammonium. A mixture of $NHBr_2$ and NH_2Cl will form in full-strength chlorinated seawater containing NH_3. At high NH_3 concentrations and longer times, NH_2Cl becomes the major component [15]. The accuracy of earlier calculations of the extent of hydrolysis of NH^+ in seawater over a range of temperatures and salinities has been confirmed [16]. Concentrations in marine salts were found to be: Ammonium absent, nitrites 0 to 0.45 ppm, nitrates 0.3 to 1.7 ppm [17].

Arsenic. Under typical seawater conditions, H_3AsO_4 is less dissociated than H_3PO_4. Significant amounts of reactive AsO_4^{3-} are ion-paired with cations other than Na^+ [18]. Determination of arsenic in sediments of the North Atlantic Ocean and the Eastern Mediterranean Sea is reported [19].

Cerium. The concentration of radioactive Ce in various water samples of Taiwan was 52.8 to 134.8×10^{-3} µCi/l, much lower than the maximum permissible limit [20].

Chromium. $^{51}Cr(III)$, a corrosion product generated in the cooling system at a nuclear power plant, is in a predominantly soluble state at pH < 6 and in a colloidal or suspended state at pH > 6 as a result of hydrolysis [21]. Speciation of chromium in municipal wastewaters and seawater of southern California is reported[22].

Cobalt. The effect of dissolved organic matter on the behavior of radiocobalt in seawater [23], as well as the interaction of cobalt with organic matter dissolved in seawater were investigated [24].

Copper. A review of the chemical speciation of copper in seawater was presented [25]. Copper chelation in seawater by a variety of organic ligands is dependent upon the stability of the model complex and the pH of the medium [26]. Copper chelation capacity of seawater was investigated by a bacterial bioassay technique[27]. Copper concentrations in surface waters off the southeastern United States coast range from 0.02 to 0.33 µg/kg [28]. The application of the rotated disc electrode to measurement of copper complex dissociation rate constants in marine coastal samples was studied [29].

Halogens. Sunlight exposure of chlorinated seawater produced a substantial conversion of the oxidative capacity in the added Cl to BrO_3. The most obvious organic products of seawater chlorination are $CHBr_3$ and to a lesser extent $CHClBr_2$ [30]. The Br^-/Cl^- ratio for surface samples from the northern Bay of Bengal was 0.00321 ± ± 0.00007 g/kg Cl and is significantly lower than for deeper layers or other regions [31]. The mass transfer coefficient and height of transfer unit were correlated with other design variables, using the laboratory for development of air-blowing technique to recover Br from seawater and on pilot plant, where Br is recovered from bittern by a conventional streaming-out technique [32]. The distribution of bromine between salt crystals and solutions of different composition and concentration was investigated [33]. Estimation of F^- in the northern Indian Ocean gives an average concentration of 1.31 ± 0.01 mg/l and a F/Cl ratio of 6.84 ± 0.01 $\times 10^{-5}$. The Ca/Cl ratio in the area is 0.02229 ± 0.000005 with an average Ca concentration of 432 ± 5 mg/l. The Mg/Cl ratio is observed to be 0.06631 ± 0.00003 with an average Mg concentration of 1281 ± 14 mg/l [34].

Hardness. Tables were constructed for the rapid calculation of equilibrium pH and equilibrant CO_2 for a water sample, which does not contain too much sulfate [35].

Helium. The helium concentration in seawater was generally found about 6×10^{-5} ml/l at depths of about 2000 m. The atmospheric concentrations were 5×10^{-6} ml/l at sealevel and 2.5×10^{-6} ml/l at about 5 km altitude [36].

Iron. The effects of the major ions in seawater on the titratable charge of goethite (α-FeOOH) and the capacity of the latter for these ions were evaluated[37].

Lead. The distribution is given of dissolved and particulate forms of naturally radioactive ^{210}Pb and ^{210}Po in ocean coastal waters, deep waters and sediment
of the continental shelf and slope in various Seas [38].

Manganese. The selectivity of ion exchangers for radiomanganese and radiozinc
depends on the presence, in their structure, of complex-forming oxygen-containing
functional groups. For the concentration of Mn and Zn from seawater, carboxyl-con-
taining polyampholytes and H_3PO_4- based preparations are recommended [39].

Mercury. Waters off the coast of Bombay at a depth 1 to 150 m contained 26 to
142 ng/l of mercury. The coastal waters of Mormugao contained 1 to 187 ng/l at a
depth of 1 to 800 m, while off the coast of Cochin the mercury concentration was
58 to 407 ng/l at depths 1 to 1200 m [40]. In Minamata Bay the total and particulate concentrations of mercury were 56 to 285 ng/l and 2.1 to 506 ng/l, respectively [41]. Most of the mercury in coastal seawater polluted with wastewater is
associated with suspended matter, especially particulate organic matter [42]. The
distribution of mercury in the Gulf Stream was investigated [43].

Nitrates. A review was presented on the nitrate removal from water supplies
by ion exchange [44].

Organics. The distribution of particulate organic carbon in the oceans and its
ecological implications were reported [45]. Petroleum hydrocarbons in the Mediterranean Sea averaged from 6.9 to 25.8 µg/l, with the Alboran Sea and the area off
Libya having the highest concentrations [46]. Simultaneous samplings and analysis
at different depths showed differences in composition of fatty acid fractions[47].
Data on dissolved free amino acids in seawater of the Tahiti area were reported[48].
Determination of nonpolar hydrocarbons in various Seas were made [49]. Benzene
and alkylbenzenes were quantitated at the parts per trillion level [50]. Organic
carbon content of seawater from over three Caribbean reefs was determined [51].
Estimations of chlorinated hydrocarbon concentrations in seawater of the German
Bight and the Western Baltic did not show correlations with the amount of particulate organic matter, distance from land and amount of rain [52]. The occurence
and temporal variations of a variety of low to medium polarity organic compounds
in the volatility range bracketed by heptane and octadecane were studied in
seawater [53].

Oxygen. The distribution of oxygen in the ocean depends on oxygen generation
during photosynthesis, oxygen consumption during biochemical oxidation of organic
matter, convective and turbulent oxygen exchange between the ocean and atmosphere
and oxygen transfer during horizontal or vertical movements [54]. Oxygen exchange
in the North Sea between sea and atmosphere and mixing within the water column was
investigated. At the surface the seawater was saturated with air and showed the
$^{18}O/^{16}O$ ratio of atmospheric oxygen. In deeper layers the oxygen was consumed and
the heavier isotope ^{18}O was enriched [55].

Phosphoric acid. The stability constants of phosphoric acid in seawater of 5
to 40‰ salinity and temperatures of 5 to 25°C were determined [56]. The three
dissociation constants of phosphoric acid have been obtained in seawater media by
means of potentiometric titrations with strong base [57].

Plutonium. The ranges of $^{239+240}Pu$ concentrations determined in the northwest-
ern Mediterranean are 0.5 to 8.5 fCi/kg for seawater, 0.3 to 4.2 pCi/kg for dried
sediments and 0.42 to 0.74 pCi/kg wet for total mussel [58].

Radionuclides. The great tendency of radiocolloids to adsorb onto finely divided hydrous oxides makes their formation of significance in seawater [59]. Measured concentrations of radionuclides, trace elements and salts in the Persian Gulf
and Caspian Sea indicate an obvious difference between the two water bodies compared with open seawater [60]. The radioactivity of advected seawaters was calculated from the changes in the salinity [61]. Variations of the concentrations of

^{90}Sr and ^{137}Cs in the Baltic Sea waters can be explained on the basis of meteorological and hydrological processes [62]. The average concentrations of thorium
isotopes in the Pacific Ocean were for ^{232}Th 0.9 ng/l, for ^{230}Th 2.1 x 10^{-2} pg/l
and for ^{228}Th 1.2 ag/l [63]. A settling model was proposed as a removal mechanism
of elements in the ocean on the basis of the behavior of insoluble radionuclides
in seawater. The settling model is applied to ^{234}Th in the surface water and the
vertical eddy diffusivity of the water is calculated as about 30 cm^2/s [64]. A review on radioactive species in seawater was presented [65]. Measurement of ^{222}Rn
concentration is a useful parameter for determining the exchange of gases between
the ocean water and atmosphere [66].

Rhenium. The average concentration of dissolved Re in ocean waters was 0.009
µg/l [67].

Strontium. The strontium concentration in the coastal water of the Baltic Sea
was 1.6 to 2.4 mg/l, with a maximum in the spring and a minimum in early autumn[68].
The maximum concentration of ^{90}Sr in the same sea was found 0.18 to 1.46 pCi/l
in surface water and decreased with depth. ^{137}Cs showed a maximum concentration
of less than 1.17 pCi/l at depths of 100 to 200 m [69]. The average concentration
of ^{90}Sr in the Baltic Sea during the calm season was 0.68 at the surface and 0.55
pCi/l at 50 m depth. In the North Sea, the surface and deep water concentrations
were 0.53 and 0.57 pCi/l, respectively. The average concentrations of stable Sr
in the surface waters of the North and Baltic Seas were 1.6 to 2.3 and 7.2 to 7.8
mg/l, respectively [70]. The concentration ratio ^{137}Cs/^{90}Sr in the Baltic Sea
varied over a broad range of 0.3 to 2.2, increasing with depth. In the North Sea
the average values of this ratio were 1.74 and 1.53 in the surface and 50 m deep
layers, respectively [71].
The mean value of the ^{90}Sr concentration in the Black Sea waters was 0.5 pCi/l
[72]. ^{90}Sr and tritium concentrations in the surface water of the North Pacific
were also determined [73]. A study was made on stable strontium content in Azov
Sea water [74].

Sulfates. The sulfur isotopic composition of ocean water sulfates was determined for samples from various depths of the Geosecs Stations II and 3 and for a Pacific Ocean surface sample [75].

Trace metals. The computer program GEOCHEM was applied to irrigation water-
brine mixtures and to Salton Sea water-brine mixtures to compute the chemical speciation of the elements Cd, Cu, Hg, Ni, Pb and Zn, along with the oxyanions of As
and B [76]. The amount of trace metals in seawater was determined for Fe 56 to 155,
Mn 2.4 to 3.6, Co 1.9 to 3.6, Ni 27 to 72, Cu 1.0 to 4.7, Zn 12 to 66, Pb trace to
36 and Cd not detected to 0.05 µg/l. The correlation coefficient between Fe and
Mn was 0.74 and between Fe and Ni 0.71 [77]. Other studies include the distribution of Mn, Fe, Zn, Cd and Cu in Baltic seawater [78], predicted trace metal concentrations in saline seep waters [79], some trace elements in the waters, marine
organisms and sediments of the Adriatic Sea [80], chemical forms of minor metallic
elements in the Ocean [81], concentrations of trace metal ions in seawater [82],
concentrations of some ions of trace metals in China coastal waters [83], trace
metal characterization of ocean water [84], and separation of naturally occuring
high molecular weight metal complexes from seawater [85]. The contents of Cu, Zn,
Pb, Ni, Cd and Hg of seawater near Ancona [86] and of Cd, Cr, Cu, Pb, Zn, Ca and
Mg in Canadian drinking water supplies were determined [87].

Patents. Toxic trace elements, e.g. mercury and lead, are removed by passing
the drinking water through a bed of zinc particles and then through a bed of magnesium, alloyed with a small amount of manganese to inhibit corrosion [88].

Tritium. Tritium concentrations were determined in the Atlantic [89], in the
Caspian Sea [90], in seawater at Aichi prefecture, Japan [91] and around Kyushu,
Japan [92]. Tritium and carbon-14 distributions in seawater from under the Ross
Ice Shelf Project ice hole were reported [93].

Uranium. Vertical profiles from the Arctic, Pacific, Antarctic, Atlantic Oceans show a U-salinity ratio of $(9.34 \pm 0.56) \times 10^{-8}$ g/g with salinity ranging from 30.3 to 36.2°/oo. The U concentration of seawater of 35°/oo salinity is $3.35 \pm \pm 0.2$ µg/l [94].

Vanadium. Vanadium content of seawater from Tokyo Bay averaged 2.43 to 2.60 µg/l during summer. Winter levels were lower [95].

Zinc. Zinc concentrations in the north east Pacific Ocean, off the coast of central California, were found to be 10 to 600 ng/l [96].

LITERATURE TO 1C.

1. J. P. Riley, R. Chester, Chemical Oceanography, Vol. 7, 2nd edition (Academic Press [1979] 524 pp.)
2. J. I. Drever, Seawater, Cycles of the Major Elements (Dowden, Hutchinson & Ross [1977] 344 pp.)
3. P. S. Meadows, J. I. Campbell, An Introduction to Marine Science (Wiley, New York [1978] 179 pp.)
4. F. E. Hermann, F. Culkin (Deep Sea Res. 25 [1978] 1265/1270)
5. M. J. Blandamer (Proc. 2nd Solvents Symp., Manchester, [1977] 24/44)
6. J. L. Bischoff, W. E. Seyfried (Am. J. Sci. 278 [1978] 838/860.- C. A. 89 [1978] 117370)
7. H. Postma (Versl. Gewone Vergad. Afd. Natuurkd., K. Ned. Akad. Wet. 85, No 9 [1976] 123/130)
8. G. M. Varshal (Prob. Anal. Khim. 5 [1977] 94/107)
9. M. Nishimura (Dojin Nyusu 6 [1977] 1/5)
10. C. Versino (Inquinamento 20, No 9 [1978] 117/123)
11. S. Tsunogai (Shizen 34, No 2 [1979] 48/58)
12. C. Versino (Inquinamento 20, No 11 [1978] 57/62.-C. A. 90 [1979] 192214)
13. S. Caschetto, R. Wollast (Marine Chem. 7 [1979] 141/155)
14. M. Stoffyn (Science 203 [1979] 651/653.- C. A. 90 [1979] 209743)
15. G. W. Inman, J. D. Johnson (Proc. Conf. Water Chlorination Environ. Impact Health Eff. 2 [1978] 235/252.- C. A. 89 [1978] 168850)
16. M. Whitfield (J. Mar. Biol. Assoc. U.K. 58 [1978] 781/786.- C. A. 89 [1978] 168744)
17. C. Cantoni, G. Beretta, M. A. Bianchi (Arch. Vet. Ital. 29, No 5-6 [1978] 166/167)
18. D. H. Lowenthal, M. E. Q. Pilson, R. H. Byrne (J. Mar. Res. 35 [1977] 653/669.- C. A. 89 [1978] 11773)
19. C. Neal, H. Elderfield, R. Chester (Marine Chem. 7 [1979] 207/219)
20. Y. M. Lin, M. F. Chang (Ho Tzu K'o Hsueh 14, No 3 [1977] 27/30.- C. A. 89 [1978] 135596)
21. E. N. Shumilin, A. A. Volkov, Yu. A. Sapozhinikov (Deposited Doc. VINITI 2725-77 [1977] 13 pp.- C. A. 90 [1979] 92117)
22. T. K. Jan, D. R. Young (J. Water Pollut. Control Fed. 50 [1978] 2327/2336.- C. A. 89 [1978] 203870)
23. Y. Kimura, Y. Honda (Kinki Daigaku Genshiryoku Kenkyusho Nempo 13 [1976] 29/43.- C. A. 89 [1978] 203940)
24. Y. Honda, Y. Kimura (Kinki Daigaku Rikogakubu Kenkyu Hokoku 13 [1978] 171/175.- C. A. 90 [1979] 109537)
25. D. C. Burrell (NTIS Rept. RLO/2229/T1-41 [1977] 43 pp.)
26. W. Rosen, P. M. Williams (Geochem. J. 12, No 1[1978]21/27.- C. A. 89 [1978] 135428)
27. P. A. Gillespie, R. F. Vaccaro (Limnol. Oceanogr. 23 [1978] 543/548.- C. A. 89 [1978] 152224)
28. H. L. Windom, R. G. Smith (Marine Chem. 7 [1979] 157/163)
29. M. S. Shuman, L. C. Michael (Environ. Sci. Technol. 12 [1978] 1069/1072)
30. J. H. Carpenter, C. A. Smith (Proc. Conf. Water Chlorination Environ. Impact Health Eff. 1977, 2 [1978] 195/207.- C. A. 89 [1978] 168849)
31. S. W. A. Naqvi, C. V. G. Reddy (Indian J. Mar. Sci. 7, No 2 [1978] 122/124.- C. A. 89 [1978] 185735)

32. A.S. Mehta, S.D. Gomkale (Proc. 4th Nat. Heat Mass Transfer Conf. [1977] 587/600.- C. A. 90 [1979] 154023)

33. M.G. Valyashko, I.K. Zherebtsova, A. Lavrova, U Bi Hao (Brom Solyanykh Otlozh Rassolakh Geokhim. Indik. Ikh Genezisa, Istor. Poisk. Priznak [1976] 381/404.- C.A. 91 [1979] 7429)

34. R. Sen Gupta, S. Naik, S.Y.S. Singbal (Mar. Chem. 6, No 2 [1978] 125/141.- C.A. 89 [1978] 64890)

35. J.A. Bardy, C. Pere (Trib. CEBEDEAU 31, No 412 [1978] 113/126.- C.A. 89 [1978] 152491)

36. E.V. Borodzich, D.V. Grichuk, Yu.N. Gurskii, V.M. Korobeinik, I.N. Yanitskii (Dokl. Akad. Nauk SSSR 243, No 5 [1978] 1239/1242.- C.A. 90 [1979] 192209)

37. L. Balistieri, J.W. Murray (A.C.S. Symp. Series No 93 [1979] 275/298.- C.A. 91 [1979] 26979)

38. D.W. Spencer, P.G. Brewer, M.P. Bacon, P.L. Sachs, C.L. Smith, A. Fleer, S. Kadar (Woods Hole Oceanogr. Inst. Rept. COO-3566-19 [1977] 74 pp.- C.A. 89 [1978] 135457)

39. B.A. Chepenko, O.T. Krylov, P.D. Novikov, V.P. Komlev (Radiokhimiya 20 [1978] 815/817.- C.A. 91 [1979] 27064)

40. S.Y.S. Singbal, S. Sanzgiri, R. Sen Gupta (Indian J. Mar. Sci. 7, No 2 [1978] 124/126.- C.A. 89 [1978] 185736)

41. M. Kumagai, H. Nishimura (Nippon Kaiyo Gakkai-Shi 34, No 2 [1978] 50/56.- C.A. 89 [1978] 203947)

42. K. Matsunaga, H. Tsujioku, S. Fukasa, K. Hasebe (Bull. Chem. Soc. Japan 51 [1978] 3519/3521.- C.A. 90 [1979] 156830)

43. P. Mukherji, D.R. Kester (Science 204 [1979] 64/66.- C.A. 91 [1979] 26985)

44. D.A. Clifford, W.J. Weber (NTIS Rept. PB-277092 [1977] 52 pp)

45. P.J. Wangersky (Int. Rev. Gesamten Hydrobiol. 63 [1978] 567/574.- C.A. 91 [1979] 26973)

46. A. Zsolnay (Mar. Chem. 7 [1979] 343/352)

47. A. Saliot, M.J. Tissier (Geochim. Org. Sediments Mar. Profonds, ORGON 1: Mer Norv. 1974 [1977] 197/208.- C.A. 89 [1978]203814)

48. J. Marchelidon, M. Fontaine, R. Taxit (Mar. Pollut. Bull. 9, No 1 [1978] 17/19.- C.A. 89 [1978] 48681)

49. N.A. Nemirovskaya, M.P. Nesterova, N.M. Anufrieva, V.G. Neiman (7th Khim.- Okeanol. Issled. 1975 [1977] 189/194.- C.A. 89 [1978] 64936)

50. R.D. Smillie, T. Sakuma, W.K. Duholke (J. Environ. Sci. Health, Part A, 13 No 2 [1978] 187/197.- C.A. 89 [1978] 185792)

51. B.L. Westrum, P.A. Meyers (Bull. Mar. Sci. 28, No 1 [1978] 153/158.- C.A. 89 [1978] 64843)

52. D. Stadler (Meeresforschung 26, No 3-4 [1978] 162/164.- C.A. 89 [1978] 203907)

53. R.P. Schwarzenbach, R.H. Bromund, P.M. Gschwend, O.C. Zafiriou (Org. Geochem, 1, No 2 [1978] 93/107.- C.A. 91 [1979] 62320)

54. V.N. Ivanenkov (7th Khim.- Okeanol. Issled. 1975 [1977] 206/209.- C.A. 89 [1978] 64938)

55. H. Foerstel, H. Zielke (Pure Appl. Geophys. 116 [1978] 486/496.- C.A. 89 [1978] 64927)

56. O. Johansson, M. Wedborg (Mar. Chem. 8 [1979] 57/69)

57. A.G. Dickson, J.P. Riley (Mar. Chem. 7 [1979] 101/109

58. C.N. Murray, R. Fukai (Estuar. Coast. Mar. Sci. 6 [1978] 145/151)

59. I. Feldman (Rochester Int. 8th Conf. Environ. Toxic. 1975 [1976] 183/190.- C.A. 89 [1978] 64765)

60. A. Mahdavi (IAEA Rept. R-1405-F [1976] 5 pp.- C.A. 89 [1978] 64895)

61. L.I. Galerkin, M.M. Domanov (Morsk. Gidrofiz. Issled. No 3 [1977] 186/192.- C.A. 90 [1979] 60896)

62. G. Kadziene, M. Lukinskiene, A. Stankaitis, D. Styro (Fiz. Atmos. 3 [1977] 207/213.- C.A. 89 [1978] 152245)

63. Y. Miyake, Y. Sugimura, T. Yasujima (Pap. Meteriol. Geophys. 29, No 2 [1978] 75/81.- C.A. 89 [1978] 185737)

64. S. Tsunogai, M. Managawa (Geochem. J. 12, No 1 [1978] 47/56.- C.A. 89 [1978] 135430)

65. M.S. Baxter, I.G. McKinley (Proc. R. Soc. Edinburgh, Sect. B, Biol. Sci. 76, No 1-3 [1978] 17/35)

66. A.D. Zemlyanoi, G.F. Batrakov, V.N. Eremeev (Morsk. Gidrofiz. Issled. No 2 [1978] 117/132.- C.A. 91 [1979] 62264)
67. T.F. Boiko, A.D. Miller (Geokhimiya No 11 [1978] 1736/1740.- C.A. 90 [1979] 141973)
68. A. Stankaitis, D. Styro (Fiz. Atmos. 2 [1976] 234/241.- C.A. 89 [1978] 152248)
69. N. Dargiene, M. Lukinskiene, D. Styro (Fiz. Atmos. 2 [1976] 241/248.- C.A. 89 [1978] 152249)
70. D. Styro, M. Lukinskiene, A. Stankaitis (Fiz. Atmos. 2 [1977] 197/201.- C.A. 89 [1978] 152243)
71. M. Aivarzi, G. Kadziene, M. Lukinskiene, A. Stankaitis, D. Styro (Fiz. Atmos. 3 [1977] 201/207.- C.A. 89 [1978] 152244)
72. S.M. Vakulovskii, I. Yu. Katrich, S.G. Malakhov, E.I. Roslyi, V.B. Chumichev, V.N. Shkuro (Tr. Inst. Eksp. Meteorol. Ser. "Zagryaz. Prir. Sred" 6 [1977] 73/76.- C.A. 89 [1978] 152250)
73. S.M. Vakulovshii, A.I. Vorontzov, I.Yu. Katrich, I.A. Koloskov, E.I. Roslyi, V.B. Chumichev (Okeanologyia 18 [1978] 244/247.- C.A. 89 [1978] 94787)
74. O.M. Aleksanyan, V.B. Stradomskii, G.S. Konovalov, A.P. Korenev (Gidrokhim. Mater. 71 [1978] 44/50.- C.A. 91 [1979] 27018)
75. C.E. Rees, W.J. Jenkins, J. Monster (Chim. Cosmochim. Acta 42 [1978] 377/381)
76. G. Sposito, A.L. Page (Univ. Calif. Rept. UCRL-13790 [1977] 121 pp)
77. A. Inoue (Kagoshima Daigaku Suisangakubu 26 [1977] 1/6.- C.A. 89 [1978] 135435)
78. K. Kremling, H. Petersen (Mar. Chem. 6 [1978] 155/170.- C.A. 89 [1978] 64891)
79. G.K. Pagenkopf (Montana State Univ. Rept. MUJWRRC 99 [1978] 20 pp.- C.A. 90 [1979] 127241)
80. L. Kosta, V. Ravnik, A.R. Byrne, J. Stirn, M. Dermelj, P. Stegnar (J. Radioanal. Chem. 44 [1978] 317/332.- C.A. [1978] 185638)
81. Y. Sugimura, Y. Suzuki, Y. Miyake (J. Oceanogr. Soc. Japan 34 [1978] 93/96)
82. H.K. Gu, M.X. Liu, X.J. Zhang, W.Y. Bao, R.X. Guo, Q. Wang, Z.W. Zeng (Hai Yang K'o Hsueh Chi K'an 14 [1978] 23/27.- C.A. 90 [1979] 156793)
83. H.K. Gu, M.X. Liu, W.Y. Bao, X.J. Zhang, Q. Wang, R.X. Guo, Z.W. Zeng (Stud. Mar. Sin. 13 [1978] 1/7.- Ocean. Abst. 79: 857)
84. T. Fujinaga, T. Kawamoto, E. Nakayama (Gendai Kagaku 86 [1978] 32/39)
85. M. Betz (Mar. Chem. 7 [1979] 165/170)
86. E. Benetti, A. Bernardini, G. Paolini. L. Mazzarini, S. Savini, E. Comunian (Inquinamento 21, No 1 [1979] 31/38.- C.A. 91 [1979] 9211)
87. J.C. Meranger, K.S. Subramanlan, C. Chalifoux (Environ. Sci Technol. 13 [1979] 707/711)
88. E.R. Du Fresne, (U.S. 4.096.064, 28 Jun 1978.- C.A. 89 [1978] 220712)
89. H.G. Ostlund, H.G. Dorsey, R. Brescher (NTIS Rept. PB-278727 [1976] 96 pp.- G.R.A. 78 No 13 [1978] 164)
90. R.L. Michel, H.E. Suess (Earth Planet. Sci. Lett. 39 [1978] 309/312.- C.A. 89 [1978] 117381)
91. K. Chaya, N. Hamamura (Radioisotopes 27, No 4 [1978] 191/193.- C.A. 89 [1978] 152276)
92. Y. Takashima, N. Momoshima, M.R. Hamidian (Mem. Fac. Sci. Kyushu Univ., Ser. C. 11 [1979] 275/280.- C.A. 90 [1979] 174369)
93. R.L. Michel, T.W. Linick, P.M. Williams (Science 203 [1979] 445/446.- C.A. 90 [1979] 209715)
94. T.L. Ku, K.G. Knauss, G.G. Mathieu (Deep Sea Res. 24 [1977] 1005/1017.- C.A. 89 [1978] 11766)
95. Y. Sato, S. Okabe (J. Fac. Mar. Sci. Technol. Tokai Univ. No. 11 [1978] 1/19.- Ocean. Abs. [1979] 60)
96. K.W. Bruland, G.A. Knauer, J.H. Martin (Nature, London, 27 [1978] 741/743.- C.A. 89 [1978] 64983)

2. PHYSICAL PROPERTIES OF SEAWATER

Activity and osmotic coefficients. Total activity coefficients of calcium in artificial seawater of various salinities were 0.395 (5‰), 0.325 (10), 0.274 (15), 0.243 (20), 0.226 (25), 0.215 (30) and 0.205 (35‰). Values found for several samples of Atlantic Ocean water agreed well with those for artificial seawater [1]. Heat of dilution and of solution data, fitted to the form of equations corresponding to that used for activity and osmotic coefficients, give the change with temperature of the activity and osmotic coefficients [2]. The activity coefficients of dilute solutions of strong electrolytes from conductance data at elevated temperatures were evaluated [3].

Conductance. A radioactive tracer method has been used to determine the transference numbers of Na^+, K^+, Ca^{2+} and Mg^{2+} cations and Cl^-, Br^-, SO_4^{2-} and HCO_3^- anions in seawater, which contribute to the overall electrical conductance by more than 99%. A proposed method for calculating seawater density is based on a nine-constituents seawater model, in which relative ionic concentrations can deviate from that of standard seawater [4].

Density. Measurements of the temperature-pressure-salinity points, at which the density of saline water is a maximum, can be made more directly by determining conditions under which compressive heating vanishes. Disagreement with previous density formulas indicate that caution should be used in extrapolating the various equations of state to these temperatures [5]. The relative densities of samples of Mediterranean Sea water collected near Gibraltar agree on the average to $\pm$ 3.9 x 10^{-6} g/cm^3 with those predicted from the equation of state of Millero et al [6]. The coefficients of the equations describing the temperature and concentration variation of the densities, viscosities and conductivities of aqueous $CaCl_2$ solutions have been calculated from the experimental values of these quantities [7]. Combining γ radiation from ^{98}Sr, ^{90}Y and ^{204}Tl with acoustic methods, made it possible to determine the movement, density and composition of salts in ocean water [8].

Dielectrics. A comprehensive and unique set of measurements of the complex-dielectric constant of sea ice is described and a set of dielectric models describing the complex-dielectric behavior of sea ice is given [9].

Electrical conductivity. The variability in the chlorinity/conductivity relation for various samples of standard seawater is significant. The measurements of density, pH, silicate and dissolved organic carbon do not suggest any clear explanation of measured discrepancy [10]. The determination of dissolved impurities in water, using an electrolytic conductivity analyzer, is described [11].

Evaporation. The apparent evaporation coefficient α of water is related to the resistance of the insoluble monolayer to the evaporation of the water on which it is spread. α is 2 x 10^{-4} both for a monolayer free surface and for a surface covered by an octadecanol monolayer in the transition region between the condensed and gaseous states [12]. Comments to this paper have been published by Hickman [13] and Palmer [14] and a reply to these comments has been given by the author [15].

Heats of solution. The variation of the heats of solution of electrolytes, comprising ions with negative hydration, is determined over the entire range of their solubility in water by the variation of the radii of the ion cavities in solution with increasing concentration, on one hand, and ionic association, on the other. For electrolytes with a cation exhibiting positive hydration, the nature of the concentration variation of the integral heats of solution is determined by the change in the radii of the cavities of the cation and anion, the change in the number of water molecules in the hydration shell of the cation undergoing electrostriction and ionic association with formation of contact ion pairs or ion pairs separated by the solvent [16].

Ion association. Recent work includes a potentiometric study of the association of sodium and sulfate ions [17] and of calcium and sulfate ions [18] in seawater type electrolyte solutions, dependence of calcite concentration on the ionic composition of water [19], ion association of chloride and sulfate with sodium, potassium, magnesium and calcium in seawater [20].

Molal volumes. The partial molal volumes of electrolytes, estimated by using the specific interaction model, were found to agree well with experimental values in NaCl solutions and seawater. For electrolytes that form ion pairs, corrections must be made [21].

Solubilities. The solubilities of NaCl, KCl, $CaCl_2$, Na_2SO_4 and K_2SO_4 were determined by a modified visual method. Previous literature data for the solubility of these salts were generally high by 0.05 to 2.0 wt.% salt [22]. The increase in the solubility of methane with increasing temperature is less for the aqueous solution of NaCl than for water [23]. A method for the determination of the solubilities of silica-containing minerals is proposed, wherein the release rate of dissolved silica is monitored from the mineral surface, using a series of extractions with seawater [24]. A solubility diagram for the multicomponent system NaCl-KCl-$MgCl_2$-$CaCl_2$-$CaSO_4$-H_2O in the range of $CaCl_2$ concentrations from 1 to 8% at various temperatures is presented [25]. The solubility of H_2S in unacidified distilled water, distilled water acidified to pH 3.0 and in acidified seawater was determined at 2 to 30° and a salinity of up to 40% [26].

Sound speed. High pressure sound speeds and PVT properties of pure water, seawater and major sea salt aqueous solutions were determined [27]. Sound absorption in seawater is dominated by chemical relaxations [28].

Specific heats. Heat capacity of aqueous NaCl and seawater at high pressures were determined using the temperature jump method [29]. Heat capacities of NaCl, KCl and NaBr aqueous solutions were reported [30].

Structure of water. The IR-spectra of water in glas-, polyimid- and cellulose acetate membranes show differences in respect to frequency and intensity from those of pure water [31].

Thermodynamic properties. A review was presented on the calculation of chemical and physical equilibria with the aid of the minimization of the free energy, the application of the solution model to the determination of activity coefficients in liquids, the determination of activity coefficients on the basis of the concept of the local composition, the calculation of the phase equilibria with the aid of the group-contribution method, the calculation of the vapor-liquid equilibrium with the aid of an equation of state that is valid for both phases, and calculation of the phase equilibria with the aid of the principle of corresponding states [32]. A correction was proposed to the formula describing the dependence of the temperature depression on the concentration of a solution within a wide range of temperatures and pressures [33]. Several formulas for calculating steam properties and programs based on them were tested on a computer and the results compared with published values [34]. In programs for calculating the thermodynamic properties of water and steam, two main approaches have been used: programs based on calculating properties as required from general formulas and programs based on interpolating as required among sets of precalculated values stored in the computer [35].

Vapor pressure. For the calculation of the saturation humidity for brine, pure water humidity charts can be used if the saturation curve is modified to reflect the effect of the salt on the water vapor pressure [36]. Equations relating to VDI-steam tables were programmed in FORTRAN-IV [37]. The temperature coefficient of the equilibrium partial pressure of CO_2 was determined [38].

Viscosity. Evaporative concentration cannot be applied to the concentrated bittern by the membrane process, because of the high viscosity of the solution due to the high contents of $CaCl_2$ and $MgCl_2$. To remove this difficulty, the viscosities of $MgCl_2$-H_2O, $CaCl_2$-H_2O and $MgCl_2$-$CaCl_2$-H_2O were measured [39]. Other work includes the effect of pressure on the viscosity of aqueous NaCl solutions in the temperature range 20 to 150°C [40] and measurement and calculation of the viscosity of mixed aqueous solutions of NaCl and KCl in the temperature range 25 to 150°C and pressure range 0 to 30 MPa [41].

LITERATURE TO 1C 2.

1. V.S. Savenko (Okeanologiya 18 [1978] 441/444.- C.A. 89 [1978] 168686)
2. L.F. Silvester, K.S. Pitzer (J. Sol. Chem. 7 [1978] 327/337)
3. N.V. Fedotov (Zh. Prikl. Khim. 51 [1978] 2352/2354.- J. Appl. Chem. USSR 51 [1979] 2236/2238)
4. A. Poisson, M. Périé, J. Périé, M. Chemla (J. Sol. Chem. 8 [1979] 377/394)
5. D.R. Caldwell (Deep Sea Res. 25 [1978] 175/181.- C.A. 89 [1978] 94777)
6. F.J. Millero, D. Means, C. Miller (Deep Sea Res. 25 [1978] 563/569.- C.A. 89 [1978] 168752)
7. Yu. I. Roman'kov, T.A. Komarova (Zh. Fiz. Khim. 53 [1979] 233/235.- Russ. J. Phys. Chem. 53 [1979] 129/131)
8. E.M. Filippov (Morsk. Gidrofiz. Issled. No 2 [1978] 165/171.- C.A. 91 [1979] 62317)
9. M.R. Vant, R.O. Ramsier, V. Makios (J. Appl. Phys. 49, Pt 1 [1978] 1264/1280.- C.A. 89 [1978] 64999)
10. A. Poisson, T. Dauphinee, C.K. Ross, F. Culkin (Oceanol. Acta 1 [1978] 425/433.- C.A. 90 [1979] 28787)
11. D. Warmoth (Kent Tech. Rev. 22 [1978] 10/13.- C.A. 89 [1978] 203894)
12. G.T. Barnes (J. Colloid Interface Sci. 65 [1978] 566/572.- C.A. 89 [1978] 118216)
13. K. Hickman (J. Colloid Interface Sci. 65 [1978] 573.- C.A. 89 [1978] 118217)
14. H.J. Palmer (J. Colloid Interface Sci. 65 [1978] 574/575.- C.A. 89 [1978] 118218)
15. G.T. Barnes (J. Colloid Interface Sci. 65 [1978] 576/577.- C.A. 89 [1978] 118219)
16. A.S. Solovkin (Zh. Fiz. Khim. 53 [1979] 443/446).
17. A.E. Kosov, O.T. Krylov, P.D. Novikov (Deposited Doc. VINITI [1976] 4479-76.- C.A. 90 [1979] 109701)
18. A.E. Kosov, O.T. Krylov, P.D. Novikov (Deposited Doc. VINITI [1976] 4480-76.- C.A. 90 [1979] 109703)
19. R. Lenkaitis, H. Laumenskas (Liet. TSR Mokslu Akad. Darb., Ser. B, No 1 [1978] 73/80.- C.A. 89 [1978] 135451)
20. K.S. Johnson, R.M. Pytkowicz (Mar. Chem. 8 [1979] 87/93)
21. F.J. Millero (Geochim. Cosmochim. Acta 41 [1977] 215/223.- E.R.A. 3 [1978] 39943)
22. R.W. Potter, M.A. Clynne (J. Res. U.S. Geol. Surv. 6 [1978] 701/705.- C.A. 90 [1979] 12982)
23. A. Yu. Namiot, V.G. Skripka, K.D. Ashmyan (Geokhimiya No 1 [1979] 147/148.- C.A. 90 [1979] 171082)
24. D.C. Hurd, C. Fraley, J.K. Fugate (A.C.S. 176th Nat. Meet. 1978, Symp. Ser. 93 [1979] 413/415.- C.A. 91 [1979] 62288)
25. A.I. Kurta, A.M. Okrepka (Zh. Prikl. Khim. 51 [1978] 1759/1764.- J. Appl. Chem. USSR 51 [1978] 1665/1669)
26. A.A. Douabul, J.P. Riley (Deep-Sea Res. 26, No 3A [1979] 259/268)
27. C.T.A. Chen (Diss. Univ. Miami, 1977 Univ. Microfilms Order No 77-28.932.- D.A. 38 [1978] 3221-B)
28. R.H. Mellen, V.P. Simmons, D;G. Browning (J. Acoust. Soc. Am. 65 [1979] 923/925.- C.A. 91 [1979] 62518)
29. K.G. Liphard, A. Jost (6th Conf. AIRAPT, High-Pressure Sci. Technol. 1977, 1 [1979] 543/547.- C.A. 90 [1979] 175580)
30. J.E. Tanner, F.W. Lamb (NTIS Rept. AD-A045707/7GA [1977] 25 pp.- G.R.A. 78, No 1 [1978] 25)
31. R. Ameis, D. Schiöberg, G. Siemann, W. Luck (Symp. Membranstofftrennprozesse in Wissenschaft und Technik, Tubingen 1977)
32. S. Peter (Ber. Bunsenges. Phys. Chem. 81 [1977] 950/959)
33. K.M. Abdullaev (Teploenergetica 241 No 5 [1977] 69/70.- Thermal Eng. 24, No 5 [1978] 60/61)
34. M.P. Burgess, G.L. Fuller, A.H. Kaiser (NTIS Rept. DPSPU-76-11-3 [1977] 49 p.- C.A. 88 [1978] 24738)
35. M. Perrin (Intern. Chem. Eng. 18 [1978] 602/610)
36. M.A. Rodgers (Chem. Eng., New York, 85, No 11 [1978] 212/214.- C.A. 89 [1978] 30498)

37. H. Siewers (GKSS Rept. 78/E/19 [1978] 26 pp)
38. F. Macintyre (Clim. Change 1 [1978] 349/354.- C.A. 90 [1979] 192221)
39. Y. Kozai, K. Minami (Nippon Kaisui Gakkaishi 31, No 1 [1977] 12/15.- C.A. 89 [1978] 152520)
40. J. Kestin, H.E. Khalifa, Y. Abe, C.E. Grimes, H. Sookiazian, W.A. Wakeham (J. Chem. Eng. Data 23 [1978] 328/336)
41. R.J. Correia, J. Kestin, H.E. Khalifa (Ber. Bunsenges. Phys. Chem. 83 [1979] 20/24)

3. ANALYTICAL CHEMISTRY OF SEAWATER

General. Standard seawater has been used for several years in measurements in the place of natural seawater. Its preparation and a high precision method is reported for the determination of its chlorinity [1]. A monograph on methods of physical and chemical analysis of fresh waters [2] and a manual on methods of chemical analysis of seawater were published [3]. U.S. Environmental Protection Agency issued a manual for the Interim Certification of Laboratories involved in analyzing public drinking water supplies criteria and procedures [4].

Ion chromatography allows direct analysis for compounds having low pK_A or PK_B, including inorganic anions, alkali metals and alkali earth metals [5]. Elements having atomic numbers up to 12 are simultaneously revealed by a proton induced X-ray emission. Sample amounts of $10^{-6}g$ or 1 ml are sufficient for this high sensitivity analysis [6]. A review on measurement methods in seawater and marine organisms is reported by Lapicque [7].

Methods of sampling and of analyzing inorganic compounds are reviewed for water and waste water [8], as well as a sampler process suitable for sampling up to 400 l of water for retention of some radionuclides in salt and fresh water [9]. A simple and reliable method for preservation of seawater samples for the determination of carbohydrates has a recovery of 95% [10].

Ion-selective electrodes can be used in water analysis [11] and in water quality analysis [12]. A specially constructed reference electrode for in situ measurements is suitable for studying the effect of hydrostatic pressure, temperature, composition and dissolved oxygen in seawater, in the magnitude of the electrode potential [13]. Methods to obtain ultra trace analysis data by portable shipboard clean laboratories, used by the U.S. Ocean Chemistry Division, are briefly reviewed [14]. The accuracy of standard methods in water analysis was evaluated for water with a salinity of 0.35 to 10g/l [15].

Acidity, alkalinity and pH. The simultaneous determination of the specific conductance and pH in natural waters by using an automated procedure, gave relative standard deviation <1% for specific conductance and for pH ≤0.06 pH units [16]. A direct method involving a single iteration (a quadratic equation solved twice) is given for the calculation of carbonate ion concentration from total CO_2 and titration alkalinity. The simplified calculation should allow wider use of existing total CO_2 and titration alkalinity data [17].

Patents. A method for adjusting the pH of a sample stream for potentiometric determination of low concentration of ion activities is based on the passive addition of acid or base to the sample stream through a membrane permeable to the reagent and impermeable to the ionic species in the sample stream [18]. The CO_3-ion content of circulating cooling water is monitored by treating a sample with excess acid to give CO_2, which is stripped from the sample into gas, the thermal conductivity of which is measured before and after stripping [19]. In an apparatus for monitoring the alkalinity of seawater fed to a desalination plant, carbon dioxide is stripped from acid-dosed seawater using hydrogen. The thermal conductivity of the gas mixture is compared with that of hydrogen to give a measure of the alkalinity of the water [20].

Aluminum. A method using semicarbazones of salicylaldehyde derivatives was used for fluorescence determination of aluminum in seawater. Determination of

0.0002 to 0.16 µg/l aluminum was effected. The analysis of seawater gave 5.5 to 5.7 µg aluminum per liter [21].

Ammonium. A method for the determination of ammonia in coastal waters, having a variety of salt contents without side effects resulting from pH differences, is described [22]. An automated procedure for the determination of ammonium in sea- water was developed and compared with the automated method for the determination of ammonia as indophenol blue. The results of the two methods are in good agreement [23]. A method for the determination of ammonia in seawater is based on the form- ation of a substituted indophenol with sodium salicylate as phenolic reagent [24].

Arsenic. A colorimetric method for Sn determination was modified to suit the determination of As in seawater. Determination sensivity was 0.25 mg As per liter seawater [25]. Another method uses a new, unique enzymatic technique by performing an oxidative arsenolysis of the enzyme. The method has a linear calibration plot for the range 0.02 to 2.0 mg As per ml [26].

Barium. Isotope dilution mass sepctrometry was used for barium determination in Pacific Ocean samples, with detection limit of 10^{-15}g of barium [27]. Direct deter- mination of barium in sea and estuarine water was reported by using graphite furnace atomic emission spectrometry. The interferences in the analysis and the techniques to overcome them are discussed [28].

Beryllium and Bismuth. A method is reported for the determination of beryllium and bismuth in spring water and seawater. The method uses coprecipitation with zirconium hydroxide and determination by atomic absorption spectrometry using a carbon tube atomizer [29].

Boron. A modified colorimetric curcumin method is described for the determination of trace amounts of boron in industrial and seawater containing various substances [30].

Cadmium. The applicability of flameless atomic absorption spectroscopy to direct determination of Cd in seawater samples is limited to a level of about 2mg per liter. Lower concentrations were investigated with an extraction method carried out with ammonium pyrrolidine dithiocarbamate [31]. An equation for stripping polarography is derived, relating the half-wave potential ligand concentration and dissociation con- stant of a complexed metal ion. A computer-assisted system for generating stripping polarographic curves is used to characterize 10^{-8} M cadmium in seawater [32]. Ulti- mate detection limit barrier in furnace atomic absorption spectroscopy was studied. As an example the determination of detection limits of cadmium in sea and fresh water is illustrated [33].

Caesium. Methods involved in routine determination of ^{134}Cs and ^{137}Cs in sea- water are described in detail [35]. The effect of the conditions for the sorption of ^{137}Cs using zirconium ferrocyanide was studied with γ-spectroscopy determina- tion. Interferences are discussed [36].

Calcium. Patents. Determination of calcium by means of an ion selective mem- brane electrode with poly(vinyl ether) matrix showed selectivity ratios for the membrane Ca-Mg ~ 1, Ca-Sr 0.5, Ca-Ba 0.3, Ca-H 10^3 to 10^4, Ca-Na and Ca-K 1.3 [34].

Carbon. Three methods for the determination of dissolved organic carbon in sea- water were compared. Samples were analysed using persulphate oxidation, high-temp- erature combustion and ultraviolet photo-oxidation. The dissolved organic carbon content of the seawater samples ranged from 0.6 to 1.6 mg/l C. The study shows that results of high-temperature oxidation and photo-oxidation procedures differ by less than 5%, whereas results with persulphate oxidation are about 15% less than those obtained with the high-temperature oxidation. The relative merits of each of the oxidation techniques for the determination of organic matter in seawater are dis- cussed [37].

Cerium. Investigations were made of the sorption of cerium-144 by using AMF, KF and KRF ion-exchange resins. Cation-exchange resin KF-11 gave especially promising results [38].

Chromium. A spectrophotometric procedure for the determination of chromium(III) and chromium(VI) in seawater involves coprexipitation and separation of chromium(III) and chromium(VI) [39]. Another method uses a selective separation of Cr(III) and Cr(VI) by using Aliquat 336 for the extraction of Cr(VI). The chromium content is measured by flameless atomic absorption spectrometry with detection limits 0.03 and 0.01 mg/l for Cr(III) and Cr(VI) respectively [40].Cr(VI) at mg/i concentration was determined by ion-exchange colorimetry with relative standard deviations about 6.5% at seawater concentration of 4.9 mg/l [41].

Cobalt. A spectrophotometric method for the determination of cobalt in seawater involves separation of cobalt from seawater by using Amberlite CG400 ion-exchanger [42]. ^{60}Co was extracted from large volumes of seawater, concentrated in a small volume of solution of pyrrolidinedithiocarbamic acid in $CHCl_3$, and determined by direct NaJ(Ti) scintillation counting. The recovery of ^{60}Co averages 99 % [43].

Copper. A rapid method for the determination of copper in desalinated water and condensates was reported. The determination of copper is made photocolorimetrically and a mathematical equation is given for the calculation of the concentration of copper [44]. The interferences of salt matrices in the determination of copper by atomic-absorption spectrometry with electrothermal atomization were studied. Interference of NaCl was suppressed by the addition of Na_2O_2. This makes possible the direct determination of copper in seawater [45]. The reversibility of copper in dilute aqueous carbonate solutions and its significance during the determination of copper by cyclic and anodic stripping voltammetry is discussed [46]. A radiometric method for the direct determination of copper in ocean water samples requires for analysis only 1.0 ml aliquots of samples containing 20 mg of copper per ml. The relative standard deviation at 20 mg Cu/ml is 2.5% [47].

Dissolved gases. In a report on measuring atmospheric tracer gases in seawater, the experimental part deals with sample containers and the sampling procedure, the degassing apparatus, and apparatus and evaluation of gas-chromatographic analyses. As results, the profiles of N_2O, tritium and fluorohydrocarbon measurements are described. The discussion deals with oceanographic tracers, a comparison of profiles and special sampling and storage characteristics. The report contains a large number of tables and diagrams of the results of the extensive measurements [48]. A method for continuous analysis of dissolved gases in seawater uses a cellulose capillary dialyzer to extract oxygen, nitrogen and CO_2. The gases are determined by a diffusion-gas chromatographic procedure [49]. A rapid and sensitive method for the determination of organohalides, in concentrations from 1 ng/l up to 760 µg/l in sea water, drinking water and industrial effluents is described. Glass capillary column gas chromatography and electron capture detection were used for the determination of the gases [50].

Gold. A highly sensitive procedure was used for the determination of gold in natural waters. The gold is preconcentrated. The determinable range is 0.01 to 5.0 µg Au per liter with a relative standard deviation of 0.022 to 0.22% [51].

Halogens. Bromine concentration in seawater brines was calculated by equations derived on the basis of a relation between initial and end concentrations of Br and the initial and end amounts of water in the evaporite basin. Examples are presented [52]. A review on methods for the separation and the colorimetric determination of bromine in seawater was presented [53]. A spectrophotometric analysis of bromine in seawater uses the reaction with phenol red. The amount of Br determined is 0.92×10^{-5} to 4×10^{-5} mg/l [54]. A procedure is suggested for the determination of the real molar absorptivity of Br in the presence of high chloride concentrations, during the spectrophotometric determination of equilibrium composition of chlorinated bromine-containing brines [55]. Direct potentiometric determination of low concentrations, (0.01-50 mg/liter), of bromide ions is possible in the presence of not more than 10,000-fold amounts of chloride ions by means of a flow-type differential cell with two identical silver bromide electrodes, the supporting electrolyte used containing ammonia [56].

Neutron methods can be used for the determination of NaCl in seawater with an accuracy up to 1%. The neutron source produced in USSR has a neutron yield of $3/10^8$ neutron per second and GE-Li radiation detector [57]. Seawater standards with salinity 5 to 40% were prepared for the calibration of salinometer based on the potentiometric determination of Cl content. The correlation between electric conductivity, salinity and Cl content were determined for Black Sea water by using these standards [58]. An anion-selective electrode having LaF_3 single crystal was used for the potentiometric determination of fluoride in seawater [59]. A method for the separation and determination of fluorine by microdiffusion with hexamethyldisiloxane was developed. The recovery for the range 5 to 500 µg F^- is claimed to be 98 to 100% [60]. A comparative study of three methods was made for the determination of iodate-iodine in seawater. In one method the iodate is determined polarographically, while in the others the iodate is determined colorimetrically as iodonium ions. The tests were conducted on a selection of open-ocean and near-shore waters with iodate concentrations ranging from 0 to 60 µg/l I. The tests indicated that the polarographic method and the colorimetric method without iondine-water give the more reliable measurement of iodate concentration [61].

Iron. Iron was determined in seawater by a rapid method converting iron in radioactive ^{59}Fe [62]. Ionic iron in small volumes of seawater was directly determined by a new radioisotope dilution method [63].

Lead. Lead (II) speciacion in seawater was determined using voltammetry at thin-mercury film electrodes [64]. It was concluded that anodic stripping of lead in organic free seawater at pH 8 gave peak currents smaller than predicted. This is due to the electrochemical inactivity of Pb(OH). The peak currents were proportional to the seawater concentration [65].

Lithium. For the determination of lithium in seawater both atomic absorption and flame emission spectrometry can be used without pretreatment of the sample. The chloride interference can be eliminated by the method of standard additions or by calibration using artificial seawater [66].

Magnesium. Conductometric titration of magnesium using NaOH saturated with Magneson 11 is applicable for magnesium concentrations up to 0.095 g/l and in the presence of Ca less than 0.835 g/l. The relative error of the method was ≤ 1% [67]. The stability constant of the fluoromagnesium complex in seawater was determined by potentiometric procedure using LaF_3 ion-selective electrode [68].

Manganese. A rapid procedure for the determination of manganese in seawater, surface water and rainwater is based on preconcentration of manganese as the pyrrolidine-dithiocarbamate complex on a layer of active carbon and determination of the manganese in the carbon concentrate by neutron activation analysis [69]. To assess exposure done by radioactivity release from light water reactors a rapid method of determining ^{54}Mn in seawater was developed [70].

Mercury. Sub-nanogram amounts of mercury in seawater were determined by plasma emission spectrometry. The detection is rapid due to a cold-vapor introduction system [71]. Flameless atomic adsorption spectrometry was used for the determination of methylmercury. Chelating resins having a selective adsorption for methylmercury were investigated for its preconcentration from seawater [72].

Published methods for the determination of mercury in water by atomic absorption were critically assessed and incorporated into a procedure, for its determination in seawater, which has been optimized for maximum reliability and simplicity under routine conditions [73]. Mercury concentrations down to 2 ng/l can be detected in a one liter sample with an aeration time of 30 minutes, using $KMnO_4$-H_2SO_4 as a trapping solution. The simplicity of the instrumentation makes it suitable for shipboard use [74].

Reducing systems were studied for the determination of mercury by cold-vapor atomic absorption procedure coupled with amalgamation on gold [75]. Chemical models of fresh and seawater were used to examine the effect of humic material on the analysis of mercury by flameless atomic absorption spectrometry [76]. Analytical problems in the mercury determination of natural waters, with emphasis to seawater, were presented. Special reference is made to errors caused prior to measurement of mercury, i.e. adsorption and loss onto the bottle wall [77].

Acid-dichromate and ultraviolet digestion are used sequentially in an automated system to extract mercury from particulate matter and oxidize organomercurials in briny waters and sediments. Mercury is then determined by reduction with tin(II) and cold-vapor atomic absorption spectrometry. The detection limit is 0.02 μg/l Hg in water. Both waters and sediments with high chloride levels can be analysed without interference at a rate of 30 samples per hour [78]. A bioassay experiment using caged mussels has suggested a total concentration of methylmercury in the seawater of a polluted estuary of 0.06 ng/dm^3. The technique may provide a useful tool for the assessment of relative concentrations of methylmercury in seawater at different locations [79].

Molybdenum. An EPR spectroscopic method was used for the determination of molybdenum in saline waters. The method is based on a 17:1 preconcentration extraction of paramagnetic $Mo(SCN)_5$ into iosamyl alcohol followed by EPR intensity measurement. The procedure is relatively rapid, requires only 10 ml of seawater, has a detection limit of 0.46 μg/l Mo, and a relative standard deviation at the 11 μg/l level of 4,7% [80]. A simple and rapid method was investigated for the direct determination by graphite-furnace atomic absorption spectrometry of traces of molybdenum in the range 0,1 to 4 ng in synthetic seawater. Samples of less than 50 μl can be analyzed directly without using a background corrector, with a precision of less than 10% [81]. Trace amounts of molybdenum in seawater were determined spectrophotometrically with 5-chloro-7-iodo-8-quinolinol [82].

Nitrates. For the determination of nitrates in seawater the Cu-Cd reduction procedure was found the most suitable with a nitrate reduction of 98.4 to 100% [83]. A new quick method having accuracy ±3% was proposed, which uses a ultraviolet light absorption resin technique [84].

Organics. Organic carbon was determined in seawater by a fast automated colorimetric method using UV destruction [85], by a high-temperature oxidation procedure which is reported as accurate and precise [86] and by the UV-induced photochemical oxidation followed by colorimetric titration or IR spectroscopy [87]. A review on special gas chromatographic procedures for the determination of organic compounds in natural waters with indication of the determination limits was presented [88]. O-dichlorobenzone and cresol in seawater were determined, after extraction, by gas chromatography with detection limits 0.01 ppm [89]. The sampling procedure for dissolved organics in seawater was studied by using silanized porous glass, which has been found easy to keep free of contamination [90]. Halogeneted hydrocarbons were determined using glass capillary gas chromatography [91] and by head space, stripping and resin adsorption gas chromatography [92]. Determination of oil and oil derivatives in seawater was presented by three modifications of the fluorescence procedure [93], by a summary report data from seven laboratories [94] and by comparison of analytical procedures by several laboratories of Baltic Sea countries [95].

Patents. Organic pollutants are determined by measuring the fluorescence intensity, after irradiation of water or seawater with UV rays [96].

Oxygen. Pt, Ag, Pb, Cd, Zn and In were evaluated as materials for the development and creation of electrochemical transducers for measuring the concentration of dissolved oxygen in seawater [97]. A series of electrolytes were studied for their use in electrochemical converters to determine the concentration of dissolved oxygen in seawater samples. Optimum parameters are recommended for determining the concentration of dissolved oxygen under automated in situ conditions at a depth of up to 2000 m [98] . Traces of dissolved oxygen in seawater were measured by using a meter of the gas transfer type [99].

Patents. A colorimetric method was described for the determination of dissolved oxygen in water [100].

Potassium. Patents. The range of measurable concentrations in determining potassium, ammonium and caesium ions by means of ion-selective electrodes is extended and the selectivity increased by adding ammonium, Cs or K tetraphenylborates to the membrane [101].

Radionuclides. The determination of ^{231}Pa in seawater and disequilibrium studies in ^{231}Pa/^{235}U activity ratios were investigated [102]. Fallout radionuclides can be concentrated from seawater by coprecipitation with barium chlorideferric ammonium alum. Recovery is pH dependent and 80% efficient at seawater pH [103]. A method for the determination of radon-222 in seawater was investigated [104]. Improved techniques for sampling, sample processing and for measuring radon and radium in seawater were described [105].

Salinity. The salinity of seawater distillates on ships was determined as a function of the water conductance by a salinometer, which is a special design for ship board measurements [106].

Sodium. In the flame-photometric determination of sodium and potassium in seawater or concentrated brines, the interfering effect of Ca, Sr and Mg was eliminated by adding to the test solutions an equal volume of a 20% AlCl$_3$ solution. The relative error in the determination of sodium and potassium was ±2.0 to 2.6 and 2.1 to 5.0%, respectively [107].

Strontium. A review of atomic absorption spectroscopy applications to strontium determination in seawater was presented [108]. Methods for determining ^{90}Sr, ^{137}Cs, ^{60}Co and ^{65}Zn from one seawater sample after precipitation of the above isotopes were described [109]. The sorption of ^{90}Sr by barium fluosilicate for the determination of ^{90}Sr in seawater by γ-spectroscopy was studied [110].

Sulfate. Sulfate can be determined with rapid and precise methods by an indirect titration procedure using barium sulfate and EDTA for precipitation followed by MgCl$_2$ titration. The standard deviation is <0.5% [111]. Dissolved SO$_4$ in seawater was determined by precipitation with ^{133}Ba in dialysis bags and counting the precipitate on a scintillation counter. The procedure permits analysis of 0.5 to 3000 μg/ml SO$_4$ in a 1.0 ml sample and is interference free in natural waters ranging from freshwater to brines of 1.5 times the salinity of seawater [112].

Sulfur. Sulfur concentrations in aqueous solutions resembling seawater were determined by X-ray fluoresence spectrometry [113].

Trace elements. Analytical techniques for speciation of trace metals in natural waters: anodic stripping voltammetry, dialysis-atomic absorption spectroscopy and ultrafiltration-atomic absorption spectroscopy, were compared [114]. Trace elements in standardized water and seawater were determined by neutron activation analysis and X-ray fluorescence. Preconcentration of the samples was made by freeze-drying and cellulose exchangers, respectively [115]. The problems in the determination of trace elements in seawater were discussed [116].
Trace elements analysis of seawater samples requires generally an extensive sample pretreatment for the concentration and separation of this element prior to the determination. Methods for the preconcentration of trace metals were studied by various authors. Thus: A preconcentration method by coprecipitation with thiooxine and 8-8'- diquinolyl-disulfide of Mn, Co, Cd, Cu, Zn and Hg was investigated [117]

Another procedure uses an extraction-chromatographic column, packed with Ftoroplast-17, for the concentration of submicrograms of Zn, Co, Cu and Hg. The metals were eluted from the column and can be used for determination [118]. Copper, nickel, and cobalt were removed by cation exchanger EDE-10p and uranium by AN-2f from seawater [119]. A new selective elution method for the extraction of zinc, cadmium, copper, nickel, cobalt and lead on chelating resin is suitable for the determination of the above elements by flameless atomic absorption spectrometry [120]. Toxic metals [Cd, Co, Cu, Mn, Pb, U and Zn] in concentrations up to 0.1 µg/l were extracted from seawater by an ion-exchange-solvent extraction procedure and then determined by atomic absorption spectrophotometry and for the uranium by fluorometry [121]. Sampling procedures to obtain uncontaminated samples for heavy metals (Cu, Cd, Zn and Ni) at the nanogram per liter level and analytical methods for their determination were described [122]. Ion-exchange membrane methods for the continuous removal of ions from water were studied. The Donnan dialysis methods were reported as superior to selvent extraction, columnal ion-exchange and chemical digestions [123].

Neutron activation. Sampling, activation and detection procedures by instrumental and radiochemical techniques of seawater samples and problems caused by high matrix activation are discussed [124]. A procedure for the pre-concentration of trace elements from seawater using ammonium pyrrolidine-1-carbodithionate and the determination of the elements by neutron activation irradiation was presented [125]. Neutron activation analysis is recommended for seawater samples in hydrophysical and oceanographic studies. The fundamentals and the methods of the neutron-activation element and salt analysis in seawater are presented [126].

Electrochemical methods. A very sensitive electrochemical method was investigated for the determination of cadmium, lead, copper and zinc in natural and waste waters. The sample is pretreated and the metals determined by anodic stripping differential pulse polarography with sensivity better than 10^{-11} M [127]. The stability of ultradiluted standard solutions and stripping coulopotentiography were studied for the determination of the lead and cadmium contents in seawater [128]. The striking and comprehensive scope of voltammetry in trace metal chemistry of aquatic ecosystems is illustrated by examples from drinking water control, investigations on rainwater, extended studies in European coastal waters and the oceans, and by fundamental work on the occurrence and behavior of toxic trace metal species in the sea [129].

Analytical procedures for the determination of zinc, cadmium, lead and copper by potentiometric stripping analysis were reported, and the results were compared with those obtained by a combined solvent extraction-atomic absorption method in laboratory and on board ship [130]. Copper, cadmium, zinc and lead were determined by anodic stripping voltammetry using preelectrolysis on a Pt electrode at -1.7 V followed by pulse polarography [131]. A simple rapid and inexpensive anodic-stripping voltammetric procedure with mercury film electrode is reported for selected natural waters. Results were well correlated with corresponding analyses performed by graphite-furnace atomic absorption sepectrometry [132]. Lead, copper and cadmium were determined in river and coastal waters by anodic stripping voltammetry. Detection limits with standard additions were 0.9 µg/l Pb, 0.3 µg/l Cd and 0.2 µg/l Cu [133]. Seawater and plankton were analysed for cadmium, lead, copper and zinc content by anodic stripping voltammetry. Cadmium concentrations were found independent of pH, lead and copper increased with increasing pH and zinc decreased by decreasing the pH [134].

Differential pulse anodic stripping voltammetry with a hanging mercury drop electrode was used for the direct and simultaneous determination of Zn, Cd, Pb, Cu, Sb and Bi at their natural levels in seawater after adjustment to pH 1 and to a 2 M chloride concentration. Relative standard deviations are about 10 to 15% [135]. By using anodic stripping voltammetry at controlled pH and after pretreatment of the sample with Chelex-100, the trace metals analysis was studied. Differences in the results of the two procedures are discussed [136]. Valenta et al. used a thin mercury film coated onto a rotating glassy carbon electrode and a method based on differential pulse anodic stripping voltammetry, for the simultaneous determination of the dissolved level of Cd, Pb and Cu in sea, inland, drinking and industrial

water in a limit of 10^{-9} g per liter [137]. The quality of the vitreous-carbon electrode on which the mercury film, thickness 20 to 100 mm, is deposited in situ is very
important in the determination of cadmium, lead and copper in the ppb and sub-pbb
ranges [138]. For the analysis of most toxic trace metals in drinking water, a
device which consists of an adapted commercial polarographic instrumentation, a
punch card program controller and automatic and standard addition facilities were
used [139]. Heavy metals, such as Zn, Cd and Pb are determined in seawater at
natural concentration by means of a new-electro-analysis technique consisting of a
combination of oscillopolarography and differential-pulse-anodic-stripping voltammetry. The determination is performed in a simple and rapid way without any preliminary treatment of the sample [140]. Organic sequestering agents in waste
waters and in seawater were oxidized by ozone, prior to the determination of lead
and cadmium content by anodic stripping voltammetry. This method has advantages to
the previous oxidation procedures by H_2O_2 and persulfate [141].

Patents. An apparatus for electrolysis or end point titration can analyze COD,
Cr^{6+}, total Cr, Mn, residual Cl, NH_4-N, NO_3-N, Se, SO_3^{2-} Sn, As, Sb, S^{2-}, CN^- and
Cl^- [142].

Spectrometry. A variety of clean sampling techniques and a dithiocarbamate extraction method coupled with atomic absorption spectrometry for the determination
of copper, cadmium, zinc and nickel in nanogram per liter concentration levels were
presented. Comparison with other extraction methods is discussed [143]. A precude
was developed for the flame atomic absorption determination of copper, lead and zinc
in mineralized waters, using preconcentration by evaporation to a dry residue. Relative deviation is less than 0.08% [144]. A dithizone extraction procedure was used
prior to flame atomic absorption determination of Cd, Zn, Cu, Ni, Co, Pb and Ag in
seawater samples. The method is sufficiently sensitive, except for lead that can be
determined only in polluted waters by this procedure [145].
Ten trace metals (V, Cr, Fe, Co, Cu, Zn, Mo, Cd and Pb) were determined in effluents and natural waters by flame atomic absorption spectrometry. Previously the
trace metals were simultaneously extracted by 2,6-dimethyl-4-heptanone. [146]. Procedures based on flameless atomic absorption spectrometry are reported for the determination of cadmium, cobalt, chromium, copper, nickel and lead in mineral waters.
The metals are separated from macrocomponents by precipitation as tetramethylene-
dithiocarbamates with Fe(III) as collector [147] . A modified standard addition
procedure was used for the determination of cadmium, lead, copper and iron in seawater by flameless atomic absorption spectroscopy. The concentrations of the above
metals were accurately determined from the sodium concentration, the sample absor-
bance versus a pure standard and the appropriate curve [148]. Elemental arsenic,
lead, cadmium,chromium and selenium, in the limits of drinking water regulations
were determined by electrodeless discharge lamps used in atomic absorption spectroscopy, which permits the determination of this element in the ppb-level without
enrichment [149].
By a careful selection of instrumental conditions it is possible to determine sub-
nanogram levels of camium, copper, cobalt, iron, manganese, nickel, lead in seawater by graphite furnace atomic absorption spectrometry. The metals are preconcentrated by a Chelex 100 resin column and separated of alkali and alkaline earth
metals by amnonium acetate elution [150]. Concentration of Cu^+, Cd^+ and Pb^+ in seawater was determined by using isotope dilution-surface ionization mass spectrometry
by means of ^{65}Cu, ^{116}Cd and ^{206}Pb [151]. Trace heavy metals (cadmium, chromium,
copper, lead,nickel and zinc) were determined in seawater by the Spectraspan dc
plasma emission spectrometer. Optimum analytical emission lines and detection limits are reported [152].

Various methods. Characteristics of gas chromatographic methods for the determination of organics, pollutants and some heavy metals in seawater are reported. The
methods are applied during the study of seawater self-purification [153]. The use
of energy-dispersive multielement X-ray analysis is described and the radionuclides
^{109}Cd and ^{241}Am X-ray tubes combined with secondary targets are compared as X-ray
sources. The water samples are treated by means of a column or a filter with chelating groups bound to cellulose. A thin layer of the cellulose powder or the filter

is analyzed with the aid of calibration graphs. Results of multielement trace analysis of standardized water, river water and seawater are presented and discussed [154]. Ion chromatography is evaluated as a technique for the separation and determination of selected anions (fluoride, chloride, bromide, and sulfate) and cations (lithium, sodium, potassium and ammonium) in geothermal well water samples. The results achieved are compared with those obtained by other methods of analysis in a round-robin test [155].

Uranium. Uranium was recovered by absorbing colloid flotation from synthetic seawater. Hydrated TiO_2 was used as collector. The mean recovery of uranium from 60 µg/l U samples containing 0 to 60 µg/l U was 91% [156]. Separation methods were investigated for separation and concentration of uranium from natural, waste, industrial and seawater. Hyphan cellulose exchanger and hydrated titanium oxide were used as sorbents [157]. Trace amounts of uranium were preconcentrated from seawater samples by Chelex-100 chelating resin and determined by neutron activation analysis [158]. Silica gel column for the concentration and separation of uranium was used. The determination was made fluorimetrically with an error less than 10% for 0.1 to 10 µg/l uranium [159]. Fluorimetry was used also after isolation of uranium by extraction with tri-n-octylphosphine oxide in Varsol. Lower content limit is 0.20 µg/l [160]. The use of selected resins provide a very high and selective recovery of uranium and allows the analysis of small volumes of seawater fluorimetrically with improved accuracy [161]. Rapid determinations of uranium are reported by spectrophotometry using Arsenazo III at 665 nm and precipitation with Al-phosphate [162] and by using a custom-built thermal-emission mass spectrometer [163]. A new method for the analysis of uranium is based on preconcentration of uranium on activated carbon, irradiation with epithermal neutrons, followed by a high resolution-γ-spectrometry of ^{239}Np [164].

Patents. Uranium is determined in natural waters and particularly in seawater by adsorption in silicagel and elution with onidizing mineral acid. The deviation at uranium concentration 1000, 100, 10, 1 and 0.1 µg/l is reported as 0,2, 0.1, 2.0, 7.0 and 10.0%, respectively [165].

Zinc. An ion-exchange colorimetric method using Dowex 1-X2 has a sensivity of $1.6 \times 10^{-6}M$. Most metals, except Cd^{2+}, Pb^2 and Bi^{3+} do not interfere when present at less than 100 times the zinc concentration [166].

LITERATURE TO 1C 3.

1. F.E. Hermann (Deep-Sea-Res. 25 [1978] 1265/1270)
2. H.L. Golterman, Methods for Physical and Chemical Analysis of Fresh Waters (Blackwell's Scientific, Oxford, [1978] 240 p)
3. S.G. Oradovskii, A.K. Prokofev et al., Manual on Methods of Chemical Analysis of Seawaters. (Gidrometeoizidat, Leningrad [1977] 108 p)
4. T.W. Stanley (NTIS Rept. PB-287118/4GA [1978] 99 p.-G.R.A. 78, No 3 [1978] 116)
5. W.E. Rich, J.A. Tillotson, R.C. Chang, in Ion Chromatogr. Anal. Environ. Pollut. [1978] 185/186 (E. Sawicki, J. Mulic, E. Wittgenstein, Editors, Ann Arbor Sci)
6. P. Baeri, E. Rimini, O. Puglisi, S. Pignataro (Chim. Ind., Milano, 60, No 6 [1978] 510/515)
7. G. Lapicque (Rev. Int. Oceanogr. Med. 49 [1978] 117/126)
8. M.S. Shuman, W.W. Fogleman (J. Water Pollut. Control Fed. 50, No 6 [1978] 1000/1021)
9. A. Nevissi, W.R. Schell (Ecol. Soc. Am. Spec. Publ. 1 [1976] 277/281)
10. H. Hirayama (Bunseki Kagaku 27, No 4 [1978] 252/255)
11. A.V. Gordievskii, E.A. Zeinalova (Probl. Anal. Khimii No 5 [1977] 135/152)
12. R.C. Thurnau (NTIS Rept. PB-285 724/1GA, [1978] 44 p.- G.R.A. 18, No 26 [1978] 96)
13. N.G. Martynova, A.Z. Khlystov, M.l. Khorsheva, V.M. Shashkov (Morsk. Gidrof. Issled. No 2 [1977] 178/183.- C.A. 89 [1978] 117425)
14. C.S. Wong (Natl. Bur. Stand., Spec. Publ. No. 464 [1977] 249/258)

36

15. V.L. Pavelko, M.N. Tarasov, I.M. Pavelko (Gidrokhim. Mater. 65 [1977] 79/88.-
    C.A. 89 [1978] 48685)
16. D.E. Erdmann, H.E. Taylor (Anal. Chim. Acta, 99, No 2 [1978] 269/271)
17. R.S. Keir (Mar. Chem. 8 [1979] 95/97)
18. A.A. Diggens (U.S. 4.131.428, 26 Dec 1978.- C.A. 90 [1979] 97077)
19. E.D. France (Brit. 1.501644, 28 Feb 1978, Addition to Brit. 1.480.728.-C.A. 89
    [1978] 94862)
20. H.M. Parr (Brit. 1.536.681, 20 Dec. 1978.- C.A. 90 [1979] 192344)
21. K. Morisige, K. Hiraki, V. Nishikawa, T. Shigematsu (Bunseki Kagaku 27, No 2
    [1978] 109/114.- C.A. 79 [1978] 152375)
22. F. Jentsch (Vom Wasser 48 [1977] 111/116)
23. P. Le Corre, P. Treguer (J. Cons. Int. Explor. Mer, 38, No 2 [1978] 147/153)
24. H. Verdouw, C.J.A. Van Echteld, E.M. Dekkers (Water Res. 12, No 6 [1978] 399/
    402)
25. A.S. Romanov, A.I. Ryabinin, L.B. Zhidkova (Gidrokhim. Mater. 70 [1977] 43/46.-
    C.A. 89 [1979] 94816)
26. S.R. Goode, R.J. Matthews (Anal. Chem. 51, NO 7 [1978] 1608/1610)
27. M. Murozumi, H. Mitobe, S. Nakamura, H. Tsubota (Bunseki Kagaku, 27, No 4
    [1978] 218/223.- C.A. 89 [1978] 135509)
28. M.S. Epstein, A.T. Zander (Anal. Chem. 51, No 7 [1979] 915/918)
29. A. Sato, N. Saitoh (Bunseki Kagaku 26 [1977] 747/751.- C.A. 88 [1978] 65795)
30. H. Kitomura, K. Okawa, Y. Kuge, S. Asada (Kogai to Taisaku 14, No 8 [1978]
    877/884.- C.A. 90 [1979] 109682)
31. S.P. Kounaves, A. Zirino (Anal. Chim. Acta 109 [1979] 327/329)
32. K.R. Sperling (Fresenius Z. Anal. Chem. 292 [1978] 113/119)
33. C.J. Rowe, M.W. Routh (Res/Dev. 28, No 11[1977] 24/30)
34. O.F. Schaefer (W. Ger. 2.700.567, 13 Jul 1978.-C.A. 89 [1978] 111795)
35. A.B. Mackenzie, M.S. Baxter, I.G. McKinley, D.S. Swan, W. Jack (J. Radioanal.
    Chem. 48, No 1 [1979] 29/47)
36. N.V. Krylov, V.G. Pitalev, A.V. Stepanov (Radiokhimiya 20, No 5 [1978] 737/741.-
    Soviet Radiochemistry, 20, No 5 [1979] 633/637)
37. R.M. Gershey, M.D. Mackinnon, P.J. le B. Williams, R.M. Moore (Mar. Chem. 7
    [1979] 289/306)
38. B.A. Chepenko, P.D. Novikov, V.P. Komlev (Radiokhimiya 19, No 1 [1977] 52/55.-
    Sov. Radioch. 19,No. 1 [1977] 42/45)
39. T. Shigematsu, S. Gohda, H. Yamazaki, Y. Nishikawa (Bull. Inst. Chem. Res.,
    Kyoto Univ., 55, No 5 [1977] 429/440)
40. G.J. de Jong, V.A.T. Brinkman (Anal. Chim. Acta 98 [1978] 243/250)
41. K. Yoshimuta, S. Ohashi (Talanta 25, No 2 [1978] 103/107)
42. T. Kiriyama, R. Kuroda (Fresenius' Z. Anal. Chem. 288 [1977] 354/356)
43. C.L. Tseng, J.M. Lo (Radiochem. Radioanal. Lett. 33, No 5/6 [1978] 315/321)
44. L.A. Kozlova (Metody Anal. Kontrolya Kach. Prod. Khim. Prom-sti No 10 [1978]
    26/27.- C.A. 91 [1979] 27039)
45. D.J. Churella, T.R. Copeland (Anal. Chem. 50, No 2 [1978] 309/314)
46. M.S. Shuman, L.C. Michael (Anal. Chem. 50, No 14 [1978] 2104/2108)
47. H.R. DuBois, G.M. Sharma (Anal. Chem. 51, No 11 [1979] 1072/1075)
48. A. Hahne (KFA Juelich, Inst. Atmosphaerische Chemie. Report No Juel-1444 [1977]
    58 p)
49. J. Navarro, A. Ballester, J. Calmet (Cron. Chem. 55 [1978] 7/11, 31.-C.A. 90
    [1979] 12049)
50. G. Eklund, B. Josefsson, C. Roos (Chromatogr. Newsl., 6, No 3 [1978] 39/41.-
    C.A. 90 [1979] 12050)
51. A.M. Plyusnin, Yu. F. Pogrebnyak, E.M. Tat'ynkina (Zh. Anal. Khim. 34, No 2
    [1979] 402/405)
52. V.S. Ogienko (Brom Solyanykh Otlozh. Rassolakh. Geokhim. Indik. Ikh Genezisa,
    Istor. Poisk. Priznak. [1976] 427/435.- C.A. 91 [1979] 7780)
53. E.R. Wright, R.A. Smith, B.G. Messik (Chem. Anal. vol 8 (Colorimetric Determin-
    ation Nonmet. [1978] 39/56.- C.A. 90 [1979] 66007)
54. A. Peron, J. Courtot-Coupez (Analysis 6, No 9 [1978] 389/394)
55. D.S. Stasinevich, A.N. Usatov (Zh. Anal. Khim. 33, No 6 [1978] 1170/1114.-
    J. Anal. Chem. USSR., 33, No 6 [1978] 909/912)

56. V.V. Bardin, A.L. Bystritskii, V.N. Tolstousov, O.F. Shartukov (Zh. Anal. Khim. 32, No 9 [1977] 1760/1766.- J. Anal. Chem. USSR, 32, No 9 [1978] 1394/1398)
57. E.M. Filippov (Morsk. Gidrofiz. Issled. No 2 [1977] 120/127.- C.A. 89 [1978] 117424)
58. A.P. Tsurikova, M.P. Nesterova, A.F. Litvinova (Khim.-Okeanol. Issled., Mater. Vses. Konf. Khim. Morei Okeanov 7th, 1975 [1977] 195/199.- C.A. 89 [1978] No 65009)
59. A.E. Kosov, O.T. Krylov, P.D. Novikov (Deposited Doc. VINITI 4477-76 [1976] 7 p.- C.A. 90 [1979] 109699)
60. M. Yoshida, Y. Makihara, T. Katsura (Nippon Kagaku Kaishi No 10 [1978] 1375/ 1379.- C.A. 89 [1978] 208604)
61. V.W. Truesdale, C.J. Smith (Mar. Chem. 7 [1979] 133/139)
62. H. Kanbe, A. Nakaoka, S. Takagi (Denryoku Chuo Kenkyusho Hokoku No 276030 [1977] 18 p.- C.A. 88 [1978] 176892)
63. G.M. Sharma, H.R. DuBois (Anal. Chem. 50 [1978] 516/521)
64. L.M. Petrie (Thesis Duke Univ., Durham N.C. [1977] 187 p.- Univ. Microfilms Int., Order No 7807626)
65. L.M. Petrie, R.W. Baier (Anal. Chem. 50 [1978] 351/357)
66. Y. Hirao, K. Fukumoto, H. Sugisak, K. Kimura (Anal. Chem. 51 [1979] 651/653)
67. A.L. Soldan, N. Saitoh (Microchim. Acta, 1 [1977] 167/171)
68. G.B. Pasovskaya (Izv. Vyssh. Uchebn. Zaved., Khim. Tekhnol., 21 No 5 [1978] 762.- C.A. 89 [1978] 152404)
69. A.E. Kosov, O.T. Krylov, P.D. Novikov, V.M. Leont'ev (Deposited Doc. VINITI 4478-76 [1976] 9 p.- C.A. 90 [1979] 127293)
70. J. Wijkstra, H.A. Van der Sloot (J. Radioanal. Chem. 46 [1978] 379/388)
71. T.R. Gilber (Anal. Lett. 10 [1977] 599/617)
72. H. Egawa, S. Tajima (U.S. Environ. Prot. Agency, Off. Res. Dev. EPA 600/3-77-083 [1977])
73. G. Dal'Pont, G.A. Major (Commonw. Scient. Ind. Res. Orgn., Div. Fish. Oceanogr. Rept., Sydney, 83 [1978] 32 p)
74. Q. Wanying, J. Zhou, J. Li, Y. Yao, L. Qiu (Oceanol. Limnol. Sin., 9, No 1 [1978] 36/42)
75. Y. Yamamoto, T. Kumamaru, A. Shiraki (Fresenius Z. Anal. Chem., 292, No 4 [1978] 273/277)
76. G.E. Millward, A. Le Bihan (Water Res. 12 [1978] 979/984)
77. K. Matsunaga, S. Konishi, M. Nishimura (Environ. Sci. Technol., 13, No 1[1979] 63/65)
78. H. Ageamin, J.A. Da Silva (Anal. Chim. Acta 104 [1979] 285/291)
79. I.N. Davies, W.C. Graham, J.M. Pirie (Mar. Chem., 7 [1979] 111/116)
80. G. Hanson, A. Szabo, N.D. Chasteen (Papers 173rd ACS National Meeting, New Orleans, [1977] Abstract No ANAL 15)
81. T. Nakahara, C.L. Chakrabati (Anal. Chim. Acta 104 [1979] 99/111)
82. N. Ohta, M. Fujita, K. Tomura (Bunseki Kagaku 28 [1979] 277/280)
83. H. Fushiwaki, K. Tanaka (Kanagawa-Ken Kogai Senta Nempo 1976, 9 [1977] 92/93.- C.A. 89 [1978] 203807)
84. L. Brown, E.G. Billinger (Water Res. 12 [1978] 223/229)
85. W. Schreurs (Hydrobiol. Bull. 12 [1978] 137/142)
86. M.D. Mackinnon (Mar. Chem. 7, No 1 [1978] 17/37)
87. A.D. Semenov, V.G. Soier (Gidrokhim. Mater., 71 [1978] 94/98.- C.A. 91 [1979] 9289)
88. S.G. Mel'kanovitskaya (Probl. Anal. Khim. 5 [1977] 176/178.- C.A. 89 [1978] 94744)
89. M. Imanaka, H. Katayama, K. Matsunaga, T. Ishida (Okayama-ken Hankyo Hoken Senta Nempo 1 [1977] 211/212.- C.A. 89 [1978] 94818)
90. J.B. Derenbach, M. Ehrhardt, C. Osterroht (Mar. Chem., 6 [1978] 351/364)
91. G. Eklund, B. Josefsson, C. Roos (J. High Resolut. Chromatogr. Chromatogr. Commun., 1, No 1 [1978] 34/40)
92. T. Okuno, M. Tsuji, T. Yamasaki, Y. Shintani (Hyogoken Kogai Kenkyusho Kenkyu Hokoku 10 [1978] 8/14.- C.A. 90 [1979] 209832)
93. G.Wennergren (Inst. Vatten Luftvardsforsk., Stockholm., Publ. B 434 [1978] 17 p)

94. R. Bonevski (Arh. Hig. Rada Toksikol., 28 [1977] 351/360.- C.A. 89 [1978] 152383)

95. S.R. Carlberg (Ambio Spec. Rep. 5 [1977] 269/277)

96. K. Hiiro, T. Tanaka, A. Kawahara, (Jap. 78.123.995, 28 Oct 1978.- C.A. 90 [1979] 61035)

97. A.Z. Khlystov, N.G. Martynova, M.I. Khorsheva (Avtom. Nauchn. Issled. Khim., Khim. Tekhnol. Mater. Vses. Shk., 9th [1978] 180/185.- C.A. 89 [1978] 168798)

98. A.Z. Khlystov, N.G. Martynova M.I. Khorsheva, V.M. Shashkov (Eleckrokhimiya 15 [1979] 605.- C.A. 90 [1979] 209855)

99. K. Sugino, K. Obata (Nippon Kaisui Gakkaishi 32 [1978] 34/40.- C.A. 89 [1978] 152422)

100. I. Kudo (Jap. 78.15.195, 10 Feb 1978.- C.A. 89 [1978] 135620)

101. A.V. Gordievskii, Yu. I. Urusov, A. Ya. Syrchenkov, A.E. Kosov, A.F. Zhukov (USSR 647.594, 15 Feb 1979.- C.A. 90 [1979] 161698)

102. L.U. Joshi, A.K. Ganguly (J. Radioanal. Chem. 41 [1977] 15/21)

103. P. Li, J. Li (Oceanol. Limnol. Sin. 9, No 1 [1978] 43/48.- Ocean. Lit. Rev.26, No 3B [1979] No 851)

104. A.D. Zemlyanov, G.F. Batrakov, V.A. Anfinogentova (Eksp. Metody Issled. Okeana [1978] 59/65.- C.A. 90 [1979] 192299)

105. R.M. Key, R.L. Brewer, J.H. Stockwell, N.L. Guinasso, D.R. Schink (Mar. Chem. 7 [1979] 251/264)

106. A. Kolodziejski, Z. Lukasik (Chemoautomatyka, No 1 [1979] 21/22.- C.A. 91 [1979] 44306)

107. S. Tsonkova, I. Kulev, A. Dyulgerova (God. Vissh. Khim.- Teknol. Inst. Burgas, Bulg. 12, Ptl [1978] 79/85.- C.A. 90 [1979] 174406)

108. S. Onuki, K. Watanuki (Kagaku Kyoiku 25 [1977] 448/454.- C.A. 89 [1978] 93004)

109. V.B. Chumichev, V.N. Shkuro (Tr.- Inst. Eksp. Meteorol., Ser. "Zagryaz. Prir. Sred." 6 [1977] 121/129.- C.A. 89 [1978] 152399)

110. N.V. Egorova, V.N. Krylov, A.V. Stepanov (Radiokhimiya 20 [1978] 742/745.- C.A. 90 [1979] 43619)

111. R.W. Howarth (Limnol. Oceanogr. 23 [1978] 1066/1069)

112. R.J. Rosenbauer, J.L. Bischoff (Limnol. Oceanog. 24 [1979] 393/396)

113. M.B. Kloster, M.P. King (J. Am. Water Works Assoc. 69 [1977] 544/546)

114. R.D. Guy, C.L. Chakrabarti (Proc. 1st Int. Conf. Heavy Met. Environ. 1 [1975] 275/294.- C.A. 89 [1978] 94825)

115. P. Burba, K.H. Lieser, V. Neitzert, H.M. Roeber (Fresenius Z. Anal. Chem. 291 [1978] 273/277)

116. T. Shigematsu (Nippon Kaisui Gakkaishi 32 [1978] 150/157.- C.A. 91 [1979] 9184)

117. M. Vircavs (Tezisy Dokl.- Konf. Molodykh Nauchn. Rab. Inst. Neorg. Khim. Akad. Nauk. Latv. SSR., 6th [1977] 10/11.- C.A. 90 [1979] 33407)

118. V. Rone (Tezisy Dokl.- Konf. Molodykh Nauchn. Rab. Inst. Neorg. Khim., Akad. Nauk. Latv. SSR., 6th [1977] 12/13 C.A. 90 [1979] 33408)

119. P.D. Novikos, O.T. Krylov, G.I. Sychkova, B.A. Chepenko (Khim.- Okeanol. Issled [Mater. Vses. Konf. Khim. Morei Okeanov], 7th [1975] B.A. Skopintsev, V.N. Ivanenkov, Editors, "Nauka", Moscow, 1977 52/56.- C.A. 89 [1978] 117430)

120. J. Lamathe (C.R. Hebd. Seances Akad. Sci., Ser. C 286 [1978] 393/396.- Anal. Chim. Acta, 104 [1979] 307/317)

121. J. Korkisch (Pure Appl. Chem. 50 [1978] 371/374)

123. J.A. Cox (NTIS Rept. PB-288918/6GA [1978].- G.R.A. 79, No 7 [1979] 93)

124. L. Greim, H.H. Schreier (GKSS Rept. 77/4/47 [1977] 42 p)

125. Y. Kusaka, H. Tsuji, Y. Tamari, T. Sagawa, S. Ohmori, S. Imai, T. Ozaki (J. Radioanal. Chem. 37 [1977] 917/926)

126. E.M. Filippov, I.A. Lamanova (Morsk. Gidrofiz. Issled. No 1 [1978] 98/110.- C.A. 90 [1979] 156844)

127. M. Branica, J. Eder-Trifunovic, M. Jurkovic, S. Kozar, T. Magjer, K. Voloder (Prehrambeno-Technol. Rev. 15, No 3 [1977] 96/97)

128. T. Fujinaga (Kagaku Zokan, Kyoto, No 78 [1978] 11/20.- C.A. 90 [1979] 156880)

129. H.W. Nuernberg (Chem. Ing. Tech. 51 [1979] 717/728)

130. D. Jagner, K. Aren (Anal. Chim. Acta 107 [1979] 29/35)

131. R. Kantin (Tethys 1975, 7, No 4 [1977] 419/425)

132. J.E. Poldoski, G.E. Glass (Anal. Chim. Acta 101, No 1 [1978] 79/88)

133. T.C. Hung et al. (Bull. Inst. Chem., Acad. Sci., Taiwan, 25 [1978] 35)
134. G. Duyckaerts, G. Gillain (Essays Anal. Chem., E. Erkki, Editor, Pergamon Pren, Oxford [1977] 417)
135. G. Gillain, G. Duyckaerts (Anal. Chim. Acta 106 [1979] 23/37)
136. D.D. Nygaard (Anal. Lett. 12, A5 [1979] 491/499)
137. P. Valenta, L. Mart, H.W. Nuernberg, M. Stoeppler (Vom Wasser 48 [1977] 89/110)
138. P. Valenta, L. Mart, H. Ruetzel (J. Electroanal. Chem. 82 [1977] 327/343)
139. P. Valenta, H. Ruetzel, P. Krumpen, K.H. Salgert, P. Klahre (Fresenius Z. Anal. Chem. 292 [1978] 120/125)
140. L. Grifone, G. Macchi (Ann. Chim., Rome, 68 [1978] 227/233)
141. R.G. Clem, A.T. Hodgson (Anal. Chem. 50, No 1 [1978] 102/110)
142. Uematsu, S. Miyake (Jap. 77.85.886, 16 Jul 1977,- C.A. 88 [1978]65836)
143. K. Bruland, R.P. Franks (Anal. Chim. Acta 105 [1979] 233/245)
144. S.V. Vratkovskaya, Yu. F. Pogrebnyak (Zh. Anal. Khim. 34 [1979] 759/763.- C.A. 91 [1979] 96382)
145. H. Armannsson (Anal. Chim. Acta 110 [1979] 21/28)
146. K.M. Bone, W.D. Hibbert (Anal. Chim. Acta 107 [1979] 219/229)
147. V. Hdnik, S. Gomiscek, B. Gorenc (Anal. Chim. Acta 98 [1978] 39/46)
148. C.P. Weisel, J.L. Fasching, S.R. Piotrowicz, R.A. Duce (Adv. Chem. Ser. 172 [1979] 134/145)
149. E. Sofzik (Vom Wasser, 50 [1978] 285/299)
150. H.M. Kingston, I.L. Barnes, T.J. Brady, T.C. Rains, M.A. Champ (Anal. Chem. 50 [1978] 2064/2070)
151. M. Murozumi, S. Nakamura, T. Igarash, H. Tsubota (Nippon Kagaku Kaishi No 4 [1978] 565/570)
152. D.D. Nygaard (Anal. Chem. 51 [1979] 881/884)
153. S.G. Oradovskii (Ambio Spec. Rep. 5 [1977] 287/289.- C.A. 89 [1978] 135533)
154. K.H. Lieger, E. Breitwieser, P. Burba, M. Roeber, R. Spatz (Microch. Acta 1 [1978] 363/373)
155. R.P. Lash, C.J. Hill (Anal. Chim. Acta, 108 [1979] 405/409) .
156. W.J. Williams, A.H. Gillam (Analyst, London, 103 [1978] 1239/1243)
157. F. Ambe, P. Burba, K.H. Lieser (Fresenius Z. Anal. Chem. 295, No 1 [1979] 13/16)
158. A. Hirose, D. Ishii (J. Radioanal. Chem. 46 [1978] 211/215)
159. A. Putral, K. Schwochau (Fresenius' Z. Anal. Chem. 291 [1978] 210/212)
160. R.J. McElhaney, J.D. Caylor, S.H. Cole, T.L. Futrell, V.M. Giles (ORNL Report Y-2111 [1978] 16 p.- ERA, 3 [1978] 33493)
161. U. Croatto, D. Sandro, L. Baracco (Ann. Chim., Rome, 68 [1978] 659/669)
162. K.H. Reinhardt, H.J. Mueller (Fresenius Z. Anal. Chem. 292 [1978] 359/361)
163. J.R. Ferguson, J.D. Caylor, E.R. Rogers, S.H. Cole (Inst. Res., Houston, Rep. Y-2073, [1977] 14 p., INIS Atomindex 10 [1979] 432201)
164. I. Kuleff, K.N. Kostadinov (J. Radioanal. Chem. 46 [1978] 365/371)
165. A. Putral, K. Schwochau (Ger. 2.720.867, 5 Oct 1978.- C.A. 90 [1979] 92213)
166. K. Yoshimura, H. Waki, S. Ohashi (Talanta, 25 [1978] 579/583)

2A. Distillation

A critical review on new developments in desalination by distillation processes, with the multistage flash evaporation process as the reference, was presented by Veenman. These developments refer to vertical tube foam evaporator, vapor compression evaporator, horizontal aluminum tube multiple effect evaporator, falling film multi-stage evaporator, direct contact condensation, multi-effect stack evaporator, multi-stage controlled flash evaporator and multi-stage flash fluidized bed evaporator. Significant factors for evaluation, such as investment costs, energy costs, and operating and maintenance costs, are summarized in three tables[1].

Other work and reviews include production of fresh water from seawater by multiple effect distillation, flash evaporation and vapor compression [2], Soviet steady-state thermal distillation units and some prospects for their development [3], methods for calculating optimum technological parameters of an automated desalting apparatus [4], selecting and designing evaporators [5] and improvement of desalination of mineralized water and purification of petroleum-containing water [6].

The future of distillation is intimately related to the availability and cost of energy and capital. Capital cost will have the most significant impact. Hence, research will be directed to reliable technology, which can utilize the least energy at the lowest possible level [7]. In a review of the history of the development of desalination processes, with emphasis on multi-stage-flash distillation, Silver, a pioneer in this development, expresses his views on the state of development, as well as on the future and the way forward for the industry [8].

Patents. The surface of seawater in a tank is heated by IR irradiation and evaporates. *The body of the seawater remains unheated [9]. Netting formed of monofilament polyamide fish net curtains is used to form thin films of the feed solution in an evaporator, separated by a thin dividing wall from the condenser [10]. A solution concentration apparatus uses mist evaporation under partial vacuum and the latent heat of condensation of a refrigerant [11]. A portable apparatus for distillation of seawater consists of an evaporation chamber, an air heater and a plate air condenser [12]. Seawater is sprayed through a nozzle on a heated metal plate and the resulting steam is condensed to obtain water [13]. An apparatus for evaporation and concentration consists of an evaporation vessel, in which the solution is naturally circulated by the density difference, and a vapor condensation vessel, provided with a liquid reservoir [14].

A multichannelled evaporator, with metal partitions separating the channels, is heated by a heating agent. Metal scrolls are placed inside the evaporator channels to increase the evaporation surface [15]. Vapor formed in an electrically heated tank is drawn by a fan to a condenser, mounted over the top of the tank [16]. A simple, single stage still with an electric heater was also described [17]. Feed solution is preheated in an conical chamber and flows to a cylindrical chamber where it is further heated until evaporation [18]. The still uses a sensor on the distillate temperature for automatic control of makeup water to the still. A flow-through or constant-flush drainage system is provided to keep mineral buildup to a minimum [19]. A vacuum is maintained in the evaporator and the condenser is cooled by a refrigeration system [20]. An electrically heated apparatus for distillation of seawater is attached to a sea wall and has operational components mounted on a float, which rises and falls with the tide [21].

A high-pressure and high-temperature liquefying coolant is vaporized in a heat-exchanger tube to prepare a gasified coolant. Hot water from the heat-exchanger is circulated between seawater and the heat-exchanger. The vapor from the seawater evaporation is condensed to softened water with the gasified coolant. The method improves thermal efficiency and requires a small apparatus [22]. A portion of the make-up seawater is withdrawn from the outlet of a seawater preheater and introduced into an evaporator having approximately the same temperature as the make-up seawater. The efficiency is increased [23].

HEAT TRANSFER

In heat transfer during evaporation of a water drop on smooth surfaces, the time-averaged heat flux and surface temperature were obtained by time-dependent measurements of the surface temperature below the drop and by the drop shape. The heat transfer was analyzed by correlating the heat flux with the difference between the time-averaged surface temperature and the saturation temperature [24, 25]. The heat-transfer rate increases and the heat-transfer coefficient decreases, as the liquid-to-gas mass transfer rate increases in a flowing gas-liquid froth. In the case of two-phase heat transfer with no mass transfer, the air introduction into the liquid increased the heat-transfer coefficients substantially and a general correlation based on dimensional analysis was proposed. In the case of two-phase heat transfer with mass transfer, the heat transfer coefficient is a function of the liquid and gas mass flux densities, the vapor pressure of the liquid and the total pressure of the system [26].

Viscous flow in the vicinity of the interline (junction of evaporating liquid film, vapor and adsorbed film on solid) significantly affects the upstream profile of a falling evaporating thin film. An evaluation of this change as a function of heat flux was made [27].

The effects of the contamination of outer surfaces of shaped heat exchanger tubes on the operating indexes of distillation units were discussed [28].

Using equations for heat transfer in a split flow heat exchanger with four tube passes, curves for logarithmic mean temperature difference correction factor and temperature efficiency, heretofore unavailable in the literature, were presented [29].

CONDENSATION

A review was presented on heat transfer in filwise condensation [30]. The governing equations for the deposition and entrainment of droplets at vapor-liquid interface during condensation inside tubes can be solved by computer aided methods to yield satisfactory equations for design predictions [31]. The mode of condensation, the nature of the vapor to be condensed and the geometry in which the condensation occurs have been reviewed [32]. A study has been made of heat transfer during vapor condensation inside vertical channels [33]. Through application of an approximate method, a closed description of the process of intensive evaporation and condensation is obtained. On the basis of experimental results, simple interpolation formulae are suggested [34].

DROPWISE CONDENSATION

The overall heat transfer is increased by a factor of 1.4 to 1.5 on Cu-Ni condenser tubes by making the outer surfaces hydrophobic. The steam condenses as drops and the heat transfer coefficient is five times that with film condensation[35]. The microscopic mechanism of dropwise condensation, results of heat transfer measurements and applications were reviewed [36]. A population balance equation was derived for dropwise condensation, which considers both the drop growth because of direct condensation and the coalescence between drops. Drop size distribution and heat flux of dropwise condensation depend strongly on the concentration of active nucleation sites on the substrate surface [37]. The effects of the solid material properties and the droplet contact angle were analyzed by solving the steady heat-conduction equation for a geometry consisting of a spherical segment droplet on a semi-infinite solid. With suitable boundary conditions at the liquid-vapor interface, some exact and approximate solutions were found and then used to obtain the overall heat flow through the droplet [38].

General dependence between the heat transfer coefficient and mean factors governing dropwise condensation were obtained, as well as a formula describing the heat transfer coefficient [39]. An experimental apparatus was constructed for determining the dropwise condensation heat transfer coefficient. Experimental data were taken from a stainless steel condensing surface [40].

The early removal of the sliding rivulets in condensing steam on a vertical
surface and breaking the rivulets into tiny fragments enhanced the heat transfer[41].

CORRUGATED AND FINNED TUBES

Boiling heat transfer on a fin in water was dependent on the horizontal or vertical position and temperature of the heat-transfer surface. The maximum boiling
increased when it was close to the transition boiling [42]. Heat transfer data
were collected on groups of horizontal corrugated tubes arranged in vertical rows
simulating tube-bundle sections. The use of heat transfer and pressure drop correlations obtained were illustrated in design examples, which indicate the advantages
of corrugated tubes [43]. Heat transfer data for in-line and staggered tube banks
indicate a great difference in performance between the two layouts. Tha data from
the in-line layouts show a strong dependence on the tube and layout geometries,
which has not been observed for staggered layouts [44].

Average heat transfer coefficient for internally finned copper tubes is 2 to 3
times as great as that for similar smooth surface tubes for a given heat flux [45].

A parametric analysis of the performance of internally finned tubes in turbulent forced convection is based on empirical heat transfer and friction correlation,
to determine the performance benefits and the preferred geometrical parameters which
allow use of minimum tube material. The best axial internal fins offer less than
10% material savings for equal pumping power and heat duty. However, the material
savings are increased to 49% using internal fins having a 30° helix angle. The
heat transfer coefficient may be increased 35 to 40% for equal pumping power and
total tubing lenght. It is possible to select internal fin geometries which provide performance mominally equal to that given by two-dimensional roughness [46].

The most promising type of profile appears to be the vertically finned tube,
with the fins on the condensing and evaporating side being mounted opposite each
other. Heat transfer measurements in a pilot plant show marked optimum Reynolds
numbers for each evaporation temperature and fin-spacing. It is concluded that
heat transfer within the film is not made by conduction only, but that a considerable amount of heat is transferred by convection. The experimental hydrodynamic
analysis reveals the existence of up to now unknown secondary flows within the
viscocapillary thin-film flow [47].

Patents. A design is described for integral finned tube for submerged boiling
applications having special outer diameter and/or inner diameter enhancement [48].

FLUIDIZED BEDS

Heat transfer data of the horizontal fluidized bed heat exchanger were compared
to the data of the previously reported vertical fluidized bed heat exchanger experiments. The first demonstration of a horizontal configuration of the liquid-
fluidized bed heat exchanger is described and the results indicate that the cost
is comparable to the conventional tube and shell heat exchanger [49]. The dynamic behavior of particulate (emulsion) phase and of bubble (void) phase at the
surface of a tube submerged in a fluidized bed were investigated by measuring the
periods, replacement frequency and time-fraction-occupancy for each phase. Average
residence times for the emulsion and the void phases varied with gas flow rate and
particle size from 0.07 to 1.13 s and from 0.08 to 0.3 s, respectively [50]. Heat
transfer coefficients measured on the surface of plain and finned tubes, located
vertically in fluidized beds, indicate that the heat transfer capability per unit
length of a finned tube with suitable fin geometry can be substantially higher than
that of a plain tube. Increases in capability up to 190% were found with some of
the finned tubes [51]. In an experimental program to determine the heat transfer
coefficients for a tube bank immersed in a fluidized bed, four limestone bed materials, two grid to tube centerline spacings and three bed material inventories
were tested. Overall bed particle size is not important in determining the heat
transfer coefficient [52].

DIRECT CONTACT HEAT TRANSFER

The flowsheet and design parameters of a 50000 gpd pilot plant were presented. The method is also adaptable to the use of solar energy, as it can operate efficiently at a maximum brine temperature of less than 70° [53]. In order to determine average and local heat transfer charateristics between continuous and dispersed phase in direct contact heat exchangers, temperature distributions in and around rising droplets in a downward moving continuous phase were measured with microthermocouples and temperatures were recorded in terms of voltages. Results obtained contradict commonly made assumptions in the theoretical treatment of direct contact heat transfer, for instance, surface temperatures of a drop are far from being uniform [54].

In a two-stage contact heat transfer plant for brine evaporation, the first stage is heated by fuel combustion products and the second by the exhaust gases of the first stage. The optimum number of the evaporative shells of the first stage is 3 to 4, instead of 30 to 40 in the stage with traditional steam heating [55].

Patents. A scheme is described for exploiting the temperature difference between deep and surface layers of the sea. A working liquid, which is immiscible with water, is evaporated by direct contact with the warm seawater and the vapor is condensed by direct contact with the cold seawater [56]. Seawater is heated by direct contact with a flue gas. The heated seawater is distilled in a two-stage unit [57].

Superheated steam is directly contacted to preheat seawater and obtain saturated steam, which are sent to the first stage evaporator [76], or blown into condensed water of the first stage [77]. Steam generated from multiple-effect evaporators is compressed and then contacted with preheated seawater, which is sent to the first stage evaporation unit [78]. Superheated steam is supplied through steam ejectors to units of a multi-effect evaporator. Waste steam is contacted with seawater and recycled to the first stage [79].

BOILING

An expression was obtained for calculating the ideal heat transfer coefficient in boiling of aqueous solutions of substances with limited solubility, when only a few experimentally determined parameters are available [58]. The kinetics of adiabatic evaporation of nonequilibrium superheated water in flash-boiling distillation units were investigated [59].

Postulating that there is a state, where the hydrodynamic condition is responsible for the critical heat flux, a theoretical presumption for the generalized correlation equation of critical heat flux was attempted with the aid of vectorial dimensional analysis. The existence of four characteristic regimes of critical heat flux were revealed [60]. Experimental data were considered, the fundamental similarity numbers were introduced and the integral laws of heat transfer and detachment effect with free convection in liquid pool-boiling were established [61]. A method was developed for calculating transition and film boiling heat transfer from vertical surfaces in inverted annular flow, based on experiments that define the flow geometry and the nature of the boiling curve [62].

Nucleate boiling in thin liquid films requires lower temperature differences for a given heat transfer rate, than nucleate boiling in general. The bursting of bubbles at the outer surface of the film generates fresh nuclei and this is a contribution to the enhanced heat transfer [63]. The surface conditions in nucleate pool boiling were represented by the number and distribution of the active nucleation sites, as well as the size and size distribution of the cavities that constitute the nucleation sites. The heat transfer rate during nucleate boiling was shown to be influenced by the surface condition through its effect on the number and distribution of the active nucleation sites and the frequency of bubble departure from each of these different size cavities, i.e. the bubble flux density [64]. The characteristics of heat transfer during nucleate boiling of an underheated liquid in a large volume at subatmospheric pressures were determined [65].

POLLUTING MATERIALS IN SEAWATER

A review of analyses for pollutants in the sea concludes that much work needs
to be done in sampling and sample preservation, as well as intercalibration between
laboratories [66]. Feed water contaminated with organic matter may cause excessive
corrosion in the boilers [67]. Pollution problems in seawater desalination by distillation were outlined [68].

Patents. Calcium and ammonia are removed by treating seawater with $MgCO_3$ slurry
while aerating, deaerating and evaporating. The obtained distillate contained
0.05 to 0.08 ppm NH_3. The water was further filtered by zeolite to decrease the
residual NH_3 to less than 0.005 ppm [69].

BIOMEDICAL ASPECTS

Studies were made to evaluate the influence of pure distillate or low-mineraliz-
ed distillate on the health of crewmen aboard a ship. Very low mineral content
does not meet hygienic requirements. The distillate could be mixed with port tap
water [70]. Toxicological aspects of scale preventative agents used in production
of drinking water by desalination have been investigated. The investigation includes sulphuric acid, polyphosphates, polyphosphonates, polyacrylates, polymaleic
acid (Belgard EV) and its sodium salt (Belgard EVN) [71].

During the last decade, it was shown that industrial chemicals, in an unexpect-
ed and unintended presence, occur everywhere including food and even the human body.
Therefore, it was internationally agreed that all new products should be subjected
to a test procedure, which includes physico-chemical, ecotoxicological and toxicological question. Evaluation procedures of the use of Belgard EV for feed water
treatment showed that no effective levels of this chemical will be present in organisms and that it is ecotoxicologically safe according to the presently used concepts [72].

STABILIZATION OF DISTILLATE

The stabilization of drinking water [73] and the quality of desalted water,
obtained from industrial distillation plants, were discussed [74].

Patents. Water with low mineral content is acidified and, after addition of
NaCl, filtered through crushed limestone or dolomite to give the desired hardness
and salt content [75].

DISPOSAL OF EFFLUENTS

The brine effluent produced by a flash distillation plant contains elevated levels of some heavy metals, which leads to increased metal concentrations in limpets
and some Fucus species. This is most evident with copper [80]. The U.S. Department of Energy proposed to use storage caverns, produced by solution mined subter-
ranean salt deposits. The environmental-biological impact on a brine disposal operation was assessed [81]. A bibliography was published with 259 abstracts for the
period 1964 to 1976, covering the disposal of wastes in the ocean [82].

Water leaving a desalination plant may have its thermal energy increased, its
chemical make up altered and its microbial biota modified. If the desalination
plant is sited in such a manner as to allow rapid dissipation of the thermal input,
the effect of the temperature change will be minimized. Signs of thermal effect
on marine ecosystems could be manifested by changes in community structure, as well
as the changes in features of individual species. The most obvious chemical changes
in desalination effluents may include increase in salinity, decrease in dissolved
oxygen, increase in dissolved organics and increase in pretreatment chemicals.
Desalination plants provide surface areas for rapid microbial proliferation. Depending upon the sequence of chemical pretreatment, it is possible that these viable

microbes will enter the ecosystem and supplement the existing biota, if conditions permit their continued growth [83].

In siting a large scale desalting plant, the possible effects of the environment to the plant must be considered, such as presence of hydrogen sulfide, ammonium ions, suspended matter, mineral pollutants, thermal pollution. On the other hand the desalting plant might affect the environment by the disposal of brine, warm effluents, corrosion products and various chemicals [84].

LITERATURE TO 2A.

1. A. W. Veenman (Desalination 27 [1978] 21/39)
2. D. Aussenac, S. Domenech, M. Enjalbert (Ing. Quim., Madrid, 9, No 101 [1977] 97/103)
3. V. B. Chernozubov (Voprosy Atom. Nauk i Tekhn. Ser. Opresnenie Solen. Vod 2, No 10 [1977] 3/6)
4. G. V. Efimov, E. I. Murakhovskaya (Tr. Vses. Teplotekhn. NII, No 13 [1978] 103/121)
5. P. Worrall (Pract. Aspects Heat Transfer, Proc. Fall Lect. Ser. N. J.-North Jersey Sect. AIChE 1976, [1978] 42/54)
6. V. F. Kovalenko (Izv. Vyssh. Uchebn. Zaved., Energ. 21, No 10 [1978] 52/61.- C. A. 90 [1979] 109707)
7. T. Kinlin (Desalination 30 [1979] 75.-Abstract only)
8. R. S. Silver (Desalination 31 [1979] 39/44)
9. E. Mousset (Belg. 861.555, 31 Mar 1978.- C. A. 78 [1978] 203968)
10. Terraqua Products Inc. (Brit. 1.487.569, 5 Oct 1977)
11. Soc. Française Whiting Fermont (French 2.351.681, 20 Jan 1978)
12. H. Merz (W. Ger. 2.737.263, 10 Mar 1979.- C. A. 90 [1979] 156957)
13. E. Shioda (Jap. 78.01.685, 9 Jan 1978.- C. A. 89 [1978] 65109)
14. S. Ueno (Jap. 78.19.974, 23 Feb 1978.- C. A. 89 [1978] 45453)
15. A. Selecki (Pol. 93.008, 15 Nov 1977.- C. A. 90 [1979] 106197.- Addition to 66.662)
16. R. McFee (U.S. 4.052.267, 4 Oct 1977)
17. G. Weiss (U.S. 4.081.331, 28 Mar 1978)
18. F. C. Kirschman, W. B. Bolte (U.S. 4.089.750, 16 May 1978)
19. F. C. Kirschman, W. B. Bolte (U.S. 4.110.170, 29 Aug 1978)
20. Auscoteng Pty. Ltd. (Brit. 1.515.683, 5 Jul 1978)
21. H. S. Green (U.S. 4.131.513, 26 Dec 1978.- C. A. 90 [1979] 192341)
22. S. Tamura, J. Ishiguro (Jap. 79.56.979, 8 May 1979.- C. A. 91 [1979] 62539)
23. K. Matsumura (Jap. 78.10.369, 30 Jan 1978.- C. A. 89 [1978] 135624)
24. I. Michiyoshi, K. Makino (Int. J. Heat Mass Transfer 21 [1978] 605/613.- C. A. 89 [1978] 217307)
25. K. Makino, Y. Isaki, M. Kurita, K. Kobayashi, K. Mori (Maizuru Kogyo Koto Semmon Gakko Kiyo 13 [1978] 1/9.- C. A. 89 [1978] 199664)
26. S. R. Ravipudi, T. M. Godbold (6th Int. Heat Transfer Conf. 1 [1978] 505/510.- C. A. 89 [1978] 217330)
27. P. C. Wayner (AIChE 71st Ann. Meet. Paper 128f, Miami Beach 1978)
28. A. P. Egorov, S. N. Filippov, S. V. Grigorenko, V. A. Koslov (Voprosy Atom. Nauki i Tekhn. Ser. Opresnenie Solen. Vod. 2, No 10 [1977] 15/17.- Ref. Zh., Khim. No 171291 [1978])
29. K. P. Singh, M. J. Holtz (AIChE Symp. Ser. 75, No 189 [1979] 219/226)
30. D. Butterworth, Two-Phase Flow Heat Transfer (Oxford Univ. Press, London 1977, 426/462)
31. E. Guevara (City Univ. London, Rept. No BLL-RM-ML-95, 1977)
32. K. J. Bell, C. B. Panchal (6th Int. Heat Transfer Conf. 6 [1978] 361/375)
33. I. P. Tairov (Nekotorye Probl. Teplo- i Massoobmena [1978] 120/122, 123/126,- Ref. Zh. Khim. No 23 I 88/89, 1978)
34. D. A. Labuntsov, A. P. Kryukov (Int. J. Heat Mass Transfer 22 [1979] 989/1002)
35. J. M. Croix (Inf. Chim. 184 [1978] 117/120.- C. A. 90 [1979] 153935)
36. I. Tanasawa (Proc. 6th Intern. Heat Transfer Conf. 6 [1978] 393/405)
37. J. R. Maa (Chem. Eng. J., Lausanne, 16 No 3 [1978] 171/176.- C. A. 90 [1979] 106353)

38. S. S. Sadhal, M. S. Plesset (Trans. ASME, J. Heat Transfer 101, No 2 [1979] 48/54)
39. R. Bairamov, L. E. Rybakova, M. Mamedov (Izv. Akad. Nauk Turkm. SSR, Ser. Fiz.-Tekh. Khim. Geol. Nauk No 3 [1978] 22/28)
40. L. R. Sharp (NTIS Rept. AD-A056395 [1978].- G.R.A. 78, No 21 [1978] 218)
41. T. G. Sunderaraman, T. Venkatram (Indian Chem. Eng. 20, No 1 [1978] 41/44.-C. A. 90 [1979] 57008)
42. M. Ouchi, T. Takeyama (Nippon Kikai Gakkai Rombunshu 44, No 377 [1978] 154/163.-C. A. 89 [1978] 131632)
43. E. H. Young, J. G. Withers, W. B. Lampert (AIChE Symp. Ser. 74, No 174 [1978] 15/19)
44. C. Weierman, J. Taborek, W.J. Marner (AIChE Symp. Ser. 74, No 174 [1978] 39/46)
45. V. G. Rifert, S. I. Chaplinskii, I. D. Iliev (Izv. Vyssh. Uchebn. Zaved., Energ. 22, No 2 [1979] 115/118.- C. A. 90 [1979] 206437)
46. R. L. Webb, M. J. Scott (AIChE Symp. Ser. 75, No 189 [1979] 314)
47. J. J. Schröder, P. Fast, W. Sander-Beuermann (Desalination 31 [1979] 19/34)
48. J. K. Thorne (U.S. 4.059.147, 22 Nov 1977)
49. C. A. Allen, O. Fukuda, E. S. Grimett, R. E. McAtee (Proc. 12th Intersoc. Energy Conv. Engg. Conf. 1 [1977] 824/831)
50. T. F. Özkaynak, J. C. Chen (AIChE Symp. Series 74, No 174 [1978] 334/343)
51. J. C. Chen, J. G. Withers (AIChE Symp. Series 74, No 174 [1978] 327/333)
52. L. P. Golan, D. C. Cherrington, R. Diener, C. E. Scarborough, S. C. Weiner (AIChE Symp. Ser. 75, No 189 [1979] 191)
53. A. Kogan (Nat. Counc. Res. Dev. Israel, Rept. NCRD 7-77 [1976] 65/73.- C. A. 90 [1979] 12070)
54. L. L. Moresco, E. Marshall (AIChE Symp. Ser. 79, No 189 [1979] 266/272)
55. E. N. Bukharin (Izv. Vyssh. Uchebn. Zaved. Energ. No 10 [1978] 68/75.- E.I.17 [1979] 33216)
56. C. S. Smith (Brit. 1.508.203, 19 Apr 1978)
57. T. Hamada, M. Arai (Jap. 78.68.679, 19 Jun 1978.- C. A. 90 [1979] 28850)
58. Z. Gropsianu, V. Jascanu, D. Kohn, A. Iovi (Bul. Stiint. Teh. Inst. Politeh. Timisoara 21, No 2 [1976] 245/249.- C. A. 88 [1978] 107288)
59. N. K. Tokmantsev, V. V. Il'yushchenko, V. B. Chernozubov (Voprosy Atom. Nauki i Tekhn. Ser. Opresnenie Solen Vod 2, No 10 [1977] 53/60.- Ref. Zh., Khim. No 17 I 292 [1978])
60. Y. Katto (Int. J. Heat Mass Transfer 21 [1978] 1527/1542)
61. S. S. Kutateladze (Int. J. Heat Mass Transfer 22 [1979] 281/299)
62. A. J. Baum, J. C. Purcupile, R. S. Dougall (ASME Ann. Meet. Paper 77-HT-82 [1977].- A. J. Baum, Diss. Carnegie-Mellon Univ. [1977].- Diss. Abs. B 38 [1977] 1836)
63. R. Mesler, G. Mailen (AIChE J. 23 [1977] 954/957)
64. M. Shoukri, R. L. Judd (Trans. ASME, J. Heat Transfer 100, No 4 [1978] 618/623)
65. A. P. Burdukov, G. G. Kuvshinov, V. E. Nakoryakov (Teploperedacha pri Kipenii i Kondensatsii 1978, 94/118.- Ref. Zh., Khim. No 22 I 130 [1978])
66. M. Marchand (Rev. Int. Oceanogr. Med. 48 [1977] 55/65.- C. A. 89 [1978] 48535)
67. I. V. Dyn'kin, B. I. Briman, Yu. M. Mikhailov (Prom. Energ. No 7 [1978] 17/19.-C. A. 89 [1978] 203935)
68. T. Goto, T. Hakuta, T. Nakahara (Tokoshi Nyusu, Kagaku Kogyo Shiryo 12 No 8 [1978] 159/176.- C. A. 89 [1978] 152136)
69. H. Tahata, R. Yokoyama, H. Choju, H. Ikeda (Jap. 78.28.255, 14 Aug 1978.-C. A. 90 [1979] 28858)
70. I. Jarnuszkiewicz, H. Swedrowska, A. Butler, M. Olszewska (Bull. Inst. Marit. Trop. Med. Gdynia 29 [1978] 211/222.- C. A. 90 [1979] 192310)
71. R. Leimgruber (Desalination 30 [1979] 615/620)
72. F. Korte (Desalination 30 [1979] 411/424)
73. P. F. Ipatov, I. S. Morozova (Tr. VNII VODGEO No 66 [1977] 26/28.- Ref. Zh., Khim. No 15 I 346 [1978])
74. A. I. Egorov, Yu. A. Rakhmanin, G. A. Ivleva (Tr. VNII VODGEO No 66 [1977] 42/44.- Ref. Zh., Khim. No 16 I 343 [1978])
75. T. Kulhavy (Czech. 168.962, 15 Apr 1977.- C. A. 89 [1978] 220711)

76. H. Teraoka (Jap. 78.05.988, 3 Mar 1978.- C. A. 89 [1978] 48760)
77. H. Teraoka (Jap. 78.05.989, 3 Mar 1978.- C. A. 89 [1978] 48759)
78. H. Teraoka (Jap. 78.05.990, 3 Mar 1978.- C. A. 89 [1978] 48758)
79. H. Teraoka (Jap. 78.13.344, 9 May 1978.- C. A. 89 [1978] 185873)
80. M. G. Romeril (U.K. Mar. Pollut. Bull. 8, No 4 [1977] 84/87.- E.R.A. 3 [1978] 7572)
81. National Oceanic and Atmospheric Administration (Rept. NOAA-78041212.- NTIS Rept. 280712 [1978] .- G.R.A. 78 [1978] p. 83)
82. R. J. Brown, Ocean waste disposal, Volume 1 (NTIS Rept. PS-78/0705 [1978] 265 pp.)
83. H. Winters, I. R. Isquith, R. Bakish (Desalination 30 [1979] 403/410)
84. J. J. Libert, A. Maurel, R. Lucas (Desalination 30 [1979] 427/435)

2B. Components of Distillation Plants

INTAKE SYSTEMS

In many areas of the Mediterranean, a residue of a particular sea-grass, Posidonia Oceanica, is often found in large quantities and cause operational problems in desalination plants. An operation oriented study deals with aspects such as concentration in seawater, distribution of sizes in residue samples, sedimentation properties, bulk densities and resistance of sea-grass cakes to seawater flow. Hence, a tailored solution of the problem can be designed [1].

Patents. An electrode, capable of generating metal ions, is used for repelling fish and shellfish from seawater intake systems [2].

CONTAINMENT VESSELS

The present and future of desalination plants using concrete evaporator shells is discussed, as well as its features [3]. A terotechnological approach is recommended for the in-service inspection of vessels, subject to significant metal loss by corrosion, as an activity within a corporate corrosion management methodology [4].

DEAERATORS

Carbon dioxide, oxygen, Freon 114 and butane were stripped from deionized water, sodium chloride solutions and freshly acidified seawater in a vacuum-spray chamber using three types of hydraulic nozzle. For both conical-spray and fan-jet nozzles, the bulk of the gas was shown to desorb from the thin liquid sheet before droplet formation occurs. A mathematical model of the diffusion in the sheet predicts well the desorption from the laminar liquid sheets formed by jet impingement nozzles, which offer the advantages of simplicity and low operating cost. An economic analysis comparing other types of industrial vacuum strippers with the proposed design showed the latter to compare very favorably in terms of estimated equipment size and cost, as well as in expected operating cost [5].

The design of the countercurrent air stream system commonly used to strip CO_2 from the water produced in the H-ion exchange section of a desalination plant was improved by using a stripping column of packings of Raschig rings with an improved design. The capacity was increased by about 100% and the residual CO_2 was reduced from 7 to about 3.5 mg/kg [6].

Patents. Dissolved oxygen is removed from water by chelating ion exchange resin containing Fe^{6+}. A hydrophilic phenolic resin having iminodiacetic acid groups is preferred [7]. Desalinated seawater from the flash-evaporation process is mixed with a decarbonating agent and cascaded down from the top of a vertical series of trays equipped with weirs [8].

DEMISTERS

A calculation of nontransported entrainment in adiabatic distillation units was presented [9]. An instrument to measure the mist concentration and the size distribution in a mulstistage flash evaporator was developed. In one case the concentration was 0.72×10^{-4} g/l and the mean diameter was 70 μm. A 20% increase in evaporation rate resulted in a factor of 3 increase in mist concentration and a factor of 2 in mist size [10].

HEAT EXCHANGERS

A review of the design of heat exchangers with special consideration of fouling and scaling effect on heat transfer [11] and the effect of scaling in heat exchangers and descaling with additives [12] were presented. Fouling of heat transfer surfaces with emphasis on deposition and reentrainment models, as well as the effects of process variables on various types of fouling were discussed [13]. A method of relating heat-exchanger fouling factors and thermal conductivity coefficients to facilitate the design of heat exchangers was presented [14]. Heat exchanger tests with moderately saline geothermal brines were reported. Chemical cleaning effectively removed the scale layer developed on tube surfaces. Corrosion of the titanium tubes was not observed [15]. A two-purpose contact heat exchanger was proposed for maximum evaporation of saline water in the separation process of $MgCl_2$ from seawater [16].

Heat exchangers for ocean thermal energy conversion (OTEC) application were developed [17]. Microfouling on heat exchanger surfaces, applicable to OTEC systems was discussed [18] and an overview was given of biofouling and corrosion tests at St. Croix in the Carribean [19]. The use of antifoulants to prevent fouling of heat exchangers was extensively tested [20].

Tests made during several years proved that very thin plastic surfaces (50 to 100 μm) may lead to overall heat transfer coefficients, which are comparable to metallic tubes in heat exchangers [21].

The Office of Saline Water accomplished a very large amount of significant work related to the design and performance of large heat exchanger bundles and enhanced heat transfer surfaces to provide basic technical and economic data for the design of seawater distillation plants. The data are scattered through a large number of reports, many of which are concerned primarily with factors other than heat transfer. Present report catalogues and organizes the heat exchanger data developed by the OSW. Some analysis as to the validity of the data is made and ranges of performance that can be expected are given [22].

Patents. Arrangements for distributing the fluids over the walls of the vertical tubes of falling-film heat exchangers are described [23].

PUMPS

Availability of desalination plants depends to a great extent from the continuing operation of seawater pumps. The effect of various parameters influencing pump operation was studied and measures to ensure safe operation are given [24].

LITERATURE TO 2B.

1. E. Gabbrielli, A. Ripasarti (Desalination 30 [1979] 127/143)
2. F. Uwano (Jap. 77.140.477, 24 Nov 1977.- C. A. 89 [1978] 131592)
3. Y. Nojiri (Civ. Eng., Japan, 16 [1977] 126/135.- E.I. 16 [1978] 70890)
4. G. W. Blackburn (Presented at Periodic Inspection of Pressurized Components London 1976)
5. S. G. Simpson, S. Lynn (AIChE J. 23 [1977] 666/673, 673/679)
6. M. M. Brando (Tsellyul. Bum. Karton No 15 [1977] 6/7.- C. A. 87 [1977] 206263)
7. T. Yoshikawa (Jap. 78.00.658, 6 Jan 1978.- C. A. 89 [1978] 135619)

8. T. Rokushi, T. Sawa, K. Izumi, S. Takahashi (Jap. 78.44.421, 29 Nov 1978)
9. V. V. Il'yushchenko (Voprosy Atom. Nauki i Tekhn. Opresnenie Solen. Vod 2, No 10 [1977] 61/69.- Ref. Zh., Khim. [1978] 17 I 295)
10. A. Gotoh, S. Toyama, K. Makino, K. Iinoya (Nippon Kaisui Gakkaishi 31, No 5 [1978] 221/227.- C. A. 89 [1978] 168826)
11. M. Kurita (Kagaku Kogaku 42, No 8 [1978] 434/438.- C. A. 89 [1978] 148591)
12. T. Suzuki (Kagaku Kogaku 42, No 8 [1978] 439/443.- C. A. 89 [1978] 148592)
13. N. Epstein (Proc. 6th Int. Heat Transfer Conf. [1978] 235/253.- C. A. 89 [1978] 199572)
14. A. Bestcherevnykh (Chem. Eng., N.Y. 85, No 4 [1978] 122.- C. A. 89 [1978] 45593)
15. G. L. Lombard (AIChE Symp. Ser. 74, No 174 [1978] 281/287.- C. A. 90 [1979] 92197)
16. E. N. Bukharkin (Izv. Vyssh. Uchebn. Zaved., Energ. 21, No 10 [1978] 68/75.- C. A. 90 [1979] 127301)
17. J. W. Michel (Rept CONF-770331-1 [1977].- C. A. 89 [1978] 45408)
18. S. M. Gerchalkov, D. S. Marszalek, F. J. Roth, B. Sallman, L. R. Udey (NTIS Rept. AD-A047302 [1977].- C. A. 89 [1978] 48734)
19. D. L. Meier (NTIS Rept. COO-4041-8 [1977].- C. A. 90 [1979] 174444)
20. L. M. Caballero Basanez, R. L. Aga Van Zeebroeck (Ing. Quim., Madrid, 8, No 92 [1976] 67/76.- C. A. 90 [1979] 56781)
21. F. Lauro (Desalination 31 [1979] 221/231)
22. Coury and Associates (NTIS Rept. PC-A05/MF A01 [1979] 98 pp.- E.R.A. 4 [1979] 31059)
23. F. Lauro (U.S. 4.106.560, 15 Aug 1978)
24. A. Le Grand, J. Gatignol (Desalination 31 [1979] 173/181)

2C. Vertical Tube Evaporators

The decrease in the temperature Δt in a multistage adiabatic water desalination plant is resulting from physicochemical and hydrodynamic parameters. The physicochemical Δt is mainly affected by the concentration of the salt water and slightly by the temperature. The hydrodynamic Δt depends on the hydraulic resistance of the condenser and separator. The Δt resulting from the pressure drop of the secondary steam is relatively high, 1.8°, and must be taken into account in projecting water desalination plants [1]. A review of vertical tube evaporator with corrosion resistant gaskets was presented [2].

The concentration factor has a double effect on specific capital cost. First, with an increase of this factor, the temperature depression increases and the heating surface must be increased. Secondly, the heat transfer coefficient from the brine to the pipe surface decreases and the condenser surface increases. A graph presents the optimum value of this factor depending on the cost of preliminary feed water treatment [3].

An experimental apparatus to study water boiling in evaporators consists of an evaporation chamber having at the bottom a heating surface of a tube bundle, which was used to show the joint effect of the process of steam formation and two-phase flow convection on heat transfer [4]. The apparatus could be used as the first stages in multi-effect desalting plants with greater efficiency [5].

A promising development in desalting is to use power plant turbine exhaust steam in conjuction with the energy-efficient vertical tube foamy flow evaporation. In this application the evaporator augments the condenser and heat rejection cycle. It also provides for zero blowdown by utilizing the power plant cooling tower blowdown as feed, and by concentrating it to a residual solid and distilled water. The steam condensate is returned to the power plant steam cycle, while the distillate is available for use as potable water, as boiler feed or as a coolant supplement [6].

FALLING LIQUID FILM

Experiments on heat-transfer coefficient and film breakdown heat flux for subcooled water films flowing down the outside of a uniformly heated vertical tube
show that when the parameter attains a constant value by increasing heat flux,
the film breakdown takes place in a thin region of the distorted film. Data on
heat-transfer coefficient and liquid droplet rate entrained by boiling were also
presented in a range of heat fluxes up to film breakdown [7].
An equation is given correlating the Nusselt, Prandtl and axial Reynolds numbers,
the weighted log-mean-temperature difference, the critical boiling temperature
difference, the unit length and inside diameter and the percentage evaporation,
using experimental heat transfer data obtained with water and organic liquids [8].
A method for calculating the heat transfer coefficient in an evaporating liquid
film makes allowance for the superheating of the liquid by friction between the
vapor and the film surface to derive a theoretical expression [9]. Local heat
transfer to a falling liquid film and a two-phase downward annular flow was studied
with allowance for the waviness of the film surface and flow resistance. It was
shown that the critical vapor content can be as high as 90% for a channel cooled
by a liquid film at atmospheric pressure [10]. Simple engineering equations were
obtained from an experimental investigation of the flow behavior and heat transfer
of a free-falling film moving along a polymer surface [11]. In an experimental
investigation of heat transfer to water in vertical tube in upflow and downflow,
the Nusselt numbers for heating in downflow were slightly higher, but, within experimental error, equivalent to those values obtained for heating in upflow [12].
The linear stability problem of a vertical falling film has been solved in terms
of a numerical solution of the Orr-Sommerfeld equation. Experimental values of
the wavelength and wave velocity were determined for water films and the results
are in reasonably good agreement with the theory. Calculated values for the wavelength, wave velocity and growth rate of the most unstable wave indicate only a
small difference between the temporal and spatial formulations for water films[13].

RISING LIQUID

In an experimental vertical evaporator, at low heat loads boiling was confined
to the upper parts of the tube bundles. As the load increased, boiling commenced
in the lower sections. The rising vapor bubbles caused tube vibration at heat
loads over 50000 W/m^2. Large discrepancies became apparent under conditions when
convective two-phase flow was significant. An improved generalized expression
was derived, in which the Nusselt number was expressed as a function of the Reynolds
and Prandtl numbers, tube geometry and the number of tubes generating vapor at any
given level [14]. The tube wall temperature, during heat transfer in the upward
spray flow of R 113 in a vertical tube, was close to the liquid saturation temperature upstream and increased suddenly to a high temperature downstream. The
location of this sudden increase in the wall temperature shifted to the upstream
side with an increase in the heat flux. The spray flow after dryout was in a
thermally nonequilibrium state in which drops and superheated steam coexisted [15].

Patents. Liquid distribution arrangements were described for uniformly feeding
the heater tubes of a falling film evaporator [16]. Preheated seawater is evaporated in a vertical tube evaporator and condensed in an ejector type condensed by
a portion of the desalted water. The hot brine collected at the bottom of the
evaporator and the desalted water are used to preheat the incoming seawater [17].
The efficiency of a multi-effect evaporator plant is improved by increasing the
surface areas in the effects, in which the heat flux is critical in terms of dryout and scale formation, relative to the areas of the evaporators in the other
effects [18]. It is suggested that at least two of the high temperature evaporators
be of the climbing-film type and the rest are falling-film type. The method eli-
minates some pumps or sprayers [19]. Vapor generated in the lowest temperature
effect is withdrawn by an ejector, driven by high pressure steam, and the vapor-
steam mixture is discharged into the condensation side of the highest temperature

effect heat transfer tubes. Solution from the lowest temperature effect is introduced onto the evaporation side of these tubes [20].

Additional perforated plates are fitted on the tube bundle of vertical tube
evaporators to collect the condensate on the outside of the tubes. This decreases
the thickness of the condensate film and results in a higher heat-transfer coefficient [21].

The steam from the evaporator recycles through the vapor-liquid separator and
ejector condenser, mixes with steam from the ejector and passes into the evaporator
to heat the evaporator tubes. The working fluid of the ejector is high-pressure
steam generated by a boiler [22].

A device is inserted in the top of the heat transfer tubes to improve the distribution and the stability of the liquid film. The device distributes the liquid
by creating a circulatory motion on the inside wall of the tubes [23].

An evaporation system using flexible tubes for the desalination of water consists
of a tower containing a series of vertical cells connected by a system of flexible,
refrigerated tubing. The temperature of each cell is slightly higher than the pre-
ceeding cell due to the heat of condensation in the water-cooled tubes. The distillate
is drawn out of the system after treatment in the cells [24].

OPERATING EXPERIENCE

A 400000 gpd vertical tube evaporator was built to recover water from a synthetic rubber manufacturing plant waste water stream. Continuous operation for extended periods was rendered impossible, due to fouling and corrosion of copper
alloy heat exchange surfaces. Corrosion was traced to presence of 2 to 5 ppm of
sulfides in the feed. Retubing with titanium eliminated corrosion, but the fouling continued [25].

LITERATURE TO 2C.

1. V. V. Savchenko (Izv. Akad. Nauk Turkm. SSR, Ser. Fiz.-Tekh., Khim. Geol. Nauk
   No 3 [1978] 29/32.- C. A. 90 [1979] 76307)
2. Y. Fukushima, A. Tani (Baruka Rebyu 21, No 3 [1977] 1/14)
3. G. K. Feiziev, F. Z. Gasanguliev (Izv. Vyssh. Uchebn. Zaved. Neft Gas No 4
   [1977] 109/112)
4. V. N. Slesarenko (Izv. Vyssh. Uchebn. Zaved. Energ. No 10 [1978] 62/67.- E. I.
   17 [1979] 33979)
5. V. Slesarenko, V. Gydakov (Desalination 30 [1979] 375/383)
6. H. H. Sephton (Desalination 31 [1979] 187/195)
7. T. Fujita, T. Ueda (Int. J. Heat Mass Transfer 21, No 2 [1978] 97/108, 109/118)
8. D. Chiang (Ta T'ung Hsueh Pao 5 [1976] 199/218.- C. A. 88 [1978] 8996)
9. Ye. I. Taubman, Yu. I. Kalishevich (Heat Transfer Sov. Res. 8, No 6 [1976]
   1/8.- E.I. 16 [1978] 18412)
10. B. G. Ganchev, A. Ye. Bokov, A. B. Musvik (Heat Transfer Sov. Res. 8, No 6
    [1976] 9/19.- E.I. 16 [1978] 18413)
11. Yu. Ye. Lukach, L. B. Radchenko, Yu. M. Tananayo, A. D. Petukhov (Heat Transfer
    Sov. Res. 8, No 6 [1976] 20/24.- E.I. 16 [1978] 18414)
12. T. E. Mullin, E. R. Gerhard (J. Heat Transfer 99 [1977] 586/589.- E.I. 16
    [1978] 18421)
13. P. W. Pierson, S. Whitaker (I.E.C. Fundamentals 16 [1977] 401/408)
14. V. N. Slesarenko, A. E. Rudakova (Izv. Vyssh. Uchebn. Zaved., Energ. 21, No 10
    [1978] 62/67.- C. A. 90 [1979] 109708)
15. Y. Koizumi, T. Ueda, H. Tanaka (Nippon Kikai Gakkai Rombunshu 44, No 377 [1978]
    191/199.- C. A. 89 [1978] 131633)
16. K. Mattern, H. Saft (French 2.340.119, 7 Oct 1977)
17. H. Satone (Jap. 78.116.272, 11 Oct 1978.- C. A. 90 [1979] 76359)
18. F. C. Wood (Brit. 1.514.480, 14 June 1978.- C. A. 90 [1979] 28851)
19. W. Nishimoto, K. Hayakawa (Jap. 78.112.279.- C. A. 90 [1979] 76358)
20. Sasakura Engineering Co. (Brit. 1.503.741, 15 Mar 1978)
21. H. G. Lankenau, A. R. Flores (U.S. 4.132.587, 2 Jan 1979.- C. A. 90 [1979] 89199)

22. H. Satone (Jap. 78.116.273, 11 Oct 1978.- C. A. 90 [1979] 76360)
23. I. Ohkochi, T. Takahashi, K. Izumi (Jap. 78.27.702, 10 Aug 1978)
24. J. Espinosa Cina (Span. 446.360, 16 Nov 1977.- C. A. 89 [1978] 113196)
25. W. C. Lang, J. H. Crozier, F. P. Drace, K. H. Pearson (NTIS Rep. No PB-265361/6.-
    EPA-600/2-76-260 [1976])

2D. Horizontal Tube Evaporators

Experimental data on horizontal tube evaporator-condenser water desalination
units agree well with the turbulent model, which covers the whole range of the
evaporating film flow rate [1]. The knowledge of the mode of liquid flow between
neighboring consecutive tubes in a horizontal bundle is important to the understanding of transport rate characteristics. Drop size, drop detachment frequency
and the distance between drop-producing sites at the bottom of the tube were studied as a function of surface tension, tube size and spatial location of the tube
in the vertical column in the bundle. Smooth and grooved tubes were examined and
the effect of dripping on the mass transfer rates is demonstrated [2].
With the aim to enhance the heat transfer rates in the horizontal-tube film-
type evaporator-condenser water desalination units, circumferentially grooved conduits on the external evaporation side were used. The theoretical analyses of
the transfer characteristics of square-edged, triangular and circular grooves show
that differently shaped grooves are advantageous at different ranges of the flow
rate. However, in the practical irrigation flow rate of 300 < Re < 1000, the
square-edged groove, with a straight or modified bottom, is the most efficient
shape. Overloading is compensated by local channeling over 2 to 3 grooves, leaving most of the grooved surface operational [3].
Experimental data were reported and relating equations are given on the heat
transfer coefficient during condensation of steam on a horizontal single tube and
on a horizontal bundle of three tubes [4]. The local heat transfer coefficient
and the pressure drop, in a uniformly heated horizontal copper tube, increase with
both heat flux and vapor quality. A correlation equation combining the heat transfer coefficient and pressure drop is proposed [5].
A two-phase system with a spray angle of over 70° was developed and installed
on a horizontal tube multiple-effect evaporator to investigate its brine distribution and heat transfer characteristics [6].
An aluminum tube multi-effect distillation plant was built at Eilat, Israel,
and began operation in May 1974, with a performance ratio of 10. The product water had less than 50 ppm salt content. Plant characteristics and proposed improvements are outlined [7].
The problem of vapor/liquid interaction and entrainment in shell-and-tube evaporator was analyzed with emphasis on the horizontal tube falling film evaporators.
For high vapor crossflow velocities, a criterion was presented for predicting the
inception of liquid entrainment by stripping. Conditions were defined under which
vapor/liquid interaction and entrainment are important [8].
The advantages of horizontal tube evaporators were emphasized, such as possibility of working at higher temperature and hence increasing water production,
possibility of use of special tubes, which improve the heat exchange efficiency
and thus reduce heat exchange surfaces, and possibility of coupling with vapor
compression apparatus [9]. A model of combined boiling and evaporation of liquid
films on horizontal tubes was developed [11].

Patents. The evaporator is made of horizontal double-tube sections. The feed
solution is introduced at the top and a concentrated product and vapor are recovered at the bottom. The concentration ratio was 0.48 to 0.55 and the overall
heat transfer coefficient was 1000 to 1600 kcal/m^{2}h°C [10].

LITERATURE TO 2D.

1. R. Semiat, D. Maron-Moalem, S. Sideman (Papers Conf. Physicochem. Hydrodyn.
   1 [1977] 81/93.- C. A. 89 [1978] 185791)
2. D. Maron-Moalem, S. Sideman, A. E. Dukler (Desalination 27 [1978] 117/127)
3. S. Sideman, A. Levin (Desalination 31 [1979] 7/18)
4. M. Furukawa, H. Sugita (Kobe Shosen Daigaku Kiyo, Dai-2-Rui 26, No 2 [1978]
   101/112.- C. A. 90 [1979] 25338)
5. D. G. Motwani, B. K. Sthapak (Proc. 4th Natl. Heat Mass Transfer Conf. [1977]
   345/352.- C. A. 90 [1979] 153967)
6. K. Matsumura, K. Yamamoto, Y. Ikenaga (Hitachi Rev. 27, No 2 [1978] 97/102)
7. S. Manor, J. Weinberg (Natl. Counc. Res. Dev., Israel, Rept. 8-76 [1977]
   49/57.- C.A. 89 [1978] 203915)
8. D. Yung, J. J. Lorenz, E. N. Genic (ASME Ann. Meet. Paper 78 WA-HT-35 [1978]
   10 pp.- E.I. 17 [1979] 33213)
9. J. C. Deronzier (Desalination 31 [1979] 115/124)
10. S. Iwata (Jap. 77.138.057, 17 Nov 1977.- C. A. 89 [1978] 8056)
11. J. J. Lorenz, D. Yung (NTIS Rept. ANL-OTEC-78-1 [1978].- E.R.A. 3 [1978] 42478)

2E. Flash Evaporation

FLASHING FLOW

A study was made on a single-stage, submerged orifice experimental flashing
unit operating on water. Correlations were obtained showing the effect of varying design parameters on the thermodynamic efficiency of the flashing process.
The most pronounced interaction is that between the flow rate, flashing range and
wire mesh [1]. Equations of two-phase flashing flow in conduits relate the pressure, quality ratio, flow rate, conduit geometry, the rate of flashing and the
distance along the conduit. Two correlations for horizontal and vertical conduits
relate the flow rate, pressure drop, initial pressure and conduit geometry for
flashing flow in conduits between flashing chambers. The developed correlations
are applied to develop polynomial equations expressing physical properties, such
as saturated temperature, vapor specific volume, enthalpy and latent heat in terms
of pressure by the least square method. The solution of the equations gives the
distribution of pressure, quality and velocity along the conduit length [2].

Flashing flow in a single-stage flash evaporator was classified into two patterns, bubble boiling and splash boiling. At 40°, bubble boiling occured when the
liquid level was increased and the liquid flow rate was decreased. When the temperature was over 50°, splash boiling was observed. Nonequilibrium temperature
difference was discontinuously changed by the transition of the flashing pattern.
The measured data in bubble boiling were 1.5 to 2 times that in splash boiling[3].

The design philosophy and considerations utilized in the construction of a
two-stage scaled-down model of a flash evaporator, as well as the instrumentation
and techniques developed or adapted for the measurement of temperature, pressure,
flow, level and salinity and for the visualization of the flow were presented [4].
The effect of the operating parameters variation on the thermodynamic efficiency
and the pressure energy losses in a scaled down flashing experimental unit, using
a new geometry for the interstage orifice, were investigated. In this type of geometry, the flow area depends on the rate of the flow, and the partial liquid recirculation and its effect on the flashing process impedes is minimized [5].

MULTI-STAGE FLASH DISTILLATION

Dynamics of a multi-stage flash evaporator, investigated on a 3000 m³/day test
MSF plant, were different, corresponding to types of input changes, i.e. step up
and step down [6]. Simulated static characteristics were developed and compared
with experimental results from the same plant. The results showed ineffective
flashing might take place at low temperature stages, when the load decreased below

a limit. The number of such stages increased with decreased load [7]. By using
a conceptual design of an MSF evaporator with a capacity of 35520 m^3/day, it was
estimated that the energy used, expressed in heavy oil consumption in 1/m^3 of fresh
water, was 1.82 for the waste-heat utilizing type and 2.54 to 5.44 for standard
MSF evaporators [8].

The thermodynamic efficiency coefficient of evaporation in stages increases with
increasing length of the evaporation chamber up to 1.6 to 2.4 m. A further increase has little effect. The efficiency coefficient decreases with increasing
the flow rate of the solution or with decreasing the equilibrium temperature of
the solution. An equation is given for the calculation of the nonequilibrium temperature for a chamber length of 2.2 m and vertical baffles [9].

Based on the overall mass and heat balances for the MSF plant, the relation
between the total flash temperature difference and the brine recirculation rate
was established. Applied to the heat input section, the basic equations provide
the dependencies for the brine recirculation, the specific energy requirements
and the temperature rise in the brine heater. Subdividing the brine heater tem-
perature-rise necessitates an additional parameter, the stage number, which combined with the total flash temperature difference, gives the heat transfer determining temperature difference for dimensioning the heat transfer surface of the
evaporator. Thereby the dependence between the process temperatures and the
plant design features is constituted [10].

Any reduction of seawater demand achieved by modifications in the heat rejection
section will raise the cold-end temperature of the evaporator. The benefits of
a reduction lie in size and cost savings for the intake structure, as well as other
plant parameters. Intake reductions to 60% of the recirculation rate will raise
either the heat input or heat transfer areas, as well as the costs for some acce-
sory components by roughly 5%. Further reductions are penalized by very rapid degradation of several plant characteristics. The best solution will depend on such
site conditions, as distance from the shore, seawater quality, shore geography and
soil, size and number of desalination streams [11].

A system of simple analog control loops has been investigated experimentally
in a 6-stage flashing test unit, which is aimed to automate the plant operation
and to improve the distillation process within any required operating condition.
The control circuit includes the control of liquid levels and temperatures in
the first and last flashing chamber, adjusting automatically the amount of steam
to the brine heater, and the brine flow rates in both the heat rejection and the
heat recovery section. Using this system, the load can be changed more easily
and quickly than by manual operation, which is an important factor for dual purpose plant coupling. Experiments have been carried out and simulated with a modified form of a digital computer code. The computer program was also applied to
an MSF plant design with a higher number of stages. One of the main results is
that, by decreasing the load, interstage blow-through can occur in the middle
stages of the plant, even if the liquid levels in the first and in the last stage
are kept constant. The presence of weirs in the flashing chambers reduces this
tendency [12].

Patents. An automatic control of the liquid level in each stage of the multi-
stage evaporator is provided [13]. Various arrangements of a horizontal cylindrical flash evaporator with two paths were outlined [14], including placement of
the condenser below the evaporator [15], centrifugal separation of drops from
vapor [16] and a long shell transversely divided into a number of flash chambers[17].
The feed stream is split in two parts, one of which is used only for cooling the
last heat removal stages [18]. An evaporator consists of a slitlike channel leading from the seawater inlet to the stream outlet, provided with alternating spiral
compartments, in which evaporation takes place, and pocketlike buffer compartments
forming liquid seals, in which liquid transport takes place [19].

A gas-liquid contacting chamber for a high temperature gas and a solution is
provided in the first stage evaporation chamber. The brine is introduced from the
top of the contacting chamber and steam from the bottom [20].

A solution is flash evaporated by temperature and pressure differences in an
apparatus consisting of slant bottom flash and slant heat transfer tubes on the

top of the apparatus. The feed enters on the top and flows down. The apparatus
prevents superheating of the feed and improves distillation [21].

A flash evaporator has a center section and two end sections, the construction
of which is described in detail. The feed enters one of the outer chambers and
the concentrated solution is transferred to the other outer chamber. The vapors
pass through the mist collectors and a heat exchanger, where they are condensed
with heating of the feed [22].

OPERATING EXPERIENCE

Germany West. The MSF Helgoland plant with a capacity of 800 m³/day is in operation since April 1972. Corrosion problems and scale deposition in the brine
heater tubes were examined. The plant was frequently shut-down during its 7 years
operation, mostly because of lack of sufficient storage capacity [23].

Japan. A distillation test plant, made of carbon steel, has operated since
14 April 1972 with a capacity to treat 1000 m³/day seawater and produce 850 m³/day
fresh water, leaving 150 m³/day concentrated brine containing 20% solid matter.
Raw seawater is pretreated to prevent scale formation and corrosion. Advantages
in using common steel are outlined [24].

Kuwait. Twenty years of desalination experience ranges from early operation
of 20 units of submerged tube type with total capacity of 9091 m³/day to the present 25 units of MSF cross tube polyphosphate dosed design with capacities ranging from 4545 to 27273 m³/day. Some of the specific technical problems encountered
are discussed and some recommendations are made for design changes and modifications to alleviate such problems in future designs [25].

Libya. The design, construction and start-up of a 1 Imp. Mgd desalination
plant of the MSF type and two 95 t/h steam boilers in Zuara were reported [26].

Mexico. Information on the history, design, operation and maintenance of the
desalination plant in Rosarito is given [27].

Netherlands. The large scale mulstistage flash seawater distillation plant at
Terneuzen produces high quality distilled water, which can directly be used in
industry. Some posttreatment is required for high-quality drinking water [28].

Oman. The Ghubrah plant is in operation since early 1976 and produces 4 Imp.
Mgd water when operated at 90°C. Belgard EV was used to allow plant operation at
temperatures above 90°. At 110°C production raised to 6 Imp.Mgd. Scale formation
was substantially less than when using polyphosphate additives [29].

U.S.S.R. A report was presented on the four-year operation of a five-stage
distillation unit of the Krasnovodsk heat and electric power plant [30].

LITERATURE TO 2E.

1. S. G. Seraq El Din, M. A. Darwish, H. T. El Dessouky (I.E.C., Process Des.
   Devel. 17, No 4 [1978] 381/388)
2. M. A. Darwish (Desalination 26 [1978] 285/297)
3. T. Sugeta, S. Toyama (Heat Transfer Jap. Res. 7, No 2 [1978] 65/73.- C. A.
   91 [1979] 41308)
4. N. Lior, R. Greif (Desalination 31 [1979] 87/99)
5. H. T. El-Dessouky, F. Mayinger (Desalination 31 [1979] 45/55)
6. M. Sato, F. Ikazaki, T. Sugeta, S. Toyama, K. Hotta, T. Tohoyama (Kagaku Kogaku Rombunshu 3, No 6 [1977] 606/611.- C. A. 88 [1978] 110289)
7. M. Sato, F. Ikazaki, S. Toyama (Kagaku Kogaku Rombunshu 4, No 4 [1978] 342/349.-
   C. A. 90 [1979] 76309)
8. Y. Nakashima, S. Toyama, M. Sato, T. Sugeta, F. Ikazaki, N. Nakazawa (Nippon
   Kaisui Gakkaishi 31, No 5 [1978] 213/220.- C. A. 89 [1978] 217890)

56

9. V. V. Savchenko, S. Seiitkurbanov (Izv. Akad. Nauk Turkm. SSR, Ser. Fiz.-Tekhn., Khim. Geol. Nauk No 3 [1978] 33/37.- C. A. 89 [1978] 203918)
10. K. Thiess, J. Hapke (Desalination 31 [1979] 101/111)
11. K. Genthner, M. El-Allawy (Desalination 31 [1979] 57/68)
12. R. Greffrath, F. Mayinger (Desalination 31 [1979] 71/83)
13. G. Pagani (Braz. 76.06.954, 30 Aug 1977.- C. A. 89 [1978] 48754)
14. R. Saari (Neth. 76.06.892, 28 Dec 1977.- C. A. 89 [1978] 181564)
15. R. Saari (French 2.355.537, 20 Jan 1978.- C. A. 90 [1979] 43638)
16. R. Saari (Jap. 78.29.281, 18 Mar 1978.- C. A. 89 [1978] 165379)
17. R. Saari (U.S. 4.105.505, 8 Aug 1978)
18. H. Risch (French 2.361.306, 14 Apr 1978)
19. M. J. Hofstede, G. Beentjes, L. Szucs, C. Tasnadi (W. Ger. 2.716.117, 26 Oct 1978.- C. A. 90 [1979] 76356)
20. T. Shioyama (Jap. 77.128.880, 28 Oct 1977.- C. A. 89 [1978] 48751)
21. H. Satone (Jap. 78.95.174, 19 Aug 1978.- C. A. 90 [1979] 73698)
22. W. R. Williamson (U.S. 4.148.693, 10 Apr 1979.- C. A. 91 [1979] 62536)
23. H. Löhrl-Thiel, A. Ejaz, J. Hapke, H. Widhalm, W. Heinrich (Desalination 31 [1979] 159/169)
24. H. Tabata, R. Yokohama (Desalination 26 [1978] 3/8)
25. A. M. S. Al-Adsani, M. A. Kalantar (Presented at the 5th Ann. Conf. of the Natl. Water Supply Improv. Assoc., San Diego, Calif. [1977])
26. F. Fioravanti, D. Falleni, G. Odone, A. De Maio (Desalination 31 [1979] 155/158)
27. E. Gomez Gomez (Desalination 30 [1979] 77/90)
28. A. A. Romeijn (Proc. Int. Water Supply Assoc. Congr. 12 [1978] S2/S8.- C. A. 90 [1979] 109711)
29. S. Chalchal (Desalination 31 [1979] 333/340)
30. L. S. Mrezhin, S. I. Golub, Yu. Sh. Kegamyan, A. N. Russkikh (Voprosy Atom. Nauki i Tekhn. Ser. Opresnenie Solen. Vod 2, No 10 [1977] 7/14.- Ref. Zh., Khim. [1978] 17 I 296)

2F. Other Distillation Processes

VAPOR COMPRESSION

A critical review was presented on the application of the heat pump principle to evaporators, operated either by steam jet thermal method or by mechanical re-compression using centrifugal or positive displacement pumps. Costs on 1974 basis were reported [1]. In trying to keep up with designed product rates, in spite of decreases in volumetric flow caused by scale or the presence of non-condensible gases, operators usually increase the evaporation temperature, increasing the vapor density and mass flow. However, the decrease in volumetric flow may suddenly shift the operating point of the compressor into the surge region. Predicted and measured performance curves were compared and rules for a safe working range were given [2]. A vapor-compression evaporator, using a seed-slurry process for scale control, was developed. Several features support the system's broad industrial application [3]. An evaporation process, using a liquid heat transfer medium and vapor compression for the concentration of scaling waste waters was presented [4]. The design of a small-capacity seawater distillation plant uses a combination of a wiped-film rotating-disk evaporator unit and a vapor compressor [5].

A new generation of desalination plants is available operating on low tempera-ture evaporation of about 50°C and using vapor compression to achieve advantageous thermal performances. Seawater is sprayed over a tube bundle and evaporated. Steam is condensed inside the tubes at a temperature difference of 1.5 to 4°C. Two kinds of compressors are used: either steam ejectors operating with medium or high pressure steam or mechanical compressors, according to local conditions. High purity water of 1 to 2 ppm is obtained. Over one hundred such plants are in service throughout the world, in particular in petroleum companies, chemical industries, etc. [6].

An analysis, from the thermodynamic point of view, of energy consumption to maintain operation of the most commonly used desalination processes includes multi-effect distillation use of a heat pump, vapor compression and multistage flash distillation. The most energy expensive process is distillation and the less expensive vapor compression [7].

Patents. In a transportable desalination plant, feedwater is preheated by the discharged brine and the distillate in separate preheaters. The preheated feed is sprayed to deaerate it and passed into the heater, then to the evaporation chamber where it is recirculated. Steam is removed and compressed and then condensed in the heater. The condensate recovers heat from the exhaust of the driving diesel engine [8]. In a water desalination apparatus, the system described has no moving parts in the primary circuits, where evaporation and condensation take place. The heat for evaporation is provided via a closed secondary circuit which includes a heat pump [9]. Fresh water is prepared from seawater by distillation, adiabatic mechanical compression and condensation of the water vapor. The waste heat is used to heat the seawater [10]. Feed solution is evaporated to produce a foamy mixture of liquid and vapor. After separation from the liquid the vapor is compressed and passed through a heat exchanger to heat the liquid in the evaporator [11].

VERTICAL MULTI-EFFECT EVAPORATORS

Patents. In a vertical multiple-effect evaporator, the number of horizontal heat-exchanger tubes in upper groups is lower than those in lower groups of tubes. A high-temperature vapor is introduced to the lowest tube group, passes through the upper heat-exchanger tube groups and is condensed by cold water sprayed outside the tubes. Vapor is generated outside the tubes [12]. The evaporators are stacked vertically in two series in a single casing. Only one pump is used for elevating the evaporation liquid [13]. Feed and heat exchange arrangements are described [14]. Ejectors extract vapor from the tubes of each effect and direct it across the outsides of the tubes of the other [15]. Evaporation and condensation trays are placed vertically, with the heater installed at the lowest evaporation chamber. Solution enters and vapor leaves the uppermost evaporation chamber[16].

VAPOR REHEAT

In the low temperature reheat distillation plant, a two-effect evaporator operating at a high vacuum is combined with a steam jet ejector-vapor compressor. Steam supplied to the ejector pumps out the vapor generated in the 2nd effect. The mixed vapor is introduced into the 1st effect and works as heating steam for the multi-effect distilling plant. Vapors from the 1st effect are used to heat the 2nd effect. A maximum brine temperature of 50°C prevails in the 1st effect. The yield of the ejectors has been recently in such a way improved, even with low pressure steam supply, that a performance ratio of 8 might be obtained. Because of low temperature operation, low potential energy such as hot water from a petro-chemical plant, flue gas waste heat or even solar energy could be used to drive the evaporators [17].

VERTICAL MSF EVAPORATOR

Patents. An evaporator consists of a bundle of tubes, each of which is an insulated pair of vertical concentric tubes. The annular space between the tubes is provided with a heater. The hot liquid drops through a series of funnel-shaped evaporation stages. Condensate rises beneath each funnel to an overflow that exchanges heat with the rising incoming liquid in the annular space [18]. A cylindrical concrete shell contains a number of flash chambers in a vertical stack with a heated liquid feed into the uppermost chamber. The chambers are connected by passages from the top of one chamber to the bottom of the next lower chamber. The pressure in each chamber is successively reduced. Vapor is counter-currently

58

contacted with cold fresh water in annular condenser chambers [19]. Several flash evaporation chambers are piled vertically. A heat transfer tube chamber is placed above the uppermost stage [20].

MULTI-STAGE FLASH FLUIDIZED BED EVAPORATOR

This newly developed multi-stage flash principle enables the application of one single standard module for the design of complete plants, covering a wide range of distillate production and/or heat consumption. The multi-stage flash fluidized bed evaporator suits the standard module philosophy very well, because of the following plant characteristics: it can operate in a once-through cycle up to relatively high maximum brine temperature, without adding any chemicals to the feed, the liquid velocities in the condenser tubes can be varied over a large range by changing the bed porosity, particle size and density of the particle material, and the compact construction of the plant as a result of the small flash chamber dimensions [21].

For more than one year the Provinciale Elektriciteits Maatschappij Nord-Holland operates on the isle of Texel a 500 m^3/day multi-stage flash fluidized bed evaporator, parallel to a 3000 m^3/day conventional multi-stage flash evaporator. Operating characteristics of both types of plants are given and the advantages of the MSF/FBE are outlined [22].

ROTATING EVAPORATORS

Patents. A design of feed solution flow arrangement of a thin film evaporator with rotor prevents back-mixing [23]. In cylindrical film evaporators, cleaning of the inside heating walls is facilitated when the axial shaft is rotatable in opposite directions and is fitted with rubbing members, which engage the heating surface with zero contact pressure, when rotated in one direction, and with a rubbing pressure when rotated in the opposite direction [24].

LITERATURE TO 2F.

1. J. W. Casten (Chem. Eng. Prog. 74, No 7 [1978] 61/67)
2. U. Fisher (Israel J. Techn. 15, No 3 [1977] 102/111)
3. Anonymous (NTIS Rept. PB-281124/8GA; W78-06994 [1977].- G.R.A. 78, No 18 [1978] 125)
4. R. Rautenbach, M. Wirtz (Chem. Ing. Tech. 51 [1979] 44/45.- Wiss. Umwelt No 1 [1978] 48/50)
5. B. W. Tleimat (NWSIA J. 5, No 2 [1978] 7/20)
6. D. Larger (Desalination 31 [1979] 125/134)
7. P. Blanc-Feraud (Desalination 30 [1979] 469/482)
8. Standard Messo Co. (Brit. 1.490.837, 2 Nov 1977)
9. J. J. Hoiss (French 2.337.693, 9 Sep 1977)
10. W. Bulang, W. Laber (W. Ger. 2.731.159, 18 Jan 1979.- C. A. 90 [1979] 156956)
11. J. E. Pottharst (Brit. 1.526.053, 23 Sep 1978)
12. M. Takada (Jap. 77.141.483, 25 Nov 1977.- C. A. 89 [1978] 91482)
13. M. Takada (Jap. 78.15.834, 27 May 1978.- C. A. 89 [1978] 165387)
14. M. Takada (French 2.348.725, 23 Dec 1977)
15. M. Takada (French 2.348.726, 23 Dec 1977)
16. H. Nishimoto (Jap. 77.131.974, 5 Nov 1977.-C. A. 88 [1978] 110348)
17. D. K. Forsyth, M. Takada (Desalination 31 [1979] 145/151)
18. H. Schreiner (W. Ger. 2.640.977, 16 Mar 1978.- C. A. 89 [1978] 165378)
19. F. L. Speed (Brit. 1.517.510, 12 Jul 1978.- C. A. 90 [1979] 61036)
20. S. Nakaya, R. Ueno, Y. Kuroda, S. Tsujita (Jap. 77.47.742, 5 Dec 1977.- C. A. 89 [1978] 61468)
21. D. G. Klaren, N. Halberg (Desalination 31 [1979] 233/240)
22. G. Spanhaak (Desalination 31 [1979] 511/519)

23. A. N. Marchenko, A. B. Tjutjunnikov (French 2.350.123, 6 Jan 1978)
24. Kansai Kagaku Kikai Seisaku K. K. (Brit. 1.506.545, 5 Apr 1978.- C. A. 89
    [1978] 131570)

2G. Combined Distillation Plants

VERTICAL TUBE AND MULTI-STAGE FLASH EVAPORATION

The desalting module at Water Factory 21 was originally designed with full-size
components and severely limited efficiency. Changes in budget allocations neces-
sitated reappraisal of the design philosophy to enable the seawater desalting facility to make a significant contribution to the total output of Water Factory
21 [1].

A VTE/MSF plant of capacity 2.25 Mgd was producing water with elevated conductance. After the demisters had been reset and the levels in each effect adjusted to 30 to 35%, instead of 50%, the conductance of the product water remained
at zero to 2 micromhos [2].

Patents. Multi-stage flash preheaters and multiple-effect evaporators are contained in a horizontal casing. Thermal energy consumption is decreased and the
efficiency increased [3].

VC - VTE - MSF COMBINED PLANT

The design of a combined vapor compression, vertical tube evaporator and multi-
stage flash distillation single purpose plant with a nominal capacity of 8 Mgd
was presented. The process uses a gas turbine powered vapor compressor that operates across four effects of a vertical tube evaporator. A multi-stage flash evaporator serves as a low-temperature brine heater. Brine recirculation is not used.
A noncondensing steam turbo-generator provides all site power requirements and
a waste heat recovery boiler provides the required steam. Two energy levels of
steam are generated, one at approximately 260°F saturated and the other at 650°F,
400 psig, supersaturated, which is used to power the backpressure steam turbine,
coupled to an electric generator. The exhaust steam at 257°F from the backpressure turbine is combined with the steam from the low pressure boiler and used as
supplemental heat unput to the first VTE effect. The vapor compressor receives
the low pressure steam from the fourth VTE effect at 226°F and increases its temperature to 343°F superheated steam, which is then desuperheated to 257°F. The
incoming seawater feed is preheated in the MSF plant to 109°F. After sulfuric
acid injection, decarbonation and deaeration, the seawater feed is further heated
through the MSF train to 209°F. This feed stream then proceeds through four feed
heaters, one between each VTE effect, and sprayed to the inside of fluted tubes
the first VTE effect. Essentially, all heat of vaporization from the steam is
transferred to the incoming brine, generating an equivalent amount of product
water. The blowdown brine from the fourth effect enters into the flash stages of
the MSF train. Similarly, the product from the vertical tube evaporator is flash-
ed down through the MSF train in order to recover its sensible heat.

The plant capital cost was estimated to 21.5 million 1979 dollars, amounting
to dollars 1.69 per gpd capacity. The production costs for water, based on dollar
2 per million BTU and 15% capital recovery area estimated to dollars 2.25 per thousand gallons [4].

Patents. The steam from a low pressure stage in a multi-stage evaporator is
heated and pressurized in a high-pressure steam condenser, using steam from a
boiler. The steam heats the first evaporator to improve efficiency [5].

LITERATURE TO 2G.

1. R. E. Bailie (Presented at the 5th Ann. Conf. of the Natl. Water Supply Improv.
   Assoc., San Diego, Calif. 1977)
2. G. Sheppard (Pure Water 5, No 1 [1976] 3/4)
3. H. Sawada (Jap. 78.13.419, 10 May 1978.- C. A. 89 [1978] 165385)
4. S. J. Senatore (Desalination 31 [1979] 135/144)
5. H. Satone (Jap. 78.116.274, 11 Oct 1978.- C. A. 90 [1979] 92218)

2H. Dual Purpose Plants

The method of allocation of steam generating costs greatly influences the specific cost of heating steam and desalted seawater. From the point of view of so-
cial economics, the cost allocation method, based on the cost of the heating energy generated by a single-purpose heat source, yields sound results. The usual criterion for steam cost evaluation, based on business administration viewpoints, is
the steam energy content. The advantages of the cost evaluation, using the geo-
metric mean of heat and exergy, in comparison with the calorific, the exergetic
and the arithmetically averaging methods are shown [1].
Dual purpose plants are to be designed to satisfy the requirements of controll-
ability, reliability, simple operation and economy. These conditions are fulfilled
not only by steam-turbine power plants, but also by gas turbine power plants [2].
A gas turbine combined with desalination plant can meet the rapidly increasing demand for both power generation and water production. The plant described consists
of three gas turbines with a 14600 kW power output, three unfired heat recovery
steam generators, one oil fired auxiliary boiler and three 4500 m³/day MSF evaporators [3]. The waste heat of gas-turbine flue gases in a power station can be
used for water desalination. For a 50 MW station, 238 t/h of water can be obtained
by cooling the flue gases [4]. Gas turbines with waste heat boilers are a suitable
method of combined heat and power generation [5].
A review of possible desalting processes in combination with electric power generation refers to possible thermal arrangements, type and unit size, possible li-
mitations and relevant design details [6]. Using well proven elements of single-
cylinder design, steam-turbines can be built for outputs up to 100 or 150 MW [7].
In combining seawater desalination with a power plant, optimization can vary
as the case may be. The most important characteristics of such a plant, as the
result of the optimization work, are the water to power ratio and the economy ratio of the desalination plant [8]. Computer simulation studies consider a steam
power station of 550 MWe and a desalination plant of 12 Mgd capacity. Each plant
has an independent cooling water intake. The study is concerned with establishing
a normal procedure for cooling water pump startup and shutdown and the system response to the hydraulic transient due to single pump trip, station backout and emer-
gency shutdown of one of the two plants. Waterhammer could occur at elevated points
of the system in one of the two plants due to improper control and operation.
Proper operating procedures to minimize the pressure surge are suggested [9].
A novel method is described in which a vertical tube evaporator generates a
secondary steam supply, which is used to power an auxiliary low-pressure turbine.
The economics of this system present considerable advantages over the conventional
dual-purpose VTE or MSF systems, particularly for low water-to-power ratios [10].

Patents. Part of the waste heat from the exhaust gases of a gas turbine is
used in a boiler to generate steam for a turbine and the other part provides hot
water for the desalination plant. Thermal efficiency is improved by combining
gas and steam turbines [11]. Seawater is heated by solar energy and passed through
multiple-stage flash evaporator [12] or a 3-effect evaporator. Vapor from the
first effect is used to generate power in a turbine before condensing to potable
water. Vapor from the second effect is used to preheat the feed water [13]. The
power station comprises a gas turbine set, a steam turbine and a waste heat boiler,
which receives the hot exhaust gases from the gas turbine. The waste heat boiler

has two sections of which the first supplies steam to the turbine and the second
supplies heat to the desalting plant [14].

The marine floating desalinator consists of a gas turbine driven generator to
provide power for driving auxiliary machinery and steam as a heat source for MSF
evaporator [15]. The distillation apparatus includes steam turbines with steam
extraction [16].

OPERATING EXPERIENCE

Libya. Combined water and electricity production plants have been installed in
Cyrenaica, Libya. The plants are running satisfactorily [17]. The Tobruk plant
consists of three condensing turbines, 33 MW each, and two back pressure turbines
15 MW each. The desalting plant produces 24000 m^3/day in four units [18].

Saudi Arabia. A brief survey is given of the combined power station and desalination plant Jeddah-2 at the Red Sea. The power plant supplies 2 × 42 MW electric power and 40000 m^3/day drinking water [19]. The major design orientations
of the combined Yanbu and Medina power and desalination plant with a capacity of
95000 m^3/day potable water and 250 MW of electrical power were reported. This
plant replaces the originally planned individual plants at Yanbu and Medina. Pipelines are conveying the product water to Yanbu town and to Medina town [20].

LITERATURE TO 2H.

1. B. Kunst, J. Hapke (Desalination 26 [1978] 309/317)
2. M. Haaland, K. Schueller (Brown Boveri Rev. 64 [1977] 532/539.- E.I. 16 [1978]
   60493)
3. M. Kato, A. Sakai, T. Inui (Congr. Int. Mach. à Combust. CIMAC, Vol. C [1977]
   2415/2437.- E.I. 16 [1978] 81122)
4. S. Ivic, S. Latifagic (Arh. Rud. Tehnol. 15, No 1-4 [1977] 63/68.- C. A. 91
   [1979] 44331)
5. K. F. Pickhardt (Energie 30 [1978] 315/318)
6. A. Spechtenhauser (Inst. Mech. Eng. Conf. Publ. [1977-3] 33/42.- E. I. 16
   [1978] 64413)
7. A. Spechtenhauser (Brown Boveri Rev. 64 [1977] 522/531.- E.I. 16 [1978] 62462)
8. K. Kuenstle, V. Janisch (Desalination 30 [1979] 555/569)
9. W. S. Hsieh, J. R. Stange (Desalination 31 [1979] 183/184)
10. I. Kamal, C. H. Hughes (Desalination 31 [1979] 533/544)
11. Brown Boveri Co. (Brit. 1.518.486, 19 Jul 1978.- C. A. 90 [1979] 106949)
12. D. B. Carson (U.S. 4.110.174, 29 Aug 1978.- C. A. 90 [1979] 192339)
13. D. B. Carson (U.S. 4.121.977, 24 Oct 1978.- C. A. 90 [1979] 174466)
14. H. Pfenninger (U.S. 4.094.747, 13 Jun 1978)
15. Hitachi Ltd. (Brit. 1.533.083, 22 Nov 1978.- C. A. 91 [1979] 44367)
16. J. Mary (French 2.383.129, 10 Nov 1978)
17. T. Michels (Brown Boveri Rev. 64 [1977] 513/521.- E.I. 16 [1978] 56983)
18. B. Boergardts (BBC Nachrichten 60 [1978] 211/218)
19. K. Zimmermann (BBC Nachrichten 59, No 5 [1977] 184/189)
20. I. M. R. Jamjoom, G. Costes (Desalination 30 [1979] 283/300)

2I. Waste Heat as Energy Source

References on waste heat utilization were compiled, covering citations from the
NTIS data base for the period 1964 to March 1978. The bibliography contains 253
abstracts, 37 of which are new entries to the previous edition [1]. A similar bibliography contains 180 abstracts of publications from the Engineering Index data
base and covers the period 1975 to 1976 [2] and the period 1977 to March 1978 containing 153 abstracts [3]. The bibliographies include studies of waste heat boilers and the use of waste heat for irrigation, sewage treatment, odor control, desalination, heating and aquaculture.

A summary was presented of the research programs in the U.S. that have studied the utilization of reject heat from both fossil and nuclear power plants [4]. Low temperature waste heat is utilized for distillation of saline and polluted water. Effluents from industry are treated where the flows of waste heat and effluent are balanced to satisfy environmental protection conditions [5]. A review of processes, facilities, operations etc. for desalination of seawater by waste heat was presented [6]. French experience in using thermal wastes from nuclear power plants for the benefit of agricultural production was reported [7].

Modern electric steel works with water-cooled electric-arc furnaces supply waste heat of temperatures suitable for conventional desalination plants. Special requirements of the desalination process, with respect to its source of energy, are compared with the energy supplied by the electric-arc furnace. Solutions to the resulting problems are indicated and various linking-up circuits are developed, with special emphasis being placed on the storage of waste heat [8]. Nuclear power plants, situated on a river bank, with a capacity of 1000 MWe require 2000 m^3/day softened cooling water. If a Δt of 10 to 15°C is available, horizontal tube evaporator with liquid spraying is more advantageous than the conventional MSF process. A study is presented of a 1500 m^3/day evaporator using hot and cold water from the condenser cooling cycle of the power plant. Product water cost is competitive to other conventional desalting processes [9].

Patents. The heat from a refrigeration compressor coolant is exchanged with seawater, which is sprayed in an aerating sprayer and the moisture condensed [10]. A automatic system controlling the operation of a desalination plant, particularly utilizing waste heat from internal-combustion engines is decribed. The automatic control system is responsive to physical parameters and is particularly useful during warm-up and cooling of the desalination systems [11].

DEHUMIDIFICATION

In an evaporation/humidification process for treating wastewater effluent, a major portion of the effluent is recovered as water of high purity suitable for recycle or reuse. Low-pressure steam generated from waste heat is sufficient to operate it. A pilot plant was operated satisfactorily [12]. The atomizing desalination process uses substantially less energy than conventional processes. Seawater is pumped to about 6 atm, mixed with compressed air and fed through nozzles producing a high velocity jet of small water droplets. Rapid vaporization and desalination occured at 30°C [13].

Patents. Moisture is extracted from humid air by an absorbant from which it is removed by a current of dry air and passed into a condenser. The vapors are condensed by heat exchange with incoming humid air [14]. A solution is evaporated by direct contact with hot air and the moisture in the air is condensed. The method can be used with waste water and the solutions from waste-gas treatment [15]. The method is also useful for salt solutions and waste-water containing salts. Scaling is absent and the abtained crystals are fine [16]. Seawater is slightly heated and partly evaporated by the passage over it of a stream of air, which is then passed through a condenser [17].

LITERATURE TO 21.

1. A. S. Hundemann (NTIS Rept. PS-78/029/5GA [1978])
2. A. S. Hundemann (NTIS Rept. PS-78/0295/2GA [1978])
3. A. S. Hundemann (NTIS Rept. PS-78/0296/0GA [1978])
4. W. F. Witzig, D. R. De Walle (Nucl. Technol. 38, No 1 [1978] 25/34)
5. F. Kranebitter (Complete Pap. Int. Top. Meet. Low Temp. Nucl. Heat [1977] 236/245.- C. A. 90 [1979] 12067)
6. Y. Nagashima (Netsu Kanri To Kogai 30, No 8 [1978] 45/50.- C. A. 90 [1979] 28690)
7. P. Balligand, P. Le Gouellec, M. Dumont, A. Granby (Nucl. Technol. 38, No [1978] 90/96)

8. R. Rautenbach, P. Monheim, D. Zebrowski (Desalination 31 [1979] 197/206)
9. P. Combaz, J. C. Deronzier, F. Lauro (Desalination 31 [1979] 207/218)
10. A. Matsumoto (Jap. 78.71.681, 26 Jun 1978.-C. A. 90 [1979] 28865)
11. H. H. Carnine, C. P. Robinson (U.S. 4.096.039, 2 Jun 1978.-C. A. 90 [1979] 43639)
12. D. G. Reininga (NTIS Rept. PB-284097 [1978] 68 pp.- C. A. 90 [1979] 192012)
13. E. Lerner, W. J. Mills (Nat. Resour. Forum 2 [1978] 399/404)
14. Maschinenfabrik Augsburg-Nürnberg A. G. (French 2.377.834, 26 Sep 1978)
15. I. Okochi, H. Yamazaki, K. Izumi, S. Takahashi (Jap. 78.119.782, 19 Oct 1978.- C. A. 90 [1979] 73924)
16. H. Yamazaki, I. Okochi, K. Izumi, S. Takahashi (Jap. 78.119.783, 19 Oct 1978.- C. A. 90 [1979] 73923)
17. Compagnie pour le Dessalement de l'Eau de Mer S.A. (French 2.379.482, 6 Oct 1978)

2K. Nuclear Energy as Heat Source

Various possibilities exist of coupling electricity, heat for district heating and desalination in multipurpose nuclear power plants [1].

A brief review is given of single-purpose desalination plants heated by nuclear reactors. Reference in made to some existing projects and to two new concepts recently developed in Israel [2]. The requirements for nuclear reactor heated seawater desalination plants, with a performance of 10000 to 80000 m^3/day, show that intergrated pressurized water reactor systems of the nuclear ship Otto Hahn, have specific advantages for generation of up to 200 MW thermal power, when compared to other reactor types being adapted for single- and dual-purpose desalination plants [3]. In a study, the multi-stage flash distillation process was selected in conjunction with a reactor rated at not less than 2100 MWth. Combined use of a condensing and a back-pressure turbine, the latter matched to distillation plant steam requirements, represents a convenient method for supplying process heat. Overall costs can be fairly allocated to the two products, using the power credit method. An economic evaluation indicates highly favorable water costs, as compared with more conventional distillation schemes based on fossil fuel [4]. The Nord-Aqua, Finland, vacuum evaporation process utilizes waste heat from cooling water of nuclear power plants at less than 30°. The vacuum is obtained by a barometric column and siphon. The pumping energy is 9 kcal/kg, or 1.5% of the heat of water evaporation [5].

Actual oil price level allows nuclear dual-purpose plants to become efficient below a thermal power of 800 MW. Two special applications of an integrated pressurized water reactor system use the waste heat of the turbine for water desalination or district heating [6]. The optimization of the turbine plant of a nuclear power station in combination with heat production is dependent upon the heat requirements, full-load equivalent operating time and the heat transport distance. The turbine can be a back-pressure turbine, a back-pressure turbine with condensing tail or a condensing turbine with heat extraction. The most attractive solution is the condensing turbine with extraction for district heating or desalination [7]. The U.S. Department of Energy program is evaluating the applications of nuclear heat to industrial heat supply, district heating, desalting and waste heat utilization. Integral pressurized water conceptual designs rated at 365 and 1200 MWth have been developed to support these applications [8].

Other work on the use of nuclear heat for industrial applications includes an assessment of a small pressurized water reactor for industrial energy [9], the use of nuclear reactors for centralized heat supply and district heating [10], experimental study of the combined utilization of nuclear power heating plants for big towns and industrial complexes [11], various papers presented at the international meeting on Low Temperature Nuclear Heat, held in Otaniemi, Finland, including economics and potential use of low temperature nuclear heat [12], heat extraction from nuclear power plants [13], front heat extraction from steam generators at the top of the steam cycle [14], economics of long-distance transmission,

storage and distribution of heat from nuclear plants with existing and newer techniques [15], nuclear wastes as a heat source [16], economics of nuclear district
heating from the utility's point of view [17] and dual-purpose nuclear power plants
for military installations [18].

Other work related to the technology of small and medium size nuclear reactors
includes a presentation on THERMOS reactors [19], the use of the THERMOS nuclear
reactor for desalting [20], arguments for the installation of small LWR nuclear
power plants in developing countries [21], problems and prospects of small and
medium power reactors [22], CAS (chaudière avancée de série) medium-size nuclear
plants [23], innovations in PHWR design, integration of nuclear power stations
into power systems and the role of small nuclear power plants in a developing
country [24] and technical and economic studies of small reactors for supply of
electricity and steam [25].

A study to develop a conceptual design and cost estimate for a 25 Mgd seawater
reverse osmosis desalting plant, operating at both Caribbean and Persian Gulf sites,
in conjunction with a 1000 MWe nuclear power plant, was made. For both sites, a
two-stage system was selected for the conceptual cost estimate and areas of potential cost reduction were indicated [26].

NUCLEAR VERSUS CONVENTIONAL POWER

Process steam from intermediate size and large reactors is cheaper than steam
produced in conventional boilers burning coal or oil at $ 2.21/GJ ($ 14 per barrel
in 1976 dollars). Small nuclear and coal-based steam supply systems may be competitive with oil under criteria that consider long-term economic inflation and
the detailed financial and tax structure of the energy user [27].

SAFETY AND ENVIRONMENT

Engineering management and manpower aspects of potential nuclear desalination
projects in non-nuclear countries were considered with emphasis on safety and
quality assurance. Detailed investigation was made of manpower and staffing requirements of dual-purpose nuclear power and desalination plants. The case of
Saudi-Arabia is of interest due to the peculiarities of a shortage in human capital and a lack of industries necessary to support desalination and nuclear energy
activities [28].

LITERATURE TO 2K.

1. P. Salminen, Y. El Mahgary, A. Ranne (Symp. on nuclear energy applications
   other than electricity production, Jülich, W. Germany 1976.- E.R.A. 3 [1978]
   12298)
2. Y. Ronen, A. Goldfield (Ann. Nucl. En. 4, No 4 [1977] 183/184.- E.R.A. 3
   [1978] 12307)
3. G. Petersen, M. Peltzer (Complete Papers Int. Top. Meet. Low Temp. Nucl. Heat
   [1977] 205/213.- C. A. 90 [1979] 12065.- Nucl. Technol. 38 [1978] 69/74.-
   E.I. 16 [1978] 32592)
4. G. Waplington, H. Fichtner (Complete Pap. Int. Top. Meet. Low Temp. Nucl. Heat
   [1977] 226/235.- C. A. 90 [1979] 11932.- Nucl. Technol. 38 [1978] 215/220.-
   E.I. 16 [1978] 43894)
5. R. Saari (Complete Pap. Int. Top. Meet. Low Temp. Nucl. Heat [1977] 214/225.-
   C. A. 90 [1979] 12006.- Nucl. Technol. 38 [1978] 209/214.- C. A. 89 [1978] 65024)
6. F. K. Boese, H. Kadella (Nucl. Technol. 38 [1978] 235/241)
7. B. Frilund, K. Knudsen (Nucl. Technol. 38 [1978] 120/125)
8. W. F. Savage, I. Spiewak (Nucl. Technol. 38 [1978] 19/24)
9. O. H. Klepper, L. C. Fuller, M. L. Myers (NTIS Rept. ORNL/TM-5881 [1977].-
   E.R.A. 3 [1978] 12299)
10. B. B. Baturov, L. A. Melent'ev, Yu. M. Bulkin, I. N. Galaktionov, G. A. Zvere-
    va, Yu. I. Tokarev, V. A. Chernyaev, E. S. Smirnova (Int. Conf. Nucl. Power
    and its Fuel Cycles, Salzburg, IAEA-CN-36/338, 6 [1977] 449/467)

11. J. Neumann, K. Barabas (Int. Conf. Nucl. Power Fuel Cycles, Salzburg, IAEA-CN-36/465, 6 [1977] 469/478)
12. D. Oesterwind (Nucl. Technol. 38 [1978] 11/18)
13. G. Deuster, P. Zenker (Nucl. Technol. 38 [1978] 35/40)
14. Ph. Aussourd (Nucl. Technol. 38 [1978] 97/103)
15. P. H. Margen (Nucl. Technol. 38 [1978] 192/203)
16. W. F. Witzig, M. E. Forster (Nucl. Technol. 38 [1978] 258/263)
17. M. Timm (Nucl. Technol. 38 [1978] 280/287)
18. G. S. Stewart, G. T. Story (Nucl. Technol. 38 [1978] 264/270)
19. G. Dupuy, M. Labrousse, B. Lerouge, J. P. Schwartz (Complete Pap. Int. Top. Meet. Low Temp. Nuclear Heat [1977] 391/399.- C. A. 89[1978] 222646.- Nucl. Technol. 38 [1978] 242/247.- E.I. 16 [1978] 43624)
20. B. Lerouge, F. Lauro (Desalination 30 [1979] 571/579)
21. B. Lezenik (Int. Conf. Nucl. Power Fuel Cycles, Salzburg, IAEA-CN-36/89, 6 [1977] 363/376)
22. A. Matin (Int. Conf. Nucl. Power Fuel Cycles, Salzburg, IAEA-CN-36/1, 6 [1977] 377/385)
23. L. Vogelweith, J. C. Lavergne, G. Martinot, A. Weiss (Int. Conf. Nucl. Power Fuel Cycles, Salzburg, IAEA-CN-36/568, 6 [1977] 387/401)
24. S. K. Mehta, A. Kakodkar, M. R. Balakrishnan, R. N. Ray, L. G. K. Murthy, B. F. Chamany, S. L. Kati (Int. Conf. Nucl. Power Fuel Cycles, Salzburg, IAEA-CN-36/394, 6 [1977] 403/417)
25. I. Spiewak, O. H. Klepper, L. C. Fuller (Int. Conf. Nucl. Power Fuel Cycles, Salzburg, IAEA-CN-36/398, 6 [1977] 419/432)
26. S. C. May, E. H. Houle, S. A. Reed, W. F. Savage (Desalination 30 [1979] 605/612)
27. I. Spiewak, O. H. Klepper (Nucl. Technol. 38 [1978] 288/294)
28. A. F. Abdul-Fattah (Thesis, Iowa State Univ. 1978.- Diss. Abs. B 39 [1978] 32/33)
    A. F. Abdul-Fattah, A. A. Husseiny (Desalination 27 [1978] 283/331)

2L. Geothermal Energy as Heat Source

Bibliographies on geothermal energy include exploration with 174 abstracts from the NTIS data base [1], corrosion and equipment with 88 abstracts from the NTIS data base [2], technology and general studies with 326 abstracts from the NTIS data base [3] and 323 abstracts from the Engineering Index data base [4]. All bibliographies cover the period May 1976 to July 1978.

Desalting tests have been conducted at the East Mesa test site, California, utilizing geothermal fluids on vertical tube evaporator, multistage distillation and a high temperature electrodialysis process. Effective scale control has been developed by the use of a threshold treatment chemical. Injection has been successfully achieved to dispose of waste fluids [5]. Both distillation plants have shown good resistance to scaling and feasible solutions have been found to various problems, including process control and scaling in process piping. The electrodialysis plant has performed well, as far as salt removal is concerned, but it does not remove boron and silica, which require other methods to reduce their concentration [6]. The degassification and desalination of mineral waters in geothermal power plant systems were presented [7].

Other work on geothermal brines includes calculation of geothermal brine thermophysical properties [8], linear polarization measurements at high temperatures in hypersaline geothermal brines [9], technology development for high-salinity geothermal resources [10], primary variables which cause some common scales in saline water systems [11], scaling in both high- and low-salinity brines [12], Coso Thermal Area geothermal corrosion studies [13], study of brine treatment [14], methods for geothermal brine treatment [15], effects of organic additives on the formation of solids from hypersaline geothermal brine [16], processing of geothermal brine effluents for injection [17], brine chemistry, combined heat and mass transfer, Vol. 1: text [18] and Vol. 2: 5 appendixes [19], geothermal direct contact heat

exchange [20], construction and operation of a heat exchange system for the study
of scale suppression from simulated geothermal brines [21], 2000-hour heat exchanger
study [22], thermodynamic study of heating with geothermal energy [23], thermosyphon
models for downhole heat exchanger applications in shallow geothermal systems [24],
heat transfer in power from hot geothermal brines [25], geothermal power from salt
domes [26], geothermal energy project Idaho [27], optimum design conditions for a
power plant at a vapor dominated geothermal resource [28], performance of a 10 MW
geothermal energy conversion test facility [29], scale formation problems in developing geothermal energy [30], corrosion susceptibilities of various metals and
alloys in synthetic geothermal brines [31], viscosity of brines from the Salton
Sea geothermal brine, Imperial Valley, California [32], operational experience at
the San Diego Gas and Electric/ERDA Niland geothermal loop experimental facility
[33] and calculation of geothermal brine thermophysical properties [34].

LITERATURE TO 2L.

1. M. F. Smith (NTIS Rept. PS-78/0664/9ENS [1978])
2. M. F. Smith (NTIS Rept. PS-78/0665/6ENS [1978])
3. M. F. Smith (NTIS Rept. PS-78/0666/4ENS [1978])
4. M. S. Smith (NTIS Rept. PS-78/0667/2EES [1978])
5. W. A. Fernelius, M. K. Fulcher (ASCE J. Environ. Eng. Div. 105, No 1 [1979]
   13/32.- C. A. 90 [1979] 156890)
6. S. H. Suemoto, T. E. Lindemuth (Presented at the 5th Ann. Conf. of the National Water Supply Improvement Assoc, San Diego, California [1977])
7. S. N. Mandzhgaladze, D. R. Ramazashvili (Metalloved. i Korroziya Met. No 5
   [1977] 141/146.- Ref. Zh., Teploenerg. [1978] 2 R 108)
8. G. L. Dittman (Univ. Calif. Rept. UCID-17406 [1977].- C. A. 88 [1978] 11620)
9. S. D. Cramer (U.S. Bur. Mines Rep. Invest. 8303 [1978].- E.I. 17 [1979] 016929)
10. A. W. Lundberg (NTIS Rept. PC-A03/MF-A01 [1977].- E.R.A. 3 [1978] 23520)
11. A. G. Collins (Proc. Conf. Scale Manag. Geotherm. Energy Dev. [1976] 115/125.-
    C. A. 89 [1978] 217877)
12. P. B. Needham, A. P. Murphy, F. X. McCawley (Proc. Conf. Scale Manag. Geotherm.
    Energy Dev. [1976] 127/144.- C. A. 89 [1978] 217878)
13. S. A. Finnegan (NTIS Rept. PC-A05/MF-A01 [1977].- E.R.A. 3 [1978] 39920)
14. S. L. Phillips, A. K. Mathur, R. E. Doebler (Rept. EPRI-ER-476 [1977].- C. A.
    89 [1978] 152139)
15. S. L. Phillips, A. K. Mathur (Univ. Calif. Berkeley, Rept. EPRI-ER-660-SR [1978]
    2.43/2.49.- E.R.A. 3 [1978] 39922)
16. J. E. Harrar, L. E. Lorensen, C. H. Otto, S. B. Deutscher, G. E. Tardiff
    (NTIS Rept. MF-A01 [1978] .- E.R.A. 3 [1978] 52347)
17. R. Quong, F. Schoepflin, N. D. Stout (NTIS Rept. PC-A02/MF-A01 [1978].- E.R.A.
    4 [1979] 510)
18 D. W. Shannon, R. A. Walter, D. L. Lessor (Rept. No EPRI-ER-635.- NTIS No
    PC-A11/MF A01 [1978] 248 pp.- C.A. 89 [1978] 152510)
19. D. W. Shannon, D. W. Faletti, D. L. Lessor, J. R. Morrey (Rept. No EPRI-ER-660-
    SR.- NTIS Rept. PC-A12/MF A01.- E.R.A. 3 [1978] 39921)
20. A. V. Sims (U.S. Dept Energy Rept. SAN/2226-1 [1976])
21. G. R. Bullard, J. S. Wilson (A.C.S. 176th Nat. Meet. Miami Beach [1978])
22. E. L. Ghormley, J. L. Stern (Rept. EPRI-ER-660-SR [1978] 2.51/2.54.- E.R.A.
    3 [1978] 39923)
23. G. M. Reistad, B. Yao, M. Gunderson (ASME, J. Eng. Power 100, No 4 [1978]
    503/510)
24. D. B. Kreitlow, G. M. Reistad, C. R. Miles, G. G. Culver (ASME, J. Heat Transfer
    100, No 4 [1978] 713/718)
25. J. S. Swearingen (Chem. Eng. Progr. 73, No 7 [1977] 83/86)
26. S. J. Altschuler (Proc. 13th Intersoc. Energy Convers. Eng. Conf., San Diego,
    Cal. [1978].- E.I. 17 [1979] 016932)
27. Anonymous (Geotherm. Energy 6, No 8 [1978] 26/28.- E.I. 17 [1979] 016930)
28. J. Pietruszkiewicz (Proc. 13th Intersoc. Energy Convers. Eng. Conf., San Diego,
    Cal. [1978].- E.I. 17 [1979] 016933)

29. W. O. Jacobson (Proc. 13th Intersoc. Energy Convers. Eng. Conf., San Diego, Cal. [1978].- E.I. 17 [1979] 016931)

30. J. D. Rimstidt, H. L. Barnes (Earth Miner. Sci 47, No 2 [1977] 9/12.-E.R.A. 3 [1978] 21447)

31. R. D. Davis, Z. A. Munir (J. Mat. Sci. 12, No 9 [1977] 1909/1913.- S. W. R. A. 11, No 9 [1978] 95)

32. A. J. Piwinskii, R. Netherton, M. Chan (NTIS Rept. UCRL-52344 [1977].- G. R. A. 78 [1978] 162)

33. G. L. Lombard (Ann. Geoth.Progr. Rept. EPRI-ER-660-SR [1978] 3.11/3.16.- E.R.A. 3 [1978] 39914)

34. G. L. Dittman (NTIS Rept. UCID-17406 [1977] 24 pp.- C. A. 88 [1978] 11620)

2M. Solar Energy as Heat Source

A monography on Distillation of water using solar energy was published [1]. A review was presented on the most important and recent studies on solar distillation [2]. Solar water desalination plants of the greenhouse type, plants of high capacity with heat pipe collectors and flat plate collectors in connection with MSF evaporation were reviewed [3]. An assessment of solar distillation plant performance, attainable with presently existing technology, provides the basis for selection of the design for a 5 Mgd solar distillation plant. For plants constructed in the future, the continued escalation of fuel oil cost will make the cost of water from solar driven and oil driven distillers equal within 15 years [4].

An analysis is reported of a solution of the linearized differential equation describing the shape of drops on both sides of an inclined plane. It was shown that the heat-transfer coefficient during drop concentration of vapor on the lower surface of the inclined plane is greater than on the upper surface [5]. In a model of an inclined plane evaporator, the density distribution of the vapor-air mixture, the flow regime in the boundary layers on the evaporating and condensing surfaces and the distribution of relative humidity were measured. A cellular structure of the movement of the vapor-air mixture and laminar flow in the boundary layer for the angles of inclination from 15 to 90° were found [6]. A methodology was presented for simulating the behavior of different types of solar stills with the aim to select the best for the existing meteorological environment [7].

The classical basin type solar still is simple and suited for small capacities. The cost of water is high. For larger yields, conventional desalination processes, using solar energy, are more appropriate and give a cost of water lower by an order of magnitude [8]. The overall efficiency of a typical basin type solar desalination plant is 30% or lower. The major design factors affecting energy utilization are basin, condensing surface and ambient air temperatures. The efficiency of a solar desalination plant can be improved by controlling radiation from the plant basin and by the reuse of the latent heat of condensation [9].

A solar still with cover made from polyethylene and an area of 50 m^2 gave an output of 3.3 to 3.6 l/m^2.day. The construction cost is about one tenth that of reinforced concrete and glass stills [10]. Annual operating data for stepped-oblique solar stills, type SOU-1000, were presented and correlated with meteorological factors. Output varies from 0.6 l/m^2.day in January to 4.75 l/m^2.day in June. The efficiency varies from 18.0% in January to 46.1% in August [11]. A regenerative solar still was developed having two glass covers. Feed water is preheated in the space between the glass covers and then fed to the still. The efficiency is considerably higher in the second half of the day [12].

In a solar MSF seawater desalination plant, the solar energy input is generated by a system of collectors. Cost estimates show that the break even point with oil fired MSF desalination plants is just about reached. It might be said that solar applications will be profitable from now on [13]. Since most solar collectors have higher collector efficiency than solar stills, a combination of a solar flat-plate collector and an MSF evaporation unit would yield higher energy efficiencies compared to a simple solar still [14].

A joint project of Mexico and West. Germany is developed to test and prove the feasibility of the solar energy utilization as the source of thermal energy required for a multistage flash evaporation plant [15]. The project consists in the design, construction, operation and evaluation of a plant integrated in three systems: MSF plant with a capacity of 10 m³/day product water, performance ratio 9 kg/1000 kcal, flat plate collectors of 518.4 m² of effective collection surface and parabolic concentrators with an effective collection surface of 160 m² and an energy storage system permitting 24 h plant operation [16]. An MSF desalting plant with a capacity of 500 m³/day, 46 stages, cross tube type, with a top brine temperature of 121°C and a specific heat consumption of about 38.5 kcal/kg of distillate, with solar energy as heat source, was designed. The best solution is to use a single field of double axis tracking collectors. A solar powered desalting plant is not only feasible, but it appears that in some countries it might also be attractive from the economical point of view [17]. Two solar conversion systems have been considered for supplying thermal energy to a low temperature multiple effect desalination unit having a nominal capacity of 30 m³/day: Hot water from flat-plate collectors and low level enthalpic heat from the cooling of photovoltaic cells for the production of electrical energy [18].

An integrated system of solar powered electrodialysis combines a photovoltaic concentrator that produced direct current electric power and collects thermal energy with an electrodialysis desalting plant, which requires less electricity as the temperature of the feed water is raised [19]. In exploring the use of novel energy sources, one reverse osmosis unit with a capacity of 2.5 m³/h was coupled to a thermodynamic solar motor and another with a capacity of 0.5 m³/h will be coupled with an aeolian generator of 4 kW [20]. Two demonstration projects are presently under performance, using wind and solar power for the energy supply of reverse osmosis desalination units, to demonstrate the reliability and the low maintenance requirements for these plants [21].

A greenhouse, combined with solar distillation of brackish or seawater for autonomous operation, without any other water supply,was developed and tested [22].

Investigations on the influence of various plastic film materials, used in constructing solar stills, on the ion composition of the distillate from various feed waters were reported [23]. The quality of distillate during desalting of brines in regenerative solar adiabatic evaporators was studied under varying loads, evaporation temperatures and salinities [24]. In experiments on closed forced circulation of a steam-air mixture in a solar desalination facility, the steam-air mixture is conveyed to the glassed-in surface and split on the glass into two flows to each end of the desalination unit. Forced circulation increases the productivity of the solar still [25].

A technical and economic assessment of a 5 Mgd solar distillation plant was presented. Under actual construction conditions, the distilled water cost is approximately $ 7 to 9 per 1000 gallons [26]. A few solar distillation plants were recently installed in India. One year's operation of a 5 m³/day plant is reported. The plant had a productivity of 1.6 to 3.6 l/m²day with an efficiency varying from 21.7 to 33.7%[27].

Patents. Means of sealing and fastening the cover on a solar still were improved [28]. Saline water, covered by a thin transparent layer of less dense and less volatile liquid, is preheated in a shallow pool, then passed to an evaporator and the vapor condensed [29]. The distillation chamber is formed by a plastic tube streched on a frame and covered with a transparent inclined plastic roof [30]. Means were proposed to increase the temperature difference between the condensation surface and the ground surface of the still [31]. Lenses were used to focus sunlight on a double-jacketed dome [32]. In an automatic solar distillation apparatus convex lenses are used and an air stream is used to promote condensation [33].

Subatmospheric pressure is maintained in the evaporation chamber by a siphon[34]. Water vapors formed in an upper reservoir flow to an underground reservoir because of vacuum produced by a fan [35]. In a multi-step solar still an aspirator reduces the pressure in the evaporator [36]. In a solar still the saline water is preheated in two stages and evaporated in a vacuum chamber. The vapors are condensed in a heat-exchange dome and deposited solids are removed by means of moving belts [37]. Seawater is heated, in a desalination and salt recovery plant, by an intermediate heat carrier and evaporated at 0.1 atm [38].

Seawater and air are forced into a solar energy-assisted humidifier to form supersaturated air. The water vapor is condensed by cooler air [39] or in a condenser located under the sea [40]. In an apparatus using solar energy, a blower is provided for the forced circulation of the vapor in the vapor circulation path [41]. In another apparatus solar energy and dryness of the air are used [42]. Seawater is heated in solar collector tubes, sprayed in an evaporating chamber, where it is contacted with hot air and then condensed by incoming seawater [43].

The seawater is evaporated, transported and condensed in heat pipes [44]. Solar heat is transferred to the evaporation unit via a heat pipe or a heat conductor rod and then condenced by incoming seawater [45].

The cover of the still consists of double-paned glass, which allows a thin film of cold seawater to flow between the glass panes and keeps them cooler than a single pane would. Seawater is preheated by the heat of condensation [46]. A similar system was also used to preheat incoming seawater and create a heat sink to condense the evaporated water [47].

For the simultaneous production of power and potable water, saline water is preheated to about 135°F and passed through a series of flash chambers. The vapor from each is passed through a separate turbine to generate power and then condensed. The warm brine rejected from the last flash chamber vaporizes a hydrocarbon, whose vapor drives another turbine [48]. A medium is heated by a solar collector and first used to drive a turbine to generate power, then in vacuum evaporation of the salt water and recycled to the solar collector [49]. The temperature of a water-vapor-containing air stream is decreased by heat exchange with a vaporizable hydrocarbon to condense potable water. The hydrocarbon vapor is introduced in a turbine to generate power. Part of the saline water in the solar radiation heat sink is passed into a low-pressure flash evaporation zone and the vapor is used to generate power in a turbine. The exit vapors are condensed to recover additional potable water [50]. The water-laden air stream can be used to evaporate a hydrocarbon, which passes through a turbine to generate power [51]. A solar distillation apparatus with automatic operation is described, which does not need energy input for the operation of pumps [52]. A large area solar still with inflated plastic cover produces distillate or is used for sprinkling plant cultures[53].

Solar energy is concentrated by a reflector to a boiler and electric power is produced in a turbogenerator. The exhaust heat is used for the distillation of seawater and electricity for the production of hydrogen and chlorine by means of electrolysis [54].

LITERATURE TO 2M.

1. R. Bairamov, S. Seiitkurbanov (Ylum, Ashkhabad, Turkm. SSR [1977] 147 pp.)
2. P. Blanco (Proc. Conf. Heliotechn. Develop. Dhahran 2 [1976] 415/423)
3. U. Heidtmann (German Solar Forum, Hamburg, [1977] 349/362)
4. Bechtel Corp. (NTIS Rept. SAND-77-8176 [1977])
5. L. E. Rybakova, M. Mamedov (Geliotekhnika 13, No 3 [1977] 45/48.- Appl. Solar En. 13, No 3 [1977] 34/36)
6. R. Bairamov (Future Energy Production Systems, Hemisphere Publishing Corp., Washington D.C., 1 [1976] 123/129)
7. P. Granier, B. N'Doye, F. Rocaries, M. Daguenet (COMPLES Bull., September [1978] 33/39)
8. H. Z. Tabor (Impact of Sci. in Soc. 28 [1978] 339/348)
9. S. M. A. Moustafa, G. H. Brusewitz (Solar Energy 22 [1979] 141/148)
10. E. Zh. Norov, B. M. Achilov, A. Odinaev (Geliotekhnika 14, No 1 [1978] 75/76.- Appl. Solar En. 14, No 1 [1978] 61/62)
11. R. A. Akhtamov, B. M. Achilov, S. Kakharov (Geliotekhnika 14, No 2 [1978] 66/68.- Appl. Solar En. 14, No 2 [1978] 47/48)
12. R. A. Akhtamov, B. M. Achilov, O. S. Kamilov, S. Kakharov (Geliotekhnika 14, No 4 [1978] 51/55.- Appl. Sol. En. 14, No 4 [1978] 41/44)
13. N. Azhari (German Solar Forum, Hamburg, [1977].- E.R.A. 3 [1979] 44354)
14. D. Singh, Y. P. Gupta (Proc. 1st Natl. Sol. Energy Conv. [1976] 171/172.- C. A. 89 [1978] 220651)
15. X. Ibarra-Herrera (Desalination 30 [1979] 31)

16. R. Manjarrez, M. Galvan (Desalination 31 [1979] 545/554)
17. S. Arazzini, G. Tomei, E. Ciccone, G. Fiorito (Desalination 31 [1979] 443/455)
18. L. Gasparini, Giuffrida, Zanelli, Mighiorini, Franzosi (Desalination 31 [1979] 457/468)
19. J. E. Lundstrom (Desalination 31 [1979] 469/488)
20. A. Maurel (Desalination 31 [1979] 489/499)
21. G. Petersen, S. Fries, J. Mohn, A. Müller (Desalination 31 [1979] 501/509)
22. P. Balligand, P. Denis, M. De Cachard, A. Gouzy, R. Berger (Desalination 31 [1979] 429/434)
23. B. M. Achilov, E. Zh. Norov, G. B. Kamaeva (Geliotekhnika 12, No 4 [1976] 90/91.- Appl. Solar En. 12, No 4 [1976] 68/69)
24. S. Seiitkurbanov, M. Khanmamedov (Izv. Akad. Nauk Turkm. SSR, Ser. Fiz.-Tekhn., Khim. Geol. Nauk No 1 [1978] 107/110.- C. A. 89 [1978] 117450)
25. S. Seiitkurbanov, L. I. Rabinovich (Geliotekhnika 14, No 2 [1978] 50/52)
26. R. L. Lessley, T. E. Lindemuth, E. Y. Lam (Greater Los Angeles Area Energy Symp., North Hollywood, Cal. 4 [1978] 8/14.- E.I. 16 [1978] 95097)
27. G. L. Natu, H. D. Goghari, S. D. Gomkale (Desalination 31 [1979] 435/441)
28. H. R. Hay (U.S. 4,055,473, 25 Oct 1977)
29. J. C. F. Courvoisier, J. L. C. Meylan, D. M. Gross, J. P. D. Fournier (U.S. 4,053,368, 11 Oct 1977)
30. G. Friese (Swiss 601,113, 30 Jun 1978.- C. A. 89 [1978] 185878)
31. Y. Midorikawa (Jap. 78.03,968, 14 Jan 1978.- C. A. 89 [1978] 48756)
32. S. Kitabayashi (Jap. 78.10,031, 11 Apr 1978.- C. A. 89 [1978] 182412)
33. Y. C. Tsai (Brit. 1,497,953, 12 Jan 1978.- C. A. 89 [1978] 48766)
34. J. Doriel (Israel 45,126, 30 Apr 1978.- C. A. 90 [1979] 192342)
35. K. Piwowarczyk (Pol. 96,878, 30 Jun 1978.- C. A. 90 [1979] 142033)
36. T. Iida (Jap. 78.50,061, 8 May 1978.- C. A. 89 [1978] 203964)
37. R. Diggs (U.S. 4,118,283, 3 Oct 1978.- C. A. 90 [1979] 92215)
38. H. Fritzen (W. Ger. 2,707,715, 24 Aug 1978.- C. A. 89 [1978] 220709)
39. T. Takeuchi (Jap. 78.32,876, 28 Mar 1978.- C. A. 89 [1978] 185870)
40. T. Takeuchi (Jap. 78.32,877, 28 Mar 1978.- C. A. 89 [1978] 185871)
41. K. Hanada (Jap. 77.129,674, 31 Oct 1977.- C. A. 89 [1978] 48752)
42. J. Billon (French 2,372,771, 30 Jun 1978.- C. A. 90 [1979] 156955)
43. H. Yanagii (Jap. 78.02,385, 11 Jan 1978.-C.A. 89 [1978] 48757)
44. K. Arayama, K. Karauchi, K. Hanada (Jap. 78.06,273, 20 Jan 1978.-C.A. 89 [1978] 48755)
45. Y. Kashiwayanagi, Y. Himono, K. Yamada (Jap. 78.43,678, 19 Apr 1978.- C. A. 89 [1978] 203963)
46. R. Hock (S. Afr. 77.06,307, 14 Jun 1978, [W. Ger. 2,650,482].- C. A. 90 [1979] 109764)
47. C. G. Currin (U.S. 4,107,000, 15 Aug 1978.- C. A. 90 [1979] 43644)
48. D. B. Carson (U.S. 4,046,639, 6 Sep 1977)
49. T. Nakahara, T. Matsuoka (Jap. 78.31,575, 24 Mar 1978.- C. A. 89 [1978] 203961)
50. J. F. Spears (U.S. 4,078,975, 14 Mar 1978.- C. A. 89 [1978] 30535)
51. J. F. Spears (U.S. 4,078,976, 14 Mar 1978.- C. A. 89 [1978] 30536)
52. A. La Rocca (U.S. 4,135,985, 23 Jan 1979.- C. A. 91 [1979] 27143)
53. R. Freilaender (W. Ger. 2,626,902, 22 Dec 1977.- E.R.A. 4 [1979] 29694)
54. H. D. Brown (U.S. 4,080,271, 21 Mar 1978)

2N. Scale Formation and Prevention

Textbooks and monographs on water treatment include Theoretical Principles of Thermal Processes [1], Reagentless Methods of Water Treatment in Power Plants [2], Engineering Technological and Chemical Methods of Cooling Water Treatment in Industry and Power Plants: Fresh Water, Seawater, Brackish Water [3], effect of the contamination of outer surfaces of shaped heat exchanger tubes on the operating indexes of distillation units [4] and model boiler studies on deposition and corrosion [5].

Scale prevention studies in the concentrating process of seawater include the effect of various additives on the formation of alkaline and CaSO₄ scale, development of methods of preventing scale formation in the salt-manufacturing process by using $Na_6(PO_3)_6$ together with HCl and in multiflash evaporation using seeding techniques [6]. Formation of scale in steam pipes during evaporation of aqueous solutions of NaCl and Na_2SO_4 results from evaporation of solution drops carried by steam [7]. A condenser tube made of titanium was compared for fouling with a 90-10 CuNi alloy tube. The biomass grew 2.4 times faster in the Ti tube than in the CuNi [8]. The amount of biological overgrowth in pipes carrying seawater for power plant cooling was proportional to time and inversely proportional to the flow velocity. The growth was effectively inhibited by cyclic heating of the water to 45 to 50° and maintaining flow velocity at over 2.9 m/s [9].

An electronic system well suited for acquiring heat-exchanger fouling data from surface installations was designed, constructed, laboratory-tested and field-proven [10]. A report was presented containing abstracts of papers, published from June 1975 to December 1978, on organisms of disease and fouling and corrosion in marine environment, as well as protective measures [11].

Microfouling in desalination is a process in which the rate of increase of the mass of surface film is dependent in large part on the specific character of the environment, most particularly the physical, chemical and biological condition of the intake seawater. The attachement of bacteria to surfaces is mediated by a glycoprotein polymer, which is active in exceedingly low concentration [12].

The methods of scale control used in MSF evaporators include sodium polyphosphate, acid dosing and organic polymers. Increasing brine temperature reduces water costs. Organic polymers are now competitive with acid dosing [13].

ALKALINE SCALE

The caharacteristics of carbonate scale deposition on heat-transfer surfaces during heating of natural waters were investigated [14]. A simple method to determine the $CaCO_3$ equilibrium and corrective treatments needed for natural waters was presented [15]. An investigation of available data on the saturation state of seawater with respect to calcium carbonate and its possible significance for scale formation on Ocean Thermal Energy Conversion (OTEC) heat exchangers has been carried out [16].

The crystallization kinetics of $CaCO_3$ and $Mg(OH)_2$ were studied in the purification of saturated NaCl solutions for the manufacture of chlorine [17]. The convective heat and mass transfer theory was used to evaluate the kinetics of $CaCO_3$ crystallization. An expression for the rate of deposition was derived as a function of the coefficient of convective mass transfer, the surface mass transfer resistance and the differential CO_3^{2-} concentration [18]. The diffusion resistance at increased temperatures, 80 to 90°, amounts to 60 to 62% of the total resistance. The initial values for calculation of the formation rate of CO_3^{2-} deposits give an adequable description of the process, because crystallization does not occur within the flow core. Calculation of the deposit formation in circulation distillers is in good agreement with practical operation results, when substantial reduction of alkalinity of the brine, due to formation of a sludge suspension, is considered [19].

Humic and fulvic acids, certain aromatic carboxylic acids and orthophosphate exhibit strong precipitation inhibition when present at low levels. Many other organic substances, e.g. amino acids, sodium stearate, EDTA, were found to have little or no effect on precipitation rate [20]. Laboratory studies on formation of alkaline scale suggest that the utilization of laboratory data to infer the actual extent of scaling in seawater distillation plants requires careful consideration [21]. The possible application of electrolytic deposition of scaling compounds from seawater in controlling alkaline scale was suggested and the feasibility of this process has been ascertained [22]. Laboratory studies on the addition of Na_2CO_3 or Na_2CO_3 + NaOH to reduce the calcium ion concentration of the seawater were presented. In the presence of sludge, the residual Ca^{2+} concentration can be reduced to 0.5 to 1.0 meq/kg if the temperature is raised to 40° with only 10% excess chemicals [23].

The precipitation of calcium carbonate has been studied in the absence and presence of hydroxyethylidene-1,1-diphosphonic acid (HEDP). At ambient temperatures, a stable, threshold inhibited solution of calcium carbonate can be made in the presence of HEDP. However, this solution is thermally unstable. Heated calcium bicarbonate solutions, containing HEDP, are stabilized by a different mechanism. The rate of CO_2 evolution, and hence of $CaCO_3$ precipitation, is reduced because of a pH rise, caused indirectly by the inhibitor [24]. Flocon antiscalant 247, a new threshold scale inhibitor, suppresses calcium carbonate and magnesium hydroxide foulings by intervention at both the bicarbonate decomposition and mineral deposition stages of the scaling sequence. Codeposition of antiscalant with scale impedes further precipitation and disperses the precipitated scales in the blow-down [25].

Patents. Scale-forming compounds are removed by preheating the feed, whereby bicarbonates are decomposed and scale compounds are precipitated [26]. NaOH solution is mixed with the seawater to recipitate calcium or magnesium and HCl is added after the precipitation to remove CO_2 [27]. NaOH is reacted with CO_2 and the product is added to seawater to precipitate $CaCO_3$ [28]. Calcium containing seawater or brine is mixed with NaOH or Na_2CO_3 and filtered to remove over 95% calcium. The filtrate contained 0.017 g/l Ca [29]. Seawater is heated to up to 50° and Na_2CO_3 is added to form $CaCO_3$ and to prevent scale formation in the desalination apparatus [30].

An electrode-ionization chamber is used in eliminating or controlling scale in water systems [31]. A part of the brine from seawater evaporation is electrolyzed to produce an alkaline solution and HCl [32]. Calcium components in seawater are removed effectively by adding NaOH and CO_2 and adjusting the pH [33, 34]. Flue gas containing CO_2 is bubbled through seawater in the presence of an alkali metal salt to remove calcium as $CaCO_3$. The treated seawater is desalinated by multiple effect evaporation without scaling [35].

SULFATE SCALE

The kinetics of calcium sulfate scale formation was studied over a wide range of temperatures. Growth in stable supersaturated $CaSO_4$ solution was completely inhibited by 7×10^{-7} M phytic acid. The seeded crystallization of $CaSO_4.0.5H_2O$ and the phase changes from α- to β-hemihydrate were investigated [36]. Anhydrite scaling occurs in exchangers with process-side inlet temperatures exceeding 118°. Scaling diagrams were given whereby accumulation of anhydrite can be predicted[37]. The solubility of $CaSO_4.2H_2O$ was measured in water and synthetic seawater. Activity coefficients, calculated from the solubility measurements, were lower than the theoretical values [38]. A mathematical determination of the saturation deficiency of water with respect to calcium sulfate was presented [39.].

Crystallization of gypsum takes place in three stages. After an initial induction period, rapid deposition of the bulk of the gypsum occurs by an auto-accelerating mechanism and, finally, equilibrium conditions are established by the slow crystallization of the residual excess. The most effective additive at 60 to 100° to prevent gypsum deposits is an amphoteric aziridine-based polyelectrolyte containing 5 to 12 phosphonomethyl groups [40]. For laboratory studies on the addition of Na_2CO_3 or Na_2CO_3 + NaOH to avoid $CaSO_4$ deposits see ref. [23].

Gypsum solubility in natural seawater was measured and scaling thresholds were evaluated graphically from solubility data. A computation scheme was used to compare these results with calculated values based on various computation models. The computer program finally developed provides read out of solubilities, as well as scaling thresholds for gypsum [41]. A simplified nomogram was presented for calculating the solubility of $CaSO_4$ between 150 and 250°, in the presence of substantial amounts of NaCl and Na_2SO_4, with an accuracy of about 10% [42].

The stability of the ion pair $CaSO_4$ was determined from measurement of the change in calcium ion activity with medium composition at constant ionic strength. The evaluation of the stability constant depends on the degree of complexation between calcium and chloride and between sodium and sulphate [43].

REMOVAL OF SCALE

Heavy deposits were formed when a hot-water boiler operating at 150° was fed with water treated with sulfuric acid. The deposits generally contained a high proportion of anhydrite, frequently approaching 50%. Scales could be removed with 5% HCl at 60 to 70°, provided they did not exceed 1200 g/m² and contained less than 30% anhydrite [44]. During the complete shut down of the Hong Kong plant, three out of six brine heaters were found fouled with calcium sulphate anhydride. Alternate reaction with sodium carbonate solution and inhibited hydrochloric acid was used for in-plant descaling process. The method has been found successful in removing most of the scale after three cycles of chemical cleaning [45]. Methods of removal of accumulated deposits in multi-stage flash distillation plants were outlined, including increased acid dosing, mechanical cleaning, sulfuric acid cleaning, sulfamic acid cleaning, hydrochloric acid cleaning [46].

Patents. An aqueous HCl solution is circulated at 85°. After rinsing several times, nitrilotriacetic acid is added to the feed [47]. The apparatus includes a cleaning rod which is moved into and out of the tubes and a cleaning solution is injected [48]. Scale was easily removed by treating with a descaling agent containing ammonium acid fluoride, oxalic acid and sulfamic acid [49]. A mixture of sodium acrylate-acrylamide copolymer and dimethylamine-ethylenediamine-epichlorohydrin copolymer was used as a descaling agent [50]. Scales deposited on metal surfaces are dissolved using aqueous solutions of 2-phosphonobutane-1,2,4-tricarboxylic acid [51].

LITERATURE TO 2N.

1. A. V. Kozhevnikov (Severo Zapaduyi Zaochnyi Politekhnicheskii Instituta, Leningrad 1977)
2. E. F. Tebenikhin (Energiya, Moscow 1977)
3. H. D. Held, W. Bujak (2nd Ed., Vulkan Verlag, Essen 1977)
4. A. P. Egorov, S. N. Filippov, S. V. Grigorenko, V. A. Kozlov (Voprosy Atom. Nauki i Tekhn. Ser. Opresnenie Solen. Vod 2, No 10 [1977] 15/17)
5. P. V. Balakrishnan, E. G. McVey (Atom. Energy Canada Ltd. Rept. No AECL-5801 [1977].- A.I. 9, No 5 [1978] 359820)
6. S. Sugita (Nippon Kaisui Gakkaishi 32, No 7 [1978] 3/33.- C. A. 89 [1978] 168827)
7. S. Pribicevic, M. Suljkanovic (Hem. Ind. 32, No 6 [1978] 394/400.- C. A. 90 [1979]12109)
8. Anonymous (Do 20 [1978] 1/6)
9. L. Genadiev (Nats. Nauchna Ses. Khim. Med. Profil [Dokl] 2nd, 2 [1977] 291/298.- C. A. 89 [1978] 65071)
10. R. W. Findley, D. L. Meier (Rept. NTIS C00-4041-6 [1977] 54 pp.- C. A. 91 [1979] 41916)
11. N. A. Ghanem, A. El-Lakany, A. F. A. Ghobashy, M. M. Abdel-Malek, M. A. Abou-Khalil (NTIS Rept. AD-A067659/3GA [1978] 81 pp)
12. H. Winters, I. R. Isquith (Desalination 30 [1979] 387/399)
13. N. M. Wade (Desalination 31 [1979] 309/320)
14. L. P. Karnaukhov, V. B. Chernozubov, L. G. Vasina (Voprosy Atom. Nauki i Tekhn. Ser. Opresnenie Solen. Vod, 2 No 10 [1977] 30/60)
15. G. Bousquet (Trib. CEBEDEAU, 31 No 410 [1978] 11/26)
16. J. W. Morse, J. de Kanel, H. L. Craig (NTIS Rept. PNL-2605 [1978] .- E.R.A. 3, No 17 [1978] 39692)
17. A. M. Pekler, V. M. Podkopov, V. A. Belayev (Kristallizatsia 2 [1976] 139/143.- C. A. 90 [1979] 170886)
18. G. Ya. Lukin (Izv. Vyssh. Uchebn. Zaved., Energ. 21, No 10 [1978] 137/141.- C. A. 90 [1979] 127302)
19. G. Ya. Lukin (Teploenergetica No 2 [1979] 45/47.- C. A. 91 [1979] 27056)
20. R. A. Berner, J. T. Westrich, R. Graber, J. Smith, C. S. Martens (Am. J. Sci. 278, No 6 [1978] 816/837.- C. A. 89 [1978] 203796)

21. P. L. Kapur, B. M. Misra (Desalination 27 [1978] 65/70)
22. M. K. V. Nair, B. M. Misra (Desalination 27 [1978] 59/64)
23. E. G. Simonyan, I. Ya. Skibinski (Vodosnabzh. Sanit. Tekh. No 4 [1978] 28/29.-
    C. A. 89 [1978] 117449)
24. K. G. Cooper, L. G. Hanlon, G. M. Smart, R. E. Talbot (Desalination 31 [1979]
    257/266)
25. S. W. Walinsky, B. J. Morton (Desalination 31 [1979] 289/298)
26. P. S. Roller (U.S. 4.054.493, 18 October 1977)
27. T. Sawa, K. Izumi, A. Yamada, S. Yoshikawa (Jap. 77.136.885, 15 Nov 1977.-
    C. A. 89 [1978] 30534)
28. T. Sawa, K. Izumi, A. Yamada, S. Takahashi (Jap. 78.14.675, 8 Feb 1978.-
    C. A. 89 [1978] 117503)
29. H. Tahata, R. Yokoyama, T. Takahara (Jap. 77.120.546, 11 Oct 1977.- C. A. 89
    [1978] 65104)
30. T. Sawa, K. Izumi, A. Yamada, S. Takahashi (Jap. 78.75.166, 4 Jul 1978.- C. A.
    89 [1978] 168919)
31. C. N. Smith (U.S. 4.127.467, 28 Nov 1978.- C. A. 90 [1979] 156974)
32. K. Izumi, A. Yamada, T. Sawa, S. Takahashi (Jap. 78.06.272, 20 Jan 1978.-
    C. A. 89 [1978] 30549)
33. T. Sawa, A. Yamada, K. Izumi, S. Takahashi (Jap. 78.60.879, 31 May 1978.-
    C. A. 89 [1978] 168917)
34. Hitachi Ltd. and Babcock Hitachi K.K. (French 2.376.079, 1 Sep 1978)
35. T. Sawa, K. Izumi, A. Yamada, S. Takahashi (Jap. 79.43.872, 6 Apr 1979.-
    C. A. 91 [1979] 62537)
36. G. H. Nancollas, A. E. Eralp, J. S. Gill (Soc. Pet. Eng. J. 18, No 2 [1978]
    133/138.- C. A. 89 [1978] 48741)
37. R. V. Comeaux (Mater. Perform. 17, No 11 [1978] 9/21.- C. A. 90 [1979] 28835)
38. C. H. Culberson, G. Latham, R. G. Bates (J. Phys. Chem. 82, No 25 [1978]
    2693/2699.- C. A. 90 [1979] 11969)
39. A. A. Govert (Tr. VNII VODGEO No 66 [1977] 33/35.- Ref. Zh., Khim. [1978]
    11 I 230)
40. L. T. Dytyuk, R. Kh. Samakaev (Neftepromysl. Delo No 10 [1978] 28/31.-
    C. A. 90 [1979] 174435)
41. M. S. Adler, J. Glater, J. W. McCutchan (J. Chem. Eng. Data 24 [1979] 187/192.-
    C. A. 91 [1979] 62511)
42. V. V. Shishchenko, M. M. Krikun (Prom. Energ. No 1 [1979] 39/40.- C. A. 90
    [1979] 192323)
43. B. Elgquist, M. Wedborg (Mar. Chem. 7 [1979] 273/280)
44. L. L. Vorobleva, E. S. Rechina, M. A. Stolov, Yu. A. Zhuravlev (Teploenergetica
    No 6 [1978] 78/80.- C. A. 89 [1978] 152508)
45. Y. K. Chan, K. K. Leung (Desalination 30 [1979] 359/371)
46. A. Brandel, J. Riebe (Desalination 31 [1979] 341/350)
47. M. Kux, D. Hartig, C. Passeck (E. Ger. 127.541, 28 Sep 1977.- C. A. 89 [1978]
    117492)
48. B. A. Troshenkin, A. Ya. Matyash (USSR 537.717, 28 Feb 1977)
49. I. Kawamura (Jap. 78.28.579, 16 Mar 1978.- C. A. 89 [1978] 65125)
50. R. M. Goodman (W. Ger. 2.804.434, 10 Aug 1978.- C. A. 89 [1978] 220717)
51. I. Sotoma, N. Yamamoto, Y. Sata (Jap. 79.20.978, 16 Feb 1979.- C. A. 91 [1979]
    44372)

2O. Treatment of Feed Water

CHLORINATION

   Relations between contact period, chlorine demand, residual chlorine and chlorine dose were established in the chlorination of seawater samples collected near
atomic power plants [1]. Nonpolar, presumably lipophilic organohalogens formed in
seawater were studied. The principal component found in all seawater samples was

CHBr$_3$. Smaller quantities of CHClBr$_2$ and traces of CHBrCl$_2$ were found. CHCl$_3$ was not present in significant amounts. CHBr$_3$ was found in all unchlorinated seawater samples analyzed [2]. To prevent biological fouling in cooling systems using seawater, a portion of seawater is electrolyzed in a cell and the chlorinated seawater is injected into the cooling system [3]. In a series of chlorination studies, the initial loss of chlorine reached a saturation level that appeared to be the true organic demand [4]. In case of deoxygenated stagnant seawater, a semilog relation was found between the residual chlorine concentration and corrosion rate. At 95° the corrosion rate was 4 fold that at 40°. The presence of a residual concentration of 0.1 ppm Cl under stagnant conditions increases the extent of corrosion by 5 to 10 % compared to the case in which chlorine is absent [5].

ACID INJECTION

A comprehensive review was published on deacidification in water treatment [6]. In a method for calculating pH values and concentrations of HCO$_3^-$ and H$_2$CO$_3$ during individual stages of a two-step chemical desalination process a system of equations was derived, which is suitable for computer processing [7]. Probe and coupon measurements in the Hong Kong MSF plant, which employs external feed deaeration and scavenging, have shown lower carbon steel corrosion rates than in the Jersey plant, where no external deaeration is provided. Both plants are acid dosed. This difference is partly due to the persistent need in the latter plant to remove fouling by acid cleaning, which has not been required in Hong Kong. Even when this additional corrosion is allowed for, the Hong Kong plant appears to enjoy a carbon steel corrosion rate significantly lower than that of the Jersey plant [8].

Patents. The amount of acid added to the decarbonation apparatus is increased or decreased according to the pH determined [9]. An automatic control of acidity for controlling water softening uses a measuring cell with noncorrosive metal electrodes [10].
Scaling is prevented by decomposing HCO$_3^-$ to CO$_2$ through addition of an acid to the feed water. Carbon dioxide, oxygen and other noncondensable gases are removed and seawater is evaporated at less than 1 atm [11].

TREATMENT BY PHOSPHATES

The calcium sulfate content of a brine from rock salt may be decreased by adding 0.001 to 0.005% alkali metal phosphates [12]. Freshly precipitated calcium phosphates had a 20 to 30 times higher solubility than the natural phosphates. The difference is attributed to gradual recrystallization with the absorption of F$^-$ from the seawater under natural conditions, leading to the formation of the less soluble fluorapatite [13].

Patents. Scale inhibitors and sludge-forming agents comprise 20 to 80 wt.% sodium tripolyphosphate, used as a 0.5 to 10% aqueous solution [14]. Crust formation is avoided by adding or forming Mg, Ca, Zn, Al or Fe hydroxide, phosphate or arsenate precipitates and adjusting the pH to 7.5 to 11 [15]. Mixtures of alkali metal, ammonium, zinc and/or organic amine polyphosphate salts showed both threshold treatment and metal corrosion inhibiting properties when added to water at 2 to 500 ppm [16]. Additives which include zinc salts and phosphoric acid are used for preventing corrosion and the formation of scale in water circulating systems [17].

ADDITIVES

Sodium CMC poly(acrylic acid) and low molecular weight polyacrylamide were effective as scale inhibitors at high pH values. Low molecular weight hydroxyethylcellulose was also effective at pH 6 to 10 [18]. Development and application for polymer scale control agents were described [19]. Laboratory and field studies were presented for inhibition of formation of new boiler scale and removal of existing scale by sulfonated styrene copolymer sludgtrol 600 [20]. Formation of scale and deposits in boilers, cooling towers and in air washing systems is controlled with a sulfonated polymer, which reduces deposits and allows removal of salts in blow-down [21].

The effects of brine circulation characteristics, retention time and supersaturation level on the degree of scale precipitation encountered in additive treated desalination plants, were analyzed by population balance models characterizing mixed flow and plug flow conditions, respectively. For identical brine retention times, the degree of bulk precipitation occuring under mixed flow conditions is considerably larger than that occuring under plug flow conditions. Similarly, the induction period in a mixed flow system is shorter than that in a plug flow system. The effects of brine retention time and additive concentration on the amount of scale precipitating in the brine were evaluated using roughly estimated values of kinetic coefficients for $CaCO_3$ retarded precipitation [22].

A safe, effective new polymeric antiscalant, the maleic acid containing polymer FLOCON 247, was tested in a laboratory single stage recycle flash evaporator. Using Atlantic Ocean seawater, with added bicarbonate, the unit is operated at 121°C for 24 h during a typical run. At the end of the run, total scale deposited in the brine heater and flash chamber were measured and magnesium/calcium ratios are determined. This procedure permits rapid generation of dose/response profiles [23].

A comparison of acid and additive treatment methods was made. The long term suitability of an additive was examined with reference to the ease of plant cleaning and the maintenance of demister efficiency, as well as the more obvious parameters, such as the maintenance of clean heat transfer surfaces throughout the plant [24]. A test evaluation of BELGARD EV was carried out in a 8500 m^3/day MSF evaporator, by comparing the plant operating performance when using sulfuric acid and BELGARD EV, as the methods of scale control. The use of this additive has reduced the specific heat consumption and substantially increased the operating period between acid cleaning. The corrosion rates of iron and copper have decreased by 84% and 75% respectively. Over the first 150 days of the trial with BELGARD EV, the mean heat consumption of the unit was 98.2% of the design for a product output of 8630 m^3/day. The mean gained output ratio was 8.46 compared with the design figure of 8.21 [25].

Patents. Addition of a mineral acid is followed by addition of a specified inhibitor to prevent alkaline scaling [26]. Prevention of scale formation is effected by addition of 0.1 to 2 ppm of a 1:1 mixture of sodium polymethacrylate[27]. The composition of an additive mixture comprises poly(meth)acrylic acid or one of its salts, a water soluble chelating agent and aminotrimethylenephosphonic acid, a hydroxyalkylidene diphosphonic acid or one of its salts [28]. Copolymer oligomers of acrylic acid with acrylonitrile or methyl acrylate with terminal bisulfate groups were used as additives [29].

Scale formation and corrosion are prevented by acrylic acid-2-hydroxyethyl methacrylate-methyl acrylate copolymer sodium salt [30]. The addition of maleinated, unsaturated fat derivatives or their alkali salts suppresses deposit formation on surfaces heated up to 140° [31]. Cyclohexane-1.2.3.4-tetracarboxylic acid is used for the prevention of scale in water-cooling systems [32].

In an additive comprising a homopolymer of acrylic acid, about 70% of the active hydrogen ions are replaced by sodium ions [33]. Low molecular weight hydrolized polyacrylamide is used as a scale inhibitor [34]. The deposition of scale is prevented by the addition of furan-maleic anhydride copolymer to the feed water [35]. Control of corrosion and scale in circulating water systems is obtained by means of partial esters of polyfunctional organic acids [36].

A copolymer obtained from acrylic acid and Me or Et ester of acrylic acid and/or methacrylic acid is used to prevent scaling [37]. Phytic acid or a phytate is added to deaerated feed water to remove scale-forming components [38]. An acrylic acid-acrylonitrile copolymer sodium salt was added to a cooling water, as a corrosion inhibitor, to prevent precipitation [39]. Isobutene-maleic anhydride copolymer and phosphonic acid or their salts are added to seawater to prevent scale formation in the evaporator [40]. Acrylic acid or its salt and poly(acrylic acid) or poly(methacrylic acid) are used to prevent scaling [41]. A descaling agent was prepared in an autoclave from a mixture of an amino acid and a sugar [42].

One component of the additive comprises a mixture of zinc and hydrolyzed polymaleic anhydride and the second component is a benzotriazole [43]. The general
formula of additives comprising sulphones or sulphoxides is given [44], as well
as additives containing diacetic acrylimine acids [45]. An additive comprises
an anionic copolymer of acrylate and acrylamide [46]. The general formula of a
phosphonomethylene amino carboxylate additive is specified [47]. A mixture of
maleic acid polymer and a non-ionic or anionic surfactant is given as additive [48].
Scale inhibiting additives contain reaction products of unsaturated fatty derivatives with maleic anhydride, maleic acid or a mixture of them with their alkaline
salts [49].

SEEDING

Patents. The MSF plant is provided with a sedimentation tank for recovering
seeds from the effluent brine and a seed mixing tank for supplying a brine with
a high seed concentration. The seed mixed brine recovered from the sedimentation
tank is blended with the seed mixed solution from the seed mixing tank. The blended solution is degassed by dispersing into the last stage flash chamber and then
introduced into a condenser. No slurry circulation pump is required [50]. An
evaporator with seed crystal separation chamber, which can be used without external pumps, is described [51].

TREATMENT BY ION EXCHANGE

Water for medium pressure steam boilers containing no Ca and Mg was prepared
by treatment with a weak-acid cation exchanger, a strong-acid cation exchanger,
a weak-base anion exchanger and a CO_2 degazifier [52], as well as with weakly basic
anion exchange resins [53]. The purification of heating steam condensate and distillate from distillation units on settling ion exchange filters was studied [54].
An apparatus and a method for partial and intensive softening of seawater by cation exchange treatment for feeding evaporators and steam generators were described. Design data for a 3 m diameter softener to treat water of 75 to 80 meq/l total hardness were presented [55]. The purification of boiler water from the Rhine
River with an electrical conductivity 1.1 µS/cm and residual silicic acid of about
0.1 mg/l SiO_2 was described [56].
The desulfation process, consisting of the make-up treatment by means of weak
basic resins characterized by a good selectivity for the sulfate ions, was fully
tested at the Bari 0.5 Mgd MSF type desalination plant, operating at 150°C top
brine temperature [57].

Patents. Acidic water is treated to remove ions causing hardness by contacting
with a weak-acid cation exchanger, the resin comprising a divinylbenzene-methacry-
lic acid copolymer in the alkali metal or NH_4^+ salt form [58]. Seawater or brine
is treated with a magnesium or calcium chelating ion-exchange resin or chelating
agent to prevent scaling in evaporators [59]. A water softening system uses an
ion-exchange resin granule bed to soften water. Saturated brine regenerates the
resin. The efficiency is improved by using a regeneration control apparatus [60].
The use of strongly basic anoin-exchange resins in the removal of silicious contaminants and dissolved organics from boiler feed water are described [61].

MAGNETIC TREATMENT

The effect of magnetic treatment of cooling water on scale deposition in a heat
exchanger depends on the magnitude of the magnetic induction. The treatment reduces the size of aragonite crystals formed upon boiling and thus reduces scale
formation [62]. In tests made with an electromagnetic generator on a laboratory
model with seawater, having a composition corresponding to the Black Sea, heat exchanger scale formation was reduced by 80% [63]. Magnetic treatment of the feed
water for hot water heating boilers and of the water circulating within the heating system can be recommended for boilers operating with heat loads of maximum 25

kW/m^2, for water of the HCO_3 grade with overall salt content not over 1000 mg/l
and total hardness of maximum 7.5 to 8.5 meq/l [64]. No carbonate depositions
were observed in turbine condenser tubes when the cooling water, containing 1900
to 2500 mg/l salts, was treated with a magnetic field [65]. A method of studying
the effect of a magnetic field on scale formation, based on determining the transmission of light through microscope slides immersed in boiling water, was discussed
with special reference to the prevention of scale formation in heat exchanger
pipes [66]. A water sample with a given analysis was subjected to the magnetic
treatment. The nucleation rate of salt crystals increased by a factor of 5 to
7 in 30 min[67].

Patents. Thermal softening was made more economical by preheating the water
at about 75 to 80° [68]. Scale formation is retarded by coating the surfaces with
a mixture of epoxy resins or polyamides and BN [69]. A magnetic-treatment unit
is placed before the lime treatment unit and is provided with a separatory filter
connected to the evaporator [70].

LITERATURE TO 20.

1. A. K. Sriraman, R. Viswanathan (Indian J. Technol. 15, No 11 [1977] 498/500.-
   C. A. 89 [1978] 135564)
2. R. M. Bean, R. G. Filey, P. W. Ryan (Proc. Conf. Water Chlorination: Environ.
   Impact Health Eff. 1977, 2 [1978] 223/233.- C. A. 89 [1978] 15227)
3. N. Kalyanasundaram, S. S. Ganti, G. P. Agrawal (Def. Sci. J. 28, No 1 [1978]
   1/4.- C. A. 90 [1979] 209861)
4. J. C. Goldman, H. L. Quinby, J. M. Capuzzo (Water Res. 13 [1979] 315/323.-
   C. A. 90 [1979] 209808)
5. Y. Ohyama (Gumma-ken Kogyo Shikenjo Nempo 1977 [1978] 98/103.- C. A, 91 [1979]
   294474)
6. B. Moergeli, J. C. Ginocchio (Wasser Luft Betr. 22, No 4 [1978] 144/148)
7. B. M. Larin (Izv. Vyssh. Uchebn. Zaved. Energ. 21, No 11 [1978] 101/105.-
   C. A. 90 [1979] 174421)
8. F. C. Wood, Y. N. Wu (Desalination 30 [1979] 347/358)
9. T. Sawa, S. Yoshikawa, M. Komai, T. Yamagata (Jap. 77.43.186, 28 Oct 1977.-
   C. A. 89 [1978] 30533)
10. A. De Chazelles (French 2.348.488, 10 Nov 1977.- C. A. 90 [1979] 8111)
11. J. E. Pottharst (Jap. 79.24.274, 23 Feb 1979.- C. A. 91 [1979] 62535)
12. A. I. Postoronko (Zh. Prikl. Khim. 51, No 2 [1978] 472.- C. A. 89 [1978] 45740)
13. V. S. Savenko (Dokl. Akad. Nauk SSR 243 [1978] 1302/1305.- C. A. 90 [1979]
    192210)
14. H. Jodko, B. Smolenski, B. Jadrzejewska (Pol. 85.468, 15 Sep 1976.- C. A. 90
    [1979] 76367)
15. Oesterreichische Studiengesellschaft für Atomenergie m.b.H., Vereinigte Edel-
    stahlwerke A.G. (Neth. 77.03.192, 28 Sep 1977.- C. A. 89 [1978] 91721)
16. G. M. Smart (Brit. 1.505.816, 30 Mar 1978.-C. A. 89 [1978] 220721)
17. G. Bohnsack, H. Kallfass, W. Radt (US 4.057.511, 8 Nov 1977)
18. R. B. Seymour, D. Powell (Chem. Ind. Dev. 12, No 3 [1978] 15/19.- C. A. 89
    [1978] 168879)
19. J. J. Schuck, I. T. Godlewski (Proc. Int. Water Conf., Eng. Soc. West. Pa 39
    [1978] 277/287)
20. D. G. Cuisia (Proc. Int. Water Conf., Eng. Soc. West. Pa 39 [1978] 289/298)
21. G. J. Helmstetter, R. A. Holzer (Proc. Int. Water Conf. Eng. Soc. West. Pa 39
    [1978] 299/308)
22. S. Steinberg, D. Hasson (Desalination 31 [1979] 267/277)
23. M. H. Auerbach, M. S. Carruthers (Desalination 31 [1979] 279/288)
24. K. G. Cooper, L. G. Hanlon, G. M. Smart, R. E. Talbot (Desalination 31 [1979]
    243/255)
25. J. Stewart, R. Bom, M. A. Finan (Desalination 31 [1979] 299/307)
26. Ciba-Geigy (UK) Ltd. (Isr. 47.088, 30 Dec 1977)
27. J. A. Gray, C. M. Hwa, D. G. Cuisia (Can. 1.033.564, 27 June 1978.- C. A. 90
    [1979] 28688)

28. Chemed Corp. (Brit. 1.491.494, 9 Nov 1977)
29. Uniroyal Inc. (Brit. 1.505.909, 5 Apr 1978)
30. M. Ii, Y. Goto, T. Suzuki, S. Kubo (W. Ger. 2.643.422, 30 Mar 1978.-C.A. 89 [1978] 117504)
31. N.B. Desai (W. Ger. 2.651.438, 18 May 1978.- C.A. 89 [1978] 48762)
32. F. Krueger, D. Palleduhn (W. Ger. 2.657.775, 22 Jun 1978.- C.A. 89 [1978] 135634)
33. T.R. Gardner, R.W. Lansford (U.S. 4.062.796, 13 Dec 1977)
34. D.S. Song, R.J. Duffy, C.R. Witschonke, A.M. Schiller, M.A. Higgings (U.S. 4.085.045, 18 Apr 1978)
35. J. Block (U.S. 4.086.146, 25 Apr 1978.- C.A. 89 [1978] 135627)
36. T.J. Suen, A.J. Begala, M. Grayson (U.S. 4.086.181, 25 Apr 1978)
37. K. Negishi, T. Oyama (Jap. 78.02.393, 11 Jan 1978.- C.A. 89 [1978] 30550)
38. A. Inoue, T. Iida (Jap. 78.05.079, 18 Jan 1978.- C.A. 89 [1978] 220719)
39. A. Maeda (Jap. 78.21.091, 27 Feb 1978,- C.A. 89 [1978] 220720)
40. T. Abiko, F. Era, N. Yamamoto (Jap. 78.95.185, 19 Aug 1978.- C.A. 89 [1978] 220710)
41. T. Takahashi, S. Kubo (Jap. 78.110.981, 28 Sep 1978.- C.A. 90 [1979] 76365)
42. Y. Kawamura (Jap. 78.113.784, 4 Oct 1978.- C.A. 90 [1978] 109775)
43. A. Harris, J. Burrows, T.I. Jones (U.S. 4.089.796 16 May 1978)
44. Ciba-Geigy A.G. (Brit. 1.526.301, 27 Sep 1978)
45. Ciba-Geigy A.G. (French 2.369.216, 30 Jan 1978)
46. A.M. Schiller, R.M. Goodman, R.E. Neff (U.S. 4.072.607, 7 Feb 1978)
47. R.S. Mitchell (U.S. 4.079.006, 14 Mar 1978)
48. W.R. Grace & Co (Brit. 1.519.512, 26 Jul 1978)
49. Grillo Werke A.G. and L. Taprogge (French 2.370.697, 14 Jul 1978)
50. S. Mutsukushi, K. Izumi, T. Takahashi, Y. Okajima, T. Sawa, M. Komai (Japan 77. 74.314, 7 Nov 1977.- C.A. 89 [1978] 48753)
51. S. Komatsu (Jap. 78.138.982, 4 Dec 1978.- C.A. 90 [1979] 170599)
52. V. Gocheci, L. Negulescu (Bul. Stiint. Teh. Inst. Politeh. Timisoara, 22, No 1 [1977] 297/302.- C.A. 89 [1978] 94844)
53. V. Gocheci, L. Negulescu (Bul. Stiint. Teh. Inst. Politeh. Timisoara, 22, No 2 [1977] 417/420.- C.A. 90 [1979] 174439)
54. I.P. Lazarev, R.N. Musikhin, S.I. Tsarenko (Voprosy Atom. Nauki i Tekhn. Ser. Opresnenie Solen. Vod, 2, No 10 [1977] 70/74.- Ref. Zh., Khim. 1978, 17 I 304)
55. K.M. Abdullaev, G.K. Feiziev, S.A. Shakhmarov, T.M. Kuliev (Engergetik No 4 [1978] 30/31.- C.A. 89 [1978] 94843)
56. J. Von Staden (VGB Kraftwerkstechnik 59, No 1 [1979] 69/71.- C.A. 90 [1979] 156912)
57. A. De Maio, G. Odone, E. Palmisano, R. Zannoni (Desalination 31 [1979] 321/331)
58. R. Kunin (U.S. 4.083.782, 11 Apr 1978.- C.A. 89 [1978] 94860)
59. H. Ueshima, T. Kudo, H. Amimoto (Jap. 78.58.150, 25 May 1978.- C.A. 89 [1978] 168915)
60. J.W. Braswell (U.S. 4.104.165, 1 Aug 1978.- C.A. 90 [1979] 156968)
61. J. Filby (U.S. 4.098.691, 4 Jul 1978)
62. B. Boyanov, R. Dimitrov, D. Khadzhistavrev (Nauchni Tr. Plovdivski Univ. 13, No 3 [1975] 303/311.- C.A. 89 [1978] 65060)
63. V.F. Ochkov, E.A. Pavlov, A.A. Kudryatsev (Tr. Mosk. Energ. Inst. 328 [1977] 88/91.- C.A. 89 [1978] 152425)
64. A.I. Shakhov, S.S. Dushkin, I.Z. Aronov, E.N. Solodovnikova (Izv. Vyssh. Uchebn. Zaved., Stroit. Arkhit. No 9 [1977] 121/126.- C.A.89 [1978] 65059)
65. M. Chirica, G. Georgescu, V. Marinescu (Energetica, 25, No 8 [1977] 289/293.- C.A. 89 [1978] 64572)
66. V.I. Morozov (Prom. Energ. No 6 [1978] 41/42.- C.A. 89 [1978] 185825)
67. K.A. Rubezhanskii, G.A. Kataev, B.P. Zhantalai, B.A. Kulikov (Khim. Prom-st, No 2 [1979] 101/103.- C.A. 91 [1979] 44343)
68. P.P. Simonov, V.V. Shishchenko, A.I. Bykov, O.P. Ostrovskii, Yu. N. Reznikov, A.N. Ostrovskii (USSR 597.643, 15 Mar 1978.- C.A. 89 [1978] 48761)
69. O.G. Maxson, G.D. Achenbach (US 4.115.606, 19 Sep 1978.- C.A. 90 [1979] 7726)
70. L.S. Sterman, V.M. Lavygin, A.V. Moshkarin, V.V. Vikhrev (USSR 594.056, 25 Feb 1978.- C.A. 89 [1978] 65114)

2P. Corrosion

A bibliography was compiled on corrosion in desalination plants. Citations from the NTIS data base refer to Federally-funded research and are primarily concerned with material properties in saline solutions, evaporators, distillation apparatus, membranes and scaling. Electrochemistry, chemical reactions, testing and performance engineering are included. The bibliography covers the period 1964 to March 1978 [1]. An updated bibliography contains 178 abstracts, one of which is a new entry, and covers the period 1964 to February 1979 [2]. Citations on corrosion in desalination plants from the Engineering Index data base cover worldwide research on corrosion in desalting processes, with special emphasis to metal pipes, heat transfer tubes, evaporators and other distillation apparatus. The report covers the period 1970 to March 1978 [3]. An updated bibliography contains 185 abstracts, 12 of which are new entries, and covers the period 1970 to February 1979 [4]. A bibliography was also published on seawater corrosion, containing citations on ships, offshore platforms, steel pilings, cables and marine equipment. Ferrous and nonferrous metals, concrete and composites are included. The bibliography contains 213 abstracts, 82 of which are new entries to the previous edition, and covers the period 1976 to January 1979 [5]. A Seawater Corrosion Handbook was published [6].

Other reviews include corrosion control in water treatment processes [7], corrosion of metals in marine environments [8] and seawater desalination and metallic materials [9].

Changes in the kinetics of the electrochemical corrosion process, during a decrease in the concentration of Cl-ions and transition to desalted water, were studied. The metals exhibit cathodic depolarization during the desalting process. Their corrosion behavior is controlled by the nature of corrosion products and by the mechanism of the intrinsic process [10]. Electrochemical techniques in corrosion studies require deaeration of the electrolyte. A fast technique for seawater deaeration is described [11]. The characteristics of seawater as a corrosive medium, methods of corrosion testing, the nature of the attack on alloys by flowing seawater and electrochemical measurements in seawater with rotating electrodes and their application in corrosion research were reviewed [12]. A simple apparatus for monitoring the corrosion of steel or other metals in seawater records potential of steel versus Zn, temperature, conductivity of seawater, current for cathodic protection, direction and speed of ocean current and salinity. The raw data from the magnetic tape can be printed out directly or can be averaged and processed in various ways [13].

Fatigue testing was conducted on four high-strength alloys at low-cycle frequency. Both seawater and potential acted to accelerate crack growth rates in the ferrous alloys, which proved to be much more sensitive to seawater and negative potential than the nonferrous alloys studied. The Ti alloy showed no measurable sensitivity to either seawater or negative potential. The high-strength steels are the most seriously affected by corrosion fatigue. The nonferrous marine alloys, the least [14]. A useful method for studying heat transfer and waterside corrosion on short lengths of evaporator tube allows simple control of the applied heat flux during steady running and of the temperature ramps during dryout transients. Waterside heat transfer coefficients are in general agreement with predictions for single phase and boiling heat transfer. The oxide layer, produced by corrosion and deposition, resulted in a small increase of outside wall temperature, corresponding to an oxide thermal conductivity of 1.9 W/m. degree [15]. The essentially antifouling nature of copper alloys is an important characteristic as condenser tube material. The need to improve resistance to inlet and impingement has led to the development of an improved copper-nickel alloy [16]. Methods of prevention of corrosion of condenser tubes are additon of Fe-ion to form an iron oxide film, cleaning to prevent clogging, monitoring of polarization and corrosion potential control [17].

Other work includes a study of the effect of velocity on corrosion of galvanic
couples in seawater using a circling-foil apparatus [18], flow effects on corrosion
of galvanic couples in seawater [19], reasons for anomalons electrokinetic properties
of corrosion products in a boiling-water reactor [20], basic criteria for evaluating
the corrosiveness of natural fresh waters [21], corrosion investigations at Panama
Canal zone [22], overgrowth and heterotrophic microflora and its effect on the
corrosion of metal samples in the Black Sea [23], studies on the mechanisms, detection and control of corrsion/erosion processes in seawater systems and assessment
of galvanic corrosion in seawater [24], Corrosion of metals in marine environments,
including recent desalination experience [25], experimental studies in dynamic conditions on the pitting corrosion resistance of some materials in natural seawater
[26] and influence of the bulk-solution-chemistry conditions on marine corrosion
fatigue crack growth rate [27].

CORROSION INHIBITION

$NaNO_2$ can be used as an inhibitor in seawater [28]. The corrosion and scale
inhibition efficiency of several agents was determined in synthetic cooling water
of varying degrees of hardness [29]. Potassium pertechnetate (KT_cO_4) was found a
very effective inhibitor in aqueous solutions. Piperidine nitrobenzoate is recommended as inhibitor in vapor phases [30]. The use of EDTA to control corrosion
in a boiler was reported [31].

Patents. Corrosion of ferrous metals in seawater is reduced with the addition
of mixtures of diethanolamine phosphate ester and zinc citrate [32]. ClO⁻ and iron
ions are added to cooling systems using seawater to prevent fouling and corrosion
[33]. An iron electrode in the Fe-ion generator forms $Fe(OH)_3$ with OH⁻ in the seawater [34]. The corrosion and scale inhibitor comprises an aminomethylenephosphonic
acid, e.g. sodium molybdate dihydrate and (hexamethylene dinitro)tetrakis[methylene-
phosphonic] acid [35].
N-hexylurocanamide was added to cooling water system as corrosion inhibitor,
which also has germicidal and antifouling effects [36]. Corrosion inhibitors for
cooling water systems contain polyphosphate, zinc salt and alkyl allyl sulfonic
acid polymer [37]. Metal corrosion in water is inhibited by lowering the dissolved
oxygen by means of N_2H_4 and/or hydroxylamine and Cu, Mn or Co [38]. Corrosion
inhibitors for water systems contain: phosphonic acids and/or their salts, zinc
compounds, N-heterocyclic compounds, isobutene-maleic anhydride copolymer and/or its
salt, and optionally hydroxycarboxylic acid and/or its salt [39]. Sodium molybdate
prevents metal corrosion in water systems [40]. Corrosion inhibitors contain polyphosphate, zinc compound and a phosphate ion precursor [41]. Polyepichlorohydrin
quaternary ammonium derivatives added to an acidic boiler or evaporator descaling
solution inhibits corrosion [42].
The formula of an additive for corrosion and scale inhibition in industrial
recirculating cooling water systems is given [43]. A corrosion inhibiting composition contains alkali metal nitrite, carbonate, hydroxide, silicate and mercapto-
benzothiazole [44]. An inhibitor includes a polyphosphate, a phosphonate, a homopolymer of maleic acid and maleic anhydride [45].

LITERATURE TO 2P.

1. M.F. Smith (NTIS Rept. PS-78/0191/3GA [1978])
2. W.E. Reed (NTIS Rept. PS-79/0133/3GA [1979])
3. M.F. Smith (NTIS Rept. PS-78/0192/1GA [1978])
4. W.E. Reed (NTIS Rept. PS-79/0134/1GA [1979])
5. W.E. Reed (NTIS Rept. PS-79/0098/8GA [1979])
6. M.M. Schumacher, Seawater Corrosion Handbook (Noyes Data Corp., Park Ridge,
   N.J. [1979] 494 pp)
7. H.S. Ramaswamy (Indian Chem. Manuf. 15, No 5 [1977] 35/39)
8. W.K. Boyd, F.W. Fink (CMIC Rept. MCIC-78-37 [1978])
9. H. Togano (Kinzoku Hyomen Gijutsu 30, No 3 [1979] 116/125)

82

10. A.V. Vedenskii, I.K. Marshakov, A.P. Karavaeva (Deposited Doc. VINITI 3225-77 [1977].- C.A. 90 [1979] 109710)
11. K.D. Efird, V.M. Putnam (Corrosion 34, No 7 [1978] 250/251)
12. F.P. Ijsseling (PT Procestech. 33, No 2 [1978] 69/82.- C.A. 89 [1978] 135296)
13. H. Arup (Proc. Ann. Offshore Technol. Conf. 11, No 3 [1979] 2129/2134.- C.A. 91 [1979] 46280)
14. T.W. Crooker, F.D. Bogar, W.R. Cares (ASTM Spec. Tech. Publ. 642, Corros.- Fatigue Technol. [1978] 189/201.- C.A. 89 [1978] 50445)
15. J.C. Ralph, L. Tomlinson, P.J.B. Silver, W. Lilley (Proc. 6th Int. Heat Transfer Conf. 4 [1978] 379/384.- C.A. 89 [1978] 187628)
16. H.T. Michels, W.W. Kirk, A.H. Tuthill (Nucl. Energy 17 [1978] 335/342)
17. K. Kawaguchi, Y. Yamaguchi, K. Onda (Chubu Denryoku Kabunshiki Kaisha Kenkyu Shiryo 57 [1975] 85/116.- C.A. 89 [1978] 48737)
18. J. Perkins, K.J. Graham, G. Storm, J. Locke, J.R. Cummings (NTIS Rept. AD-A048065 [1977].- C.A. 89 [1978] 65076)
19. J. Perkins, K.J. Graham, G.A. Storm, G. Leumer, R.P. Schack (Corrosion 35, No 1 [1979] 23/32.- C.A. 90 [1979] 209896)
20. A.A. Gromoglasov, A. Yu. Mikhailov (Teploenergetika No 6 [1978] 81/82.- C.A. 89 [1978] 152509)
21. L.S. Alekseev, Yu. P. Belichenko (Energetik No 5 [1978] 23/24.- C.A. 89 [1978] 117481)
22. M.A. Pelensky, J.J. Jaworski, A. Gallacio (ASTM Spec. Tech. Publ. 646, Atmos. Factors Affecting Corros. Eng. Met. [1978] 58/73.- C.A. 89 [1978] 184420)
23. M.N. Lebedeva, A.I. Shtevneva (Biopovrezhdeniya Mater. Zashch. Nikh [1978] 135/139.- C.A. 90 [1979] 125992)
24. G.C. Booth, J.C. Rowlands, B. Angell, J.F.G. Conde (Proc. 5th Inter-Nav. Corros. Conf. [1976] 42-1/23, 42 + 43 - 1/4.- C.A. 90 [1979] 59557)
25. W.K. Boyd, F.W. Fink (MCIC Rept. No 78-37 [1978].- E.I. 16 [1978] 64186)
26. O. Radovici, G. Beizadea, O. Ioachimescu (Lucr. Simp. Biodeterior. Clim. 7, No 1 [1977] 112/119.- C.A. 89 [1978] 206327)
27. F.D. Bogar, T.W. Crooker (J. Test. Eval. 7, No 3 [1979] 155/159.- C.A. 91 [1979] 43251)
28. K. Yano, T. Sato (Bosei Kanri, 21 No 12 [1977] 34/36.- C.A. 89 [1978] 201962)
29. S. Ivascanu, E. Ludosan, C. Cimpoeru (Rev. Chim., Bucharest, 28, No 5 [1977] 463/469.- C.A. 87 [1977] 189226)
30. V.I. Spitzin, J.L. Rosenfeld (Chem. Tech., Leipzig, 30, No 9 [1978] 437/441)
31. V. Cavazzana, C. Della Rocca (Termotecnica 32, No 1 [1978] 13/20.- C.A. 89 [1978] 94846)
32. M. Crambes, H. Granette, P. Pivette, P. Haicour (W. Ger. 2.731.711, 19 Jan 1978.- C.A. 89 [1978] 133805.- French 2.358.473, 10 Feb 1978.- C.A. 89 [1978] 219158)
33. T. Inoue, T. Shoda, I. Saguchi (Jap. 78.108.075, 20 Sep 1978.- C.A. 90 [1979] 43652)
34. T. Inoue, T. Shoda, I. Saguchi (Jap. 78.108.076, 20 Sep 1978.- C.A. 90 [1979] 43653)
35. R.J. Lipinski (Belg. 865.608, 17 Jul 1978.- C.A. 90 [1979] 76373)
36. S. Yamanaka, T. Saito, H. Ei, S. Inazuka, K. Mitsugi (Jap. 78.28.863, 17 Aug 1978.- C.A. 90 [1979] 76372)
37. Y. Yonekura (Jap. 78.39.858, 24 Oct 1978.- C.A. 90 [1979] 76374)
38. S. Shiga (Jap. 78.58.940, 27 May 1978.- C.A. 89 [1978] 203979)
39. N. Yamamoto, F. Era, H. Sata (Jap. 78.81.440, 18 Jul 1978.- C.A. 89 [1978] 220722)
40. K. Takayasu, N. Sato (Jap. 78.75.139, 4 Jul 1978.- C.A. 89 [1978] 185890)
41. M. Kamata, S. Kanada, S. Katayama (Jap. 78.124.138, 30 Oct 1978.- C.A. 90 [1979] 76375)
42. K. Katsura, T. Tsutsui, H. Tsuchida (Jap. 77.138.031, 17 Nov 1977.- C.A. 89 [1978] 48777)
43. T.J. Suen, A.J. Begala (U.S. 4.092.244, 30 May 1978)
44. C.M. Hwa (U.S. 4.098.720, 4 Jul 1978)
45. D.R. Sexsmith (U.S. 4.105.581, 8 Aug 1978)

2Q. Materials of Construction

CARBON STEEL AND IRON BASE ALLOYS

Ferrous materials are affected by residual chlorine in saline water feed of desalination plants. In stagnant Cl test solution a semilogarithmically increased corrosion rate was observed with increasing dissolved chlorine content [1]. The corrosion resistance of ferritic steels was much greater than that of Cu-Ni alloys or the austenitic stainless steels. No pitting corrosion was observed after 180 days in 3% NaCl solution at 28 to 101°. A low carbon content further increases their corrosion resistance [2]. Nonalloyed or low-alloyed steel corrosion behavior was studied in simulated seawater desalination at ambient temperature to 120° and varying pH and oxygen concentrations [3]. The corrosion rates in artificial seawater were considerably lower for a 6% Cr-steel than for an unalloyed steel C15, a base melt containing C 0.05, Si 0.16, Mn 0.48, Al 0.036, Cu 0.023% and the rest iron, and a 5.5% Ni steel. Transition from uniform corrosion to pitting was observed with increasing oxygen content from 10 to 20 ppb. Sulfide impurities in seawater became corrosion stimulators only at pH <7, but were highly active even in the ppm range. Addition of 1 ppm Cu-ion provided considerable stimulation also [4]. A relation between the corrosion rate of mild steel and the concentration of H_2S in seawater was presented. A proportional rate of corrosion to the concentration of H_2S was found until the steel surface was covered by sulfide film [5].

Immersion tests made in aerated seawater to determine the potential for complete protection of base mild steel showed that this is achieved at -790 mV. The corrosion rate is critically dependent on the direction and extend to polarization. An increased degree of protection was attained with potentials more negative than the -790 mV [6]. In determining fatigue crack growth characteristics of a structural steel in air, tests were in agreement for 38 and 76 mm plates. Effect of seawater on crack growth rate was not observed [7].

Other work includes effects of specimen thickness, saline environment and electrochemical potential on fatigue crack growth [8], effect of the composition of oxidative argon mixtures on the corrosion resistance of welded steel joints in seawater [9], kinetics of the steady-state electrode potential of steel specimens in seawater during low-cycle loading [10], threshold corrosion fatigue crack growth in steels [11], new method for monitoring corrosion fatigue properties [12], potential dependence of the corrosion fatigue of high strength sheet piling steel in salt water [13], pitting corrosion and scaling of carbon steels in geothermal brines [14] and influence of cathodic polarization upon fatigue of notched structural steel in seawater [15].

The corrosion rate at 20° was decreased from 0.25 in untreated seawater to 0.15 g/m^2h in stabilized seawater by using marble filtration. Further improvement in the corrosion protection was achieved by introducing 100 mg/l ANPO and N-(p-ethoxyphenyl)sulfopyridinium chloride following stabilization [16].

The welding operation did not cause any significant effects on the corrosion behavior of mild steel and stainless steel, either of plates of stainless steel welded together or of samples of stainless steel welded to mild steel, except for the expected production of a galvanic couple in the latter case. There was a general tendency for all-mild-steel welded samples to undergo enhanced corrosion at the weld [17]. Carbon steel has been widely used for flash chamber construction, but corrosion problems in this material have created a need for improved materials. A theoretical study has been made of the effect of oxygen content, flow rate and temperature on the corrosion of carbon steel in seawater over the range of conditions likely to be met in a mult-stage flash distillation plant. Data derived from this study were compared with measured data. The use of stainless steels and 90/10 cupro-nickel for flash chamber construction was also considered [18].

Patents. A low alloy steel, to be used in desalination plants, contains C 0.01 to 0.1, Cr 3.0 to 4.0, Nb 0.5 to 1.0, Cu 0.5 to 1.0, Mn 0.3 to 0.6, Ni 0 to up to

half the Cu content, S less than 0.017 and P less than 0.015%. In addition the
steel contains Ti, V and/or Al [19]. A steel of similar composition showed also
good corrosion resistance especially in seawater [20]. Welds in steel containing
C 0.1, Si 0.16, Mn 0.45, P 0.016, S 0.005, Al 0.009 and Ni 0.69% had little or no
corrosion after 3 months immersion in seawater containing 15 ppm oxygen [21]. A
similar composition of welds is also reported [22]. No corrosion was observed of
low alloy steels of given composition after 96 h immersion in synthetic seawater
saturated with H_2S [23].

STAINLESS STEELS

Biological patina on stainless steels in natural seawater increased the corrosion potential and localized attack [24]. Tests confirmed the easy passivation of
W2A and W4A stainless steels in seawater at 20 to 60°. Pitting inceased with temperature and agitation making seawater applications risky [25]. The use of stainless
steel in seawater showed extremely variable results, due to the potential for chloride pitting or crevice corrosion in the commercial stainless grades [26]. Results
of a five-year test on stress corrosion cracking behavior of wrought Fe-Cr-Ni alloys
exposed to a marine atmosphere were reported [27]. Exposure tests in seawater and
desalination loop experiments showed that stainless steel can be used in seawater
under desalination conditions, provided that precautions are taken in design, fabrication and operation [28]. Corrosion resistance of several types of Cr-Ni steels
in various salt media showed that 29% NaCl solution at 125° was the most aggressive
[29]. Methods of studying corrosion resistance were described and some stainless
steel grades resisting to seawater corrosion were given [30].
Other work includes corrosion potential and critical potential for pitting of
Inconel-600, type 304-SS and Zircaloy-2 stainless steels in seawater [31], effect
of heat treatment on corrosion resistance in martensitic stainless steel welded
joints [32] and possible use of stainless steels in seawater exchangers [33].

Patents. A stainless steel containing C 0.004, Si 0.41, Mn 0.62, P 0.011, S
0.007, Cu 0.16, Ni 5.06, Cr 21.41, Mo 1.99 had a corrosion rate in seawater less
than 5 $g/m^2.h$ [34].

COPPER-BASE ALLOYS

The corrosion resistance of experimental copper-base alloys in seawater [35] and
the characteristics of corrosion of copper in desalinated water under dynamic conditions were studied [36]. Various copper-base alloy tubes and Admiralty brass
were tested in synthetic seawater with Na_2S additive and benzimidazole-2-thiol
inhibitor. The inhibitor protected the Admiralty brass corrosion and blocked a
corrosion process already in action [37]. Admiralty brass was coupled to various
copper-base alloys and the corrosion behavior in flowing synthetic seawater was
studied, both in uninhibited condition and in the presence of benzimidazole-2-thiol.
The corrosion rate of the Admiralty brass increased by a factor of over 2 when
coupled to cupronickel alloys, considerably increased when coupled to CuZn22Al2 and
CuAl7 and slightly decreased when coupled to Cu. The inhibitor had a satisfactory
efficiency both in free corrosion and in the coupled condition [38]. The effects
of velocity on the corrosion of galvanic couples of 70/30 CuNi/plain C steel and
Monel/plain C steel in synthetic seawater were expressed by a model which considers
both hydrodynamic and electrochemical boundary layer effects [39].
Corrosion of Cu-Ni-Fe alloys in hot seawater environments is very dependent upon
the initial film formed under specific seawater conditions. Bicarbonate alkalinity
dissolved oxygen and pH are critical species controlling corrosion. A corrosion
mechanism is postulated to show how seawater conditions effect the structure and
composition of corrosion product films [40]. A review was presented on the use of
Cu-Ni alloys in the form of welded tubes, plated or medium thick and thick plates
for the construction of seawater desalination plant [41].
Rotating cylinder experiments were conducted to rapidly determine the relative
corrosiveness of seawater containing sulfide, oxygen or sulfide oxidation products.
Progress of preliminary experiments is summarized [42]. A study on the effect of

hydrostatic pressure on the corrosion of copper in seawater, under constant dissolved oxygen content, showed that the weight loss of copper increased with increase
in pressure and reached a maximum at a pressure of 150 atm. The increased pressure
has no influence on the anodic dissolution process for copper, but accelerates the
cathodic process [43]. The effects of hydrodynamic variables on the corrosion rate
of 90/10 Cu-Ni were studied in seawater and general concepts of turbulence that
affect corrosion were discussed [44]. The effects of hydrodynamic variables were
also studied in single metal exposures and in galvanic couples with Pt with particular emphasis on the determination of variable parameters of fluid dynamics and on
the correlation of nondimensional hydrodynamic and mass transfer parameters with
experimentally determined corrosion rates [45]. The effect of turbulent flow on the
corrosion behavior of 90/10 Cu-Ni were also determined in synthetic seawater by
direct weight loss and by several electrochemical methods. The contribution of
convective diffusion was considered dominant over the velocity range studied and
the rate of eddy diffusivity was described [46]. The corrosion of 90/10 and 70/30
Cu-Ni alloys in flowing seawater was studied as a function of oxygen concentration
using the linear polarization, a.c. impendance, and potential step methods for
measuring the polarization resistance. The high nickel alloy was found to be more
corrosion resistant than the 90/10 Cu-Ni alloy, provided that oxygen content is not
more than 6.60 mg/l [47]. Horse shoe corrosion is the result of the formation of
an active-passive cell. The pH of the seawater is the determining factor and development of the corrosion is strongly influenced by the local hydrodynamic conditions
[48].

Compositions, engineering properties, standard available forms and high temperature properties, especially maximum allowable stresses and commonly used copper
alloys, were presented [49]. A review was also presented on nickel-containing
materials for marine applications, including Cu-Ni alloys and stainless steel [50].
A case history on copper alloy tube failure in seawater condensers was described [51].
There are four different types of fouling on heat exchangers using seawater as
coolant and several operational parameters affect the fouling, i.e. temperature,
flow rate, oxygen concentration, material of exchanger and combination of factors [52].

By application of a characterization technique, consisting of electrochemical
and visual/microscopical methods, qualitative and quantitative information about
the protective properties of layers on CuNi10Fe in seawater can be obtained. After
about four weeks the system seems to be stabilized [53].

The decomposition of plant and animal macroorganisms in seawater leads to a
solution of low pH, high sulfide and high total organic content, which can accelerate corrosion of copper base alloys [54]. High Cl$^-$ waters have an increased corro-
sivity toward Admiralty brass. A mathematical relation was developed between
aggressive Cl$^-$ concentration and the inhibitor combination of benzotriazole and
chromate [55].

Patents. Copper alloys containing Al 2.0 to less than 6.0, Ni 1.0 to 12.0, Fe
0.1 to less than 2.0 and Mn 0.1 to 3.0% are resistant to concentrated seawater. The
pitting depth and corrosion loss in a synthetic seawater during 300 h at 8 m/s were
0.08 mm and 1.38 mg/cm^2, respectively [56].

ALUMINUM

A search of the existing literature and unpublished data on the corrosion performance of aluminum alloys in seawater was conducted and quantitative data were computer catalogued [57]. Pitting and crevice corrosion proved to be the types of attack
that predominated. The catalogued data are the results of many static tests conducted under natural conditions of marine fouling. Aluminum alloys 5052 and Alclad
3003 can be used for heat exchanger tubes [58]. The adsorption isotherms for chloride on a corroding aluminum surface were measured by using ^{36}Cl as a radioactive
tracer. Addition of nitrate or sulfate delayed but did not prevent the uptake of
chloride. It was concluded that a corroding aluminum surface has a variety of adsorption sites with differnet adsorption properties. Only a minority of these sites are
potential or active sites for pitting corrosion [59].

In studying the galvanic corrosion behavior of 5086 aluminum alloy, when coupled with more noble metals and immersed in aerated synthetic seawater, galvanic current density measurements, potentiodynamic polarization determinations and optical and electron microscopic observations were made [60]. A study was undertaken to examine the corrosion behavior of 5456-H117 aluminum in high velocity seawater. The results show that both the rate and mode of corrosion are velocity dependent. The corrosion rate increased significantly with increasing velocity. The observed morphological changes suggest that the basic mechanism for high velocity corrosion of 5456-H117 aluminum involves film disruption/removal leading to enhanced micro-pitting about intermetallic particles [61]. Stress corrosion tests in marine atmosphere and seawater on smooth and precracked specimens of various aluminum alloy plates showed that threshold stress intensity in both environments was independent of plate thickness and was inversely proportional to yield strength. The fracture surfaces and metallographic cross sections showed crack blunting, branching, un-failed ligaments and mechanical fracturing in the precracked specimens [62]. In a condensed summary of pertinent literature regarding the corrosion of aluminum in water, corrosion resistance and pitting were primarily considered, as well as the effects of deposits and the various corrosive substances, including living organisms [63].

TITANIUM

The performance of titanium and copper-nickel tubing under all the various forms of chemical and mechanical attack, to which they are subjected in desalting plant operation were compared. In discussing the factors influencing the choice of material for a plant, the uncertainties in the future prices of both materials are emphasized [64]. The use of titanium tubes as heat transfer tubes for large scale multistage flash seawater desalination plant was reviewed [65]. Titanium is comparable to other corrosion resistant alloys and corresponds with stainless steel in thermal properties. Corrosion resistance is due to the oxide film developed on titanium and extends to all kinds of natural waters, including brackish water and steam at high velocities [66]. Titanium tubes are completely resistant to corrosion, whereas under the same conditions condenser tubes of austenitic and austenoferritic Cr-Ni stainless steels containing up to 4% Mo may undergo corrosion pitting in flowing seawater [67].

Investigation of corrosion cracking of titanium in deaerated 6% NaCl solution, simulating conditions in an MSF desalination plant showed: Titanium is affected by corrosion cracking at temperatures over 80°C. Coating with PdO/TiO_2 is effective against corrosion attack. Couples with iron or zinc pick up hydrogen at temperatures over 80°. Hydrogen is not picked up by not coupled titanium [68].

Choice of Cu, Ni or Ti alloys for uses subject to corrosion by seawater was discussed and protective measures were considered [69]. Titanium heat transfer tubes are gradually beginning to be used in MSF plants. However, there is a possibility of crevice corrosion and hydrogen absorption. An investigation of some corrosion problems includes crevice corrosion of titanium tube, galvanic corrosion of copper-alloy tube-plate equipped with titanium tubes and hydrogen absorption of titanium tubes [70].

CONCRETE

The post-tensioned high-tensile steel wire embedded in cement grout or in high-strength concrete was completely protected from corrosion, even under conditions of alternate immersion in simulated seawater. Some rusting was observed when the grout or concrete contained 2 to 5% $CaCl_2$ [71]. Metal-fiber reinforced concrete is a structural material with several unique advantages for offshore applications [72]. Development of seawater corrosion resistant steel bars for offshore concrete structures was reported [73]. A prototype of three MSF stages of a 30000 m^3/day desalination plant in full size has been constructed and tested to temperatures up to 120°C. Concrete performance was satisfactory [74].

Patents. A cement containing C2S 55, C2S 20, C3A 1.6 and C4AF 15.2% resists sulfates, has strong resistance to cracking in aqueous salts and provides better protection for steel [75].

NON METALLICS

An evaporator design was presented providing non-metallic reinforced thermosetting materials for all parts except heat exchanger tubing. The proposed plant capacity is 1 Mgd, assuming the use of a polymer additive treatment at 212°F top brine temperature. Pumps except shafts, are also fabricated from a compressed molding. Cost estimates indicate this evaporator to be competitive with the metallic plants being fabricated today [76].

PROTECTIVE COATINGS

Protective paint "Silikatsink-2" was used to increase the corrosive fatigue strength of steels in seawater [77]. A system based on zinc silicate paint is recommended for protection of metallic structures in atmospheric conditions with periodic wetting with salt water [78]. An accelerated testing procedure was used to determine the effect of average zinc dust particle size on the electrochemical behavior of a zinc-rich chlorinated rubber primer applied to mild steel. A particle size of 2.5 μm provided the best protection [79]. In experiments with bare and painted steel, high Reynolds flow effects were not found to be harmful as long as the painted to bare steel ratio was high [80]. The zinc dust content necessary to ensure protection of mild steel was 80 wt.% in the dried film [81]. By using mixtures of zinc and zinc oxide, the zinc content could be replaced by up to 43% of zinc oxide, on a total solids content of 93% in the dried film, without unduly affecting the protection [82]. A review was presented on protection of steel against corrosion by means of coatings, including properties, advantages, disadvantages and applications in tabulated form [83]. Protection of steel piles in seawater by non-metallic coatings was reported [84]. Best corrosion protection in vapor of synthetic Caspian Sea water was observed for the coating based on synthetic rubber with the addition of Akor-1 as inhibitor [85].

CATHODIC PROTECTION

The combination of the two protective systems, organic coating and cathodic, complement each other but it is important that the potential applied in cathodically painted steel not be too high or the paint will be damaged [86]. The hydrogen absorption during cathodic protection by samples prepared from high-strength steel was greater in moving than in quiet seawater [87]. A breaking point was observed on the anodic polarization curve in artificial seawater of a titanium electrode coated with RuO_2, which was due to the marked increase in chlorine gas generation. In seawater, the RuO_2 layer was corroded at high current densities [88]. A special cathodic protection of the inside of mild steel tubes was obtained by applying a layer of zinc powder or zinc grains. The entire inside part of the tube was polarized to about -900 mV. The current generated by the dispersed anodes protected the entire inside surface of the tube [89]. A short report was given on the use of an aluminum alloy for the cathodic protection of marine structures [90]. Electrodeposition of calcium carbonate on structures immersed in seawater and subjected to cathodic protection was investigated [91]. Electrochemical characteristics and corrosive wear of graphite anodes containing up to 50 wt.% of lead powder were determined in 3% NaCl solution at 20°. During anodic charge wear of the anodes decreased when the lead content increased to 10% [92].

Patents. A zinc sacrificial anode, partly coated with polystyrene, installed on a heat exchanger in a refrigerator system prevented tube corrosion for one year. Without coating, the sacrificial anode was consumed and tube corrosion was significant [93]. An aluminum alloy containing Zn 4.5, In 0.02, Ca 0.02, Mg 2 and miscellaneous metals 0.01% was used as sacrificial anode [94]. A zinc sacrificial anode contains Al 0.05 to 0.5, Sn 0.01 to 0.5, Mn 0.005 to 0.04, Ti and/or B 0.006 to 0.05% [95].

LITERATURE TO 2Q.

1. H.Nakauchi, Y. Oyama, K. Osato, H. Togano (Boshoku Gijutsu 26 [1977] 629/636.- C.A. 89 [1978] 133600)
2. P.G. Kulkarni, S.N. Bagchi, K. Balaramamoorthy (Tool Alloy Steels 12, No 5 [1978] 157/159.- C.A. 89 [1978] 218890)
3. R. Manner, E. Heitz (Werkst. Korros. 29 [1978] 559/567.- C.A. 89 [1978] 185828)
4. E. Heitz, R. Manner (Werkst. Korros. 29 [1978] 783/791.- C.A. 90 [1979] 125493)
5. H. Sasaki, T. Tanaki, T. Nakahara, Y. Kanda (Boshoku Gijutsu 26 [1977] 229/236.- C.A. 89 [1978] 133602)
6. B.T. Moore, P.J. Knuckey (Corros. Australas. 3, No 3 [1978] 4/5.- C.A. 89 [1978] 187891)
7. R. Johnson, I. Bretherton (U.K. At. Anergy Auth. North Div. Rept. ND-R-151(S).- C.A. 90 [1979] 42077)
8. A.M. Sullivan, T.W. Crooker (Proc. Int. Conf. Fract. Mech. Technol. 1 [1977] 677/698.- C.A. 90 [1979] 27113)
9. S.T. Rimskii, V.G. Svetsinskii, E.P. Los (Avtom. Svarka No 4 [1978] 55/59.- C.A. 89 [1978] 115300)
10. A.V. Kobzaruk, G.V. Karpenko (Fiz.-Khim. Mekh. Mater. 14, No 1 [1978] 75/80.- C.A. 89 [1978] 119541)
11. L.K.L. Tu, B.B. Seth (J. Test. Eval. 6, No 1 [1978] 66/74.- C.A. 89 [1978] 218861)
12. E. Maahn, H. Noppenau (Proc. Dan. 1st Conf. High Perform. Mater. Processes [1978] 157/170,-C.A.89 [1978] 170816)
13. R.K. Poepperling, W. Schwenk , G. Vogt (Werkst. Korros. 20 [1978] 445/451.- C.A. 89 [1978] 154531)
14. A. Goldberg, L.B. Owen (Corrosion, Houston, 35, No 3 [1979] 114/124.- C.A. 90 [1979] 176891)
15. W.C. Hooper, W.H. Hartt (Corrosion, Houston, 34 [1978] 320/323.- C.A.89 [1978] 187893)
16. A.D. Kengerli, A. Ya. Abdullaev, T.M. Mursakulova (Azerb. Neft. Khoz. No 9 [1978] 48/49, 60.- C.A. 91 [1979] 25049)
17. T. Hodgkiess, W.T. Hanbury, M. Arndt, N. Eid (Desalination 31 [1979] 399/410)
18. J.W. Oldfield, B. Todd (Desalination 31 [1979] 365/383)
19. Vereinigte Oesterreichische Eisen- und Stahlwerke-Alpine Montan A.G. (W. Ger. 2.624.515, 15 Dec 1977.- C.A. 90 [1979] 10469)
20. Vereinigle Oesterreichische Eisen- und Stahlwerke-Alpine Montan A.G. (Belg. 844769, 16 Nov 1976.- C.A. 90 [1979] 172579)
21. T. Kurusu, C. Kuno (Jap. 78.100.120, 1 Sep 1978.- C.A. 90 [1979] 27211)
22. T. Kurusu, C. Kuno (Jap. 78.100.121, 1 Sep 1978.- C.A. 89 [1978] 219164)
23. Y. Inagaki, T. Nishimura, M. Kodama, M. Tanimura (Jap. 79.11.017, 26 Jan 1979.- C.A. 90 [1979] 172599)
24. A. Mollica, A. Trevis (Acciaio Inossid. 44, No 4 [1977] 3/13.- C.A.[1978] 183816)
25. O. Landauer, C. Mateescu, D. Geana, O. Iulian (Bul. Inst. Politch. "Gheorghe Gheorghiu-Dej" Bucuresti, Ser. Chim.-Metal. 39, No 3 [1979] 39/46.- C.A. 89 [1978] 119531)
26. H.E. Deverell, J.R. Maurer (Mater. Perform. 17, No 3 [1978] 15/20.- C.A. 89 [1978] 119538)
27. K.L. Money, W.W. Kirk (Mater. Perform. 17, No 7 [1978] 28/36)
28. R. Droin (Mater. Tech., Paris, 66, No 9/10 [1978] 283/288.- C.A. 90 [1979] 156889)
29. K. Krzysztofowicz, K. Ksiazek, N. Zajdek (Ochr. Koroz. 21, No 2 [1978] 37/42.- C.A. 89 [1978] 218863)
30. J.L. Crolet (Aciers Spec. No 42 [1978] 21/27.- Rev. Metall., Paris, 75, No 11 [1978] 633/639.- C.A. 90 [1979] 78319)
31. M. Shimada (Denryoku Chno Kenkyusho Hokoku [1977] 276024.- C.A. 89 [1978] 187875)
32. J. Paszota (Pr. Nauk Uniw. Slask. Katowicach 214 [1978] 14/26.- C.A. 90 [1979] 172437)
33. R. Droin (Rev. Metall. 75 [1978] 655/661.- Aciers Spec. 45 [1979] 3/9)
34. M. Kowaka, H. Nagano (Jap. 78,25,214, 8 Mar 1978.- C.A. 89 [1978] 133744)

35. A.A. D'yakov, T.D. D'yakova (Voprosy Atom. Nauki i Tekhn. Ser. Opresnenie Solen. Vod, 2, No 10 [1977] 84/88.- Ref. Zh., Khim. 17 I 293 [1978])
36. O.P. Kuznetsov, N.I. Borovin (Tr. VN11 VODGEO No 66 [1977] 5/7.- Ref. Zh., Teploenerg. 5 R 163, [1978])
37. F. Zucchi, G. Brunoro, G. Trabanelli (Werkst. Korros. 28, No 12 [1977] 834/837.- C.A. 89 [1978] 184405)
38. F. Zucchi, G. Brunoro, M. Zucchini (Mater. Chem. 3, No 2 [1978] 91/104.- C.A. 89 [1978] 167808)
39. G.A. Storm (NTIS Rept. AD-A048669 [1977].- C.A. 89 [1978] 183819)
40. R.W. Ross (Corrosion '77, Int. Corros. Forum, Prot. Perf. Mater., Paper 94, San Francisco, Calif. 1977)
41. P. Bronzini de Caraffa (Sondage Tech. Connexes 31, No 7-8 [1977] 290/297.- E.I. 16, No 3 [1978] 22654)
42. B.C. Syrett, S.S. Wing, D.D. Macdonald (NTIS Rept. AD-A058 052/2GA [1978].- G.R.A. 78, No 24 [1978] 157)
43. E.D. Mor, A.M. Beccaria (Br. Corros. J. 13, No 3 [1978] 142/146.- C.A. 90 [1979] 30975)
44. R.P. Schack (NTIS Rept. AD-A056 325/4GA [1978].- C.A. 90 [1979] 42737)
45. G.H. Leumer (NTIS Rept. AD-A056381/7GA [1978].- C.A. 90 [1979] 30979)
46. G. Leumer, R.P. Schack, K.J. Graham, J. Perkins (NTIS Rept. AD-AO 59646/OGA [1978].- C.A. 90 [1979] 158965)
47. D.D. Macdonald, B.C. Syrett, S.S. Wing (Corrosion, Houston, 34, No 9 [1978] 289/301-NTIS Rept. AD-A051 833/2GA.- C.A. 89 [1978] 187892)
48. G. Bianchi, G. Fiori, P. Longhi, F. Mazza (Corrosion, Houston, 34, No 11 [1978] 396/406.- E.I. 17 [1979] 015436)
49. W.S. Lyman, A. Cohen (Chem. Engng. 85, No 6 [1978] 99/102)
50. B. Todd (Anti-Corros. Methods Mater. 25, No 10 [1978] 4/7)
51. J.M. Schluter (Mater. Perform. 17, No 2 [1978] 25/28)
52. Anonymous (Do 19 [1978] 1/6)
53. F.P. Ijsseling. J.M. Krougman (Euro. Congr. Met. Corros., London [1977] 181/188.- E.I. 16 [1978] 72740)
54. K.D. Efird, T.S. Lee (Corrosion 35 [1979] 79/83)
55. G.T. Farley, F.G. Vogt (Proc. Int. Water Conf., Eng. Soc. West. Pa. 39 [1978] 352/355)
56. T. Hiramatsu, M. Nonaka, Y. Monju (Jap. 78.41.096, 31 Oct 1978.- C.A. 90 [1979] 75541)
57. R.A. Bonewitz (Proc. 4th Ann. Conf. Ocean Thermal Energy Convers., New Orleans [1977] VII. 37/40.- E.R.A. 3 [1978] 47165)
58. R.H. Wagner, R.A. Bonewitz (NTIS Rept. PNL-2606 [1978].- E.R.A. 3 [1978] 39693)
59. A. Berzins, R.T. Lowson, K.J. Mirams (Aust. J. Chem. 30 [1977] 1891/1903.- E.R.A. 3 [1978] 43312)
60. J. Perkins, J.S. Locke, K.J. Graham (NTIS Rept. AD-A050917 [1978])
61. G.A. Gehring (NTIS Rept. AD-A058814/5GA [1978].- G.R.A. 79, No 2 [1979] 157)
62. R.C. Dorward, K.R. Hasse (Corrosion, Houston, 34, No 11 [1978] 386/395.- C.A. 90 [1979] 75484)
63. C.R. Schmitt, J.M. Schreyer, J.M. Googin (NTIS Rept. Y-DA-7955 [1977])
64. R.B. Cox (Pure Water, 5, No 2 [1977] 5/9)
65. Y. Watanabe (Chitaniumu Jirukoniumu, 24, No 3 [1976] 113/129, 25, No 1 [1977] 5/18)
66. L.C. Covington, R.W. Schutz, I.A. Franson (Chem. Engng. Progr. 74, No 3 [1978] 67/69)
67. G. Herbsleb (Arch. Eizenhuettnewes. 49, No 11 [1978] 545/547.- C.A. 90 [1979] 109741)
68. T. Fukutsuka, K. Shimogohri, H. Sato, F. Kamikubo, T. Hara (R + D Kobe Seiko Gihi 28, No 3 [1978] 76/80.- C.A. 90 [1979] 43634)
69. H. Richter (Werkst. Korros. 28 [1977] 671/676.- C.A. 89 [1978] 10509)
70. T. Fukuzuka, K. Shimogori, H. Satoh, F. Kamikubo (Desalination 31 [1979] 389/398)
71. N.S. Rengaswamy, S. Chandrasekharan, K.S. Rajagopalan (Indian Concr. J. 51, No 11 [1977] 342/346.- C.A. 89 [1978] 167327)
72. R.G. Rider, R.H. Heidersbach (NTIS Rept. PB-290333/4GA [1978])
73. H. Shimada, H. Okada, H. Nishi (Proc. Ann. Offshore Technol. Conf. 10, No 2 [1978] 1243/1250.- C.A. 89 [1978] 201934)

74. A. Bernard, R. Lacroix, M. Izabel (Desalination 31 [1979] 385)
75. Aktieselskapet Norcem (Belg. 856.653, 9 Jan 1978.- C.A. 89 [1978] 219844)
76. O.J. Morin, R.A. Johnson (Desalination 31 [1979] 355/363)
77. V.A. Orlov, Yu. G. Ozhiganov, O.F. Shevebenko (Sov. Mater. Sci. 13, No 5 [1977]
    545/547.- E.I. 17 [1979]005959)
78. B.I. Borovskii, G.F. Afanas'ev (Lakokras. Mater. Ikh Primen. No 6 [1978] 24/26.-
    C.A. 90 [1979] 73367)
79. H.H. Chua, B.V. Johnson, T.K. Ross (Corrosion Sci. 18, No 6 [1978] 505/510.-
    O.A. 26 [1979] 79:74)
80. B.V. Johnson, T.K. Ross (Corrosion Sci. 18, No 6 [1978] 511/518.- O.A. 26 [1979]
    79:77)
81. R. Pedram, T.K. Ross (Corrosion Sci., 18, No 6 [1978] 519/522.- O.A. 26 [1979]
    79:78)
82. H.H. Chua, T.K. Ross (Corrosion Sci., 18, No 6 [1978] 523/525.- O.A. 26 [1979]
    79:75)
83. K.A. van Oeteren (VDI-Z. 120, No 23 [1978] 1089/1099)
84. E. Escalante, W.P. Iverson (Mater. Perform. 17, No 10 [1978] 9/15)
85. V.I. Nikitin, A.M. Gvozd, E.A. Demirchyan, T. Ya. Karpova, I.E. Zibozov,
    L.V. Demidova (Tr. Tsentr. Nauchno-Issled. Prvektno-Konst. Kotloturbinnyi Inst.
    141 [1977] 74/81.- C.A. 91 [1979] 24874)
86. S. Wiktorek (Prepr. Pap. Ann. Conf. Australas. Corros. Assoc., Paper L2 [1977].-
    C.A. 90 [1979] 30968)
87. S.M. Beloglazov, L.K. Yagunova (Korroz. Zashch. Met. 3 [1979] 31/36.- C.A. 90
    [1979] 30980)
88. M. Takasu, E. Sato (Boshoku Gijutsu 26, No 9 [1977] 499/502.- C.A. 89 [1978]
    137479)
89. H. Nakauchi, K. Osato, H. Hogano (Boshoku Gijutsu 26, No 12 [1977] 711/720.-
    E.I. 16 [1978] 34256)
90. T. Murai, Y. Tamura, C. Miura (Chemsa 5, No 1 [1979] 11.- C.A. 90 [1979] 143465)
91. J. Ledion, M. Moreau (Mater. Tech., Paris, 65 [1977] 751/753, 66 [1978] 9/13,
    51/54.- C.A. 89 [1978] 187895, 137474.- E.I. 17 [1979] 5958)
92. J. Kubicki, B. Szczygiel (Ochr. Koroz. 22 No 4 [1979] 98/102.- C.A. 91 [1979]
    29507)
93. T. Ozaki, T. Sasaki, Y. Ishikawa (Jap. 78.06.240, 20 Jan 1978.- C.A. 89 [1978]
    45465)
94. T. Murai, Y. Tamura (Jap. 78.100.115, 1 Sep 1978.- C.A. 90 [1979] 14049)
95. Y. Tamura, T. Yoshikawa (Jap. 78.100.125, 1 Sep 1978.- C.A. 90 [1979] 31123)

3A. Ion Exchange

Development of modern techniques made it possible to use ion exchangers not only for the preparation of demineralized water, but also for the recuperation of materials from waste water and solutions. The future of such separation processes may be in the combination of reverse osmosis and ion exchange [1]. Preparation, technology and utilization of ion exchangers were reviewed, with emphasis on separation processes, adsorption and catalysis with ion exchangers, as well as binding of heavy metals for soil improvement and ion exchanger use as carriers for nutrients in aquaculture [2].

A review was presented on the mechanism of ion exchange and its impact on water quality [3]. Several process arrangements for ion exchange and adsorption were described, including large-scale swimming-pool systems for water treatment, mixed bed deionization, cyclic multi-bed systems, continuous ion exchange, reverse-flow regeneration, combination fixed-bed and fluidized-bed systems for water softening and the Sirotherm or thermal-cycling process for desalination. Schematic flow diagrams are provided for each of the process configurations [4]. A method was proposed for calculating the amount of ion-exchange resin in mono-ionic form for correction of multicomponent solutions. An experimental study of the K^+-Mg^{2+}-Ca^{2+} and Na^+-K^+-Mg^{2+}-Ca^{2+} systems demonstrated that correction of a multicomponent solution by this method may be carried out for all ions, the correction errors being 5 to 10% [5].

Other work includes the design, operation and maintenance of a demineralized water plant [6], renovation of brackish and waste water through ion exchange technology [7] and protecting demineralizers from organic fouling [8].

<u>ION EXCHANGE TECHNOLOGY</u>

A method for calculating the performance of settling filters is based on the theory of dynamic sorption of ions on granular particles and requires a knowledge of the ion exchange velocity constant, resin exchange capacity, grain size and the kinetic characteristics of the filter. The theoretcial and experimental results for a horizontal settling filter containing H-ion type cation resin are in good agreement [9].

Precipitation can accompany ion exchange when ions released from the exchanger produce a supersaturated solution with oppositely charged ions present in the contacting solution which has tended to complicate several processes having potential practical interest or to render them infeasible. The ion-exchange step accompanied by precipitation is carried out by upflow in a fluidized bed and the precipitate is separated from the exchanger particles by hydraulic classification. This idea has been reduced to practice in the example of softening the high-calcium, high-sulfate feed for a desalting operation, the softening step being carried out downflow in the fixed-bed mode. The results of bench-scale development indicate that continued cyclic ion exchange can be sustained solely with the reject brine from the desalting process, that calcium removal in the order of 98% and magnesium removal in the order of 95% can be achieved, and that successful performance is obtainable over a wide range of operating conditions [10]. Separation of an ion-exchange desalination filter into two columns reduced reagent consumption during the regeneration process [11].

In the Lewatit-WS-process the ion exchanger comprises a fluidized bed contacting a fixed bed contacting a bed of inert resin. This system provides efficient water treatment. The system is regenerated by passing regenerant through countercurrently, the fixed bed is moved onto the fluidized bed, thus immobilizing it and providing rapid, efficient regeneration. The rinsing volume of treated water and rinse time are both small. The bed movement prevents the formation of channels [12]. Countercurrent regeneration in the Asahi process, packed bed systems and deep bed systems are more effective and have increased capacity. Outage time and effluent disposal problems are reduced, especially when operated at high flow rates which cause high pressure drops across the resin beds [13].

Other work includes countercurrent ion-exchange columns of high specific capacity and their use for solving certain problems [14] and control of the productive capacity of automated apparatus for the chemical desalination of water [15].

Patents. In a countercurrent fixed-bed ion-exchange apparatus, the water flows upwardly and regeneration is carried out downwardly [16]. Ion exchange resin beads are added at the top and removed from the bottom without stopping for regeneration [17]. Regenerating solution is passed through the fixed bed in the opposite direction to that of the feed solution. Resin particles are withdrawn from one end of the bed, passed through the wash zone and returned to the other end of the bed [18]. Via a backwash column provided at the top of the desalination column, the ion-exchange resin is supplied to the desalination column. The resin is introduced as a slurry into the backwash column and withdrawn from the bottom of the desalination column [19].

Water is treated with H-type weakly acidic cation exchange resin to remove HCO_3 and then treated with mixed resin bed [20]. A mineral acid is added to water to decarbonate. The method prevents $CaCO_3$ deposits [21]. The cation-exchange resin is regenerated by NaCl solution in the first stage and a H_2SO_4 solution in the second stage [22].

Water is desalinated with mixed ion exchange resins. The anion exchanger is regenerated with NaOH and N_2H_4 solutions [23]. Regenerated resins are mixed and rinsed with water until they are reused [24]. The mixed ion-exchangers are separated and regenerated by a method that decreases the consumption of regeneration agent [25]. Ion-exchange resin beads containing weak cation exchange groups and weak anion exchange groups are used for the desalination of a solution containing strongly acidic salts. The resin is heated and treated with H_2CO_3 for regeneration [26]. In the cyclic operation of a mixed bed, the transfer schedule for the mixture of anion- and cation-exchange materials between the treatment and the separation columns is described. Chemicals which assist in the separation of the two resins are added at an intermediate point in the cycle [27].

SIROTHERM PROCESS

Thermally regenerable ion exchange resins were used to partially demineralize water containing up to 3000 mg/l of dissolved salts. The resins are regenerated by hot water. They were used for treatment of brackish underground and surface waters and for effluent treatement [28]. Heat regenerable ion-exchangers were applied in Japan for desalination of water containing 300 to ·3000 ppm of salts [29].

A new periodic operation in which a thermally regenerable ion-exchange resin in a basket is alternatively immersed in cold and hot reservoirs is developed to desalt water. A graphical solution for mass balance equations is presented together with analytical solutions for special cases [30] A review on the production of desalted water by heat-regenerative ion exchange resins covers the method, the use of Amberlite XD-2, a batch system and continuous one- and two-bed systems and applications [31].

Patents. Basic ion exchange resins that can be thermally regenerated are manufactured by treating diallylamine polymers with organic dihalo compounds to reduce the tendency of the polymers to swell [32]. Waters containing salts are passed upwardly through columns packed with heat regeneration-type ion-exchange resins to absorb the salts. Hot water is passed downward through the columns [33]. A hybrid resin comprises two polymer phases, one functioning for anion-exchange and the other for cation-exchange. The regeneration by elution is made with a liquid at a higher temperature than the operating temperature [34]. A thermally regenerable ion exchange resin containing carboxyl and amino groups was preloaded with a solution containing Na_2HPO_4 and NaCl. The preloaded resin was used to treat solution containing 1870 ppm $CaCl_2$ [35].

LITERATURE TO 3A.

1. K. Marquardt (Chem. Tech.. Heidelberg, 7 [1978] 535/539)
2. G.Kühne (Fortschr. Verfahrenstech. 16 [1978] 235/255)
3. M.J. Semmens (Proc. Public Water Supply Eng. Conf. 19 [1977] 115/127)
4. T. Vermeulen (Chem. Eng. Prog. 73, No 10 [1977] 57/61)
5. V.A. Bychkova, V.S. Soldatov, A.M. Vitkovskaya (Vestsi Akad. Navuk BSSR, Ser. Khim. Navuk No 1 [1978] 19/22. - C.A. 89 [1978] 131613)
6. D.M. Butala, P.H. Patel (Indian Chem. Manuf. 15, No 11 [1977] 37/45)
7. A.K. Sengupta (Indian Chem. Manuf. 16, No 7 [1978] 43/48)
8. J.D. McWilliam (Chem. Eng., N.Y., 85, No 12 [1978] 80/84)
9. N.G. Skvortsov,G.M. Kolosova (Zh. Fiz. Khim. 52 [1978] 1763/1765.-C.A. 89 [1978] 185741)
10. G. Klein, T.J. Jarvis, T. Vermeulen (Recent Dev. Sep. Sci. 5 [1979] 185/198)
11. G.K. Feiziev, V.M. Baladzhanov (Energetik No 3 [1978] 31)
12. F.B. Martinola (Proc. Int. Water Conf., Eng. Soc. West. Pa. 39 [1978] 29/39)
13. A.W. Gillis, K. Smith (Proc. Int. Water Conf., Eng. Soc. West. Pa. 39 [1978] 43/59)
14. V.I. Gorshkov, G.A. Medvedev, D.M. Murav'ev, N.B. Ferapontov (Teoriya i Praktika Sorbtsion Protsessov No 12 [1978] 83/91. - Ref. Zh., Khim. 12 I 146)
15. G.V. Efimov, E.I. Murakhovskaya, E.A. Perova, E.V. Titarenko (Tr. Vses. Teplotekhn. NII [1978] 121/133.- Ref. Zh., Khim. [1978] 17 I 302)
16. T. Sasaki (Jap. 78.03.755, 9 Feb 1978.- C.A. 89 [1978] 65111)
17. K. Komatsu, N. Sasaki (Jap. 78.19.976, 23 Feb 1978.- C.A. 89 [1978] 65120)
18. Asahi Chemical Co. (Brit. 1.505.384, 22 Mar 1978)
19. K. Chonan, K. Fujiwara, H. Kashima (Jap. 77.123.382, 17 Oct 1977. - C.A. 88 [1978] 110347)
20. I. Moriyasu, Y. Arai, M.Fujita (Jap. 77.134.876, 11 Nov 1977.-C.A. 89 [1978] 30548)
21. Y. Arai, M. Fujita (Jap. 77.134. 877, 11 Nov 1977.- C.A. 89 [1978] 117501)
22. I. Moriyasu, Y. Arai (Jap. 77.146.782, 6 Dec 1977.- C.A. 89 [1978] 65117)
23. Crane Co. (Neth. 76.02.976, 26 Sep 1977.- C.A. 89 [1978] 65102.- French 2.348. 157, 16 Dec 1977)
24. H. Takeuchi, M. Saito (Jap. 78. 04.776, 17 Jan 1978.- C.A.89 [1978] 117502)
25. K. Chonan, K. Fujiwara (Jap. 78.67.682, 16 Jun 1978.- C.A. 89 [1978] 185877)
26. H. Shimizu (Jap. 78.114.780, 6 Oct 1978.- C.A. 90 [1979] 61037)
27. K. Marquardt (U.S. 4.088.563, 6 May 1978)
28. B.A. Bolto, N.H. Pilkington, D.E. Weiss, P.M. Sharples, G.K. Stephens, K.O. Wade (Progr. Water Technol. 9 [1978] 911/922)
29. T. Maeda, T. Hirano (Nippon Kaisui Gakkaishi 31, No 1 [1977] 20/30)
30. S. Goto, N. Sato, H. Teshima (Sep. Sci. Technol. 14, No 3 [1979] 209/217)
31. T. Iwatsuka (Nenryo Oyobi Nensho 45, No 3 [1978] 211/222)
32. B.A. Bolto, K.H. Eppinger (W. Ger. 2.755.504, 15 Jun 1978.- C.A. 89 [1978] 111236)
33. T. Maeda, Y. Kaminami, R. Sugimoto, K. Sasaki (Jap. 78.69.449, 20 Jun 1978.- C.A. 89 [1978] 220708)
34. J.H. Barrett, D.H. Clemens (U.S. 4.087.357, 2 May 1978)
35. D.E. Weiss, B.A. Bolto, G.K. Stephens (Jap 78. 129.186, 10 Nov 1978.-C.A.90 [1979]192346)

3B. Membrane Processes

Under the general term of membrane processes separation methods are included, in which the desalination effect is due to the action of a membrane. Electrodialysis uses ion selective membranes and a direct current as driving force. In all other membrane processes the driving force is a pressure differential. Reverse osmosis makes use of semi-permeable membranes. Hyperfiltration is synonymous with reverse osmosis. Ultrafiltration differs from reverse osmosis in the characteristics of the membranes used, in the size of the separated particles and necessity of lower pressures. In a newer term these processes are also named as membrane filtration. Piezodialysis is a pressure driven process which uses ion exchange membranes.

Potable water can be produced by means of membrane filtration, as well as waste water purification at reasonably low cost. Newly developed composite membranes are characterized by a higher water permeability with equal salt retention. Poisonous materials can be retained with special membranes. The advantages of membrane filtration are low operating temperature, flexible capacity modification, recuperation of valuable materials and low energy requirement [1]. Membrane technology with the existing know-how and the characteristics of the developed membranes offers interesting economic and technical solutions to problems of production, recycling and environmental protection [2].

The proceedings of the 13th Annual Desalination Conference on desalination and waste water treatment by means of membranes, held in Jerusalem 8 to 12 January 1978, were edited by S. Gairon [3].

Other published work refers to the principles, practice and prospects of membrane separation processes [4], transport models in osmotic membranes [5], modern ion exchange and reverse osmosis processes in fresh and waste water treatment [6], advances in membrane processes technology, reverse osmosis and ultrafiltration [7], membrane desalination technology in the world [8], developments in membranes, module and systems [9], future aspects of membrane processes [10], current status of separation technology with membranes [11].

LITERATURE TO 3B.

1. C.A. Smolders (Polytechn. T. Ed. Processtechn. 33 [1978] 362/367)
2. W.J. Brenner (Chem. Techn. 7 [1978] 221)
3. S. Gairon, editor (Nat. Coun. for R & D, Publ. No NCRD 8-78 [1978] 463pp)
4. P. Meares (Interdiscip. Sci. Rev. 2 [1977] 327/336)
5. G. Mossa (Quad. Ist. Ric. Acque 22 [1977] 3/22)
6. K. Marquardt (Chem. Tech., Heidelberg, 7 [1978] 535/539)
7. J.W. Carter (ALTECH 27 [1978] 69/94)
8. T. Yamabe (Maku 3 [1978] 116/124)
9. S. Kimura (Kagaku Kogaku 42 [1978] 487/488)
10. H. Ohya (Kagaku Kogaku 42 [1978] 480/484)
11. E. Staude (Fette, Seifen, Anstrichm. 81, No 2 [1979] 75/83)

3C. Ion Exchange Membranes

Methods of preparing cation- and anion-exchange membranes with low areal resistances were developed for use in lower cost desalination of seawater. Membranes
of both types were prepared having the desired resistance of 1 Ω.cm or less.
Preparation involved using reinforcements to strengthen polymer compositions. In
1 N sodium choloride the best experimental cation-exchange membranes that could be
made consistently had transference numbers of 0.76 to 0.79, compared to commercial
membranes with a range of about 0.85 to 0.87. The best experimental anion-exchange
membranes had transference numbers of 0.90, the same as commercially available
membranes [1].

Cationic membranes of acrylic acid-ethylene graft copolymer were prepared by
γ-irradiation [2].

Copolymer anion-exchange membranes were prepared from aminated films of polyethylene and chloromethylated diphenyl ether. The nature of the amine affected
the properties of the membranes. Characteristics varied, according to preparation
from 1.54 to 4.9 meq./g ion exchange capacity, 30 to 70 Ω.cm electrical resistivity
and 1.4×10^{-14} to 5.8×10^{-14} water permeability [3]. Anion exchangers, based
on reaction products of epichlorohydrin polymer with di- and polyamines from, on
mixing with high-density polyethylene solutions as binders, gave anion-exchange
membranes which have an ion exchange capacity 3.2 to 5.1 meq/g and specific
resistivity 80 to 200 Ω.cm [4]. Ion exchange membranes with interpenetrating
networks, having optimum mechanical strengths and increased adsorption capacity
for Cu^{2+} ions, were prepared from a 40-100 polyethylenimine epichlorohydrin rubber
mixture. The membranes with a high degree of crosslinking had specific ion-
selective transfer properties in diffusion dialysis [5]. UV irradiation and subsequent graft copolymerization of polyethylene film with vinyl monomers give
homogeneous ion exchange membranes with low electrical resistivity and increased
ion exchange capacity [6].

Strongly alkaline anion exchanger membranes were prepared by grafting hexa-
fluoro-propene-vinylidene fluoride copolymer with β-(diethylamino)ethyl methacrylate and alkylating the product. The membranes had ion exchange capacity
1.2 meq/l, electrical resistivity 50 Ω.cm and tensile strength 90 kg/cm^2 [7].
Transmission electron microscopy has been used for the recognition of the structural heterogeneity and the distribution of ionogenic groups in the cation exchange
membranes. The investigations were performed on the fluorinated membranes (NA-
FION, MRF) and on the earlier known membranes containing poly(styrenesulfonic
acid) [8]. Electrodialytic transport properties of cation exchange membranes are
changed by adhesion of polyethylenimine on the membrane. The relative Ca-Na ion
transport numbers decreased rapidly when the conformation of polyethylenimine
was compacted by adding salt to the solution adjusting the pH to a high value
where the imine was scarcely protonated, and adding water soluble neutral polymers and multivalent anions [9]. Interpolymer ion exchange membranes were prepared from a compatible casting solution that contained poly(styrene sodium sulfonate), poly(vinyl methyl ether-alt-maleic anhydride) and poly(vinyl alcohol).
Crosslinking of the films was accomplished through the formation of ester linkages that were stable in aqueous environments. Membrane properties (water content, capacity, concentration potential, equivalent conductivity) were measured
over a wide range of membrane compositions. Rejection and hydraulic permeability
data were reported as a function of membrane composition [10].

Frictional coefficients between membrane matrix and permeants were evaluated
for the ThO_2 membrane-DMF permeant system. The degrees of coupling and the

maximum value of the efficiencies of electrokinetic energy conversion for both modes of conversion, electroosmotic flow and streaming potential, were calculated from these coefficients. The values of frictional coefficients were also used to characterize the membrane, i.e. to determine the equivalent pore radii and the number of pores [11].

The film thickness for the cation exchange membrane electrolyte solution system was measured by electrodialysis of NaCl and $CaCl_2$ solutions. All the changes in the geometry of the cell, in the kind and thickness of the membrane, and in the electrolyte concentration had little effect on the thickness of the film that adhered to the membrane. The thickness decreased with an increase in the stirring rate [12]. An anion exchange membrane having a backbone of polyethylene and a sidechain of sulfonamide amine and the water sorbed on this membrane have been studied by infrared spectroscopy and thermogravimetric measurements. The various groups of this membrane, as well as the changes occurring during chemical treatment, have been identified by these techniques [13].

A membrane model was considered, in which the membrane surface has a distribution of discrete electrical dipoles with axes parallel to each other and perpendicular to the surface. The ion penetration process through porous membranes with surface electrical field and the interaction of solutes and membrane material with polar solvents is described. A correlation exists between ion hydration energies and ion fluxes through the membrane [14].

Other work on ion exchange membranes includes production technology and standardized characteristics of ion-exchange membranes [15], membranes for electrodialysis [16], swelling of MK-40 cation exchange membranes and their behavior during electrolysis of concentrated solutions [17], effect of the nature of the ion exchanger on the physicochemical properties of bipolar ion exchange membranes [18], effect of the structure of MPFS ion exchange membranes on ion mobility [19], study of the physicochemical properties of new types of ion exchange materials in relation to their use in electrodialysis [20], determination of humidity in the vapor phase by use of ion exchange membrane [21], Nernstian response of Neosepta membranes for common cations in electrolyte solutions [22], ionic diffusion and ion clustering in a perfluorosulfonate ion exchange membrane [23], bromine diffusion through Nafion perfluorinated ion exchange membranes [24], the interdiffusion of counterions in a cation exchange membrane [25], permselectivities of amphoteric ion-exchange composite membranes [186], osmotic pressure of polystyrene sulfonic acid solutions [187], and determination of the osmotic stability of ion-exchange materials [188].

Patents. A large number of patents, especially Japanese, refers to the preparation and improvement of properties of ion exchange membranes. Only a short description of these patents is given, which include fluoropolymer cation-exchange membranes [26,27,28], fluororesin cation exchange membrane with carrier [29], perfluoropolymer film membranes [30], fluorine-containing cation-exchange resin membrane [31], styrene-tetrafluoroethylene copolymer cation exchange membrane [32], membranes of fluoropolymers containing SO_2Cl pendant groups [33], modified fluoropolymer cation exchange membranes [34], cation exchange membrane from fluoropolymers [35], sulfonic acid containing perfluorocarbon copolymer membrane [36], perfluorocarbonsulfonic acid polymer membrane [37], fluorocarbon membrane with excellent selective permeability [38], fluor-containing cation exchange membrane [39], fluoropolymer cation exchange membranes [40], fluorocarbon cation exchange membrane [41], sulfonated fluorocopolymer cation exchange memrane [42], fluorocarbon cation exchange membranes having phosphorus exchange groups [43], fluoropolymer ion exchange membrane [44], fluoropolymer-based cation exchange membranes [45], sulfonated trifluorostyrene polymer ion exchange membrane [46], fluoropolymer cation exchange membranes [47], fluoropolymer double-layer cation exchange membrane [48], fluoropolymer ion exchange membrane [49], fluoropolymer complex cation exchange membrane [50], fluoro-copolymers containing ion-exchange groups and their crosslinked products [51], fluorine containing cation exchange membrane [52], fluoropolymer containing cation-exchange membrane [53], fluorocarbon cation-exchange membrane [54], fluoro-copolymer cation exchange membrane [55], cation exchange membrane by polymerizing lactone ring containing unsaturated monomer in the presence of fluorinated polymer [56], fluorocarbon cation exchange membrane [57], and binding a cation exchange membrane based on fluorinated polymers [58].

Other ion-exchange membranes include sufonated polyethylene [59], sulfonated crosslinked cation exchange membranes [60], use of a fluoropolymer membrane of the sulfonic acid type for electrolysis [61], ethylene-vinyl acetate copolymer membranes [62], treatment of polyethylene film [63], homogeneous cation exchange membranes of SO_3H-type [64], or of a polymer containing SO_3H [65], improvement of fluoropolymer resin cation exchange membrane [66, 67] , aromatic polymer membranes [68], membranes prepared from allylsulfonic acid-acrylic ester copolymers [69], fluoropolymer cation exchange membranes containing CO_2H-groups [70], radiochemical grafting of styrene derivatives on polyethylene films [71], graft hydrocarbon copolymer ion-exchange membranes [72], anion exchange membranes with quaternization of the polymers with amines [73], ion-exchange membranes by aminating a membrane containing aromatic halomethyl groups [74], ion-exchange membranes from sulfonated divinylbenzene styrene copolymer [75], fibrous ion-exchange membrane comprising a crosslinked vinyl polymer [76], cation-exchange film consisting mainly of a polybutadiene chain [77], cation exchange carbon film [78], ion-exchange membranes based upon polyphenylene sulfide [79], membrane consisting of polyquaternary amine ion exchange polymer network interpenetrating the chains of thermoplastic matrix polymer [80]. Ion exchange membranes useful as an electrolysis diaphragm are swelled with a hydrophilic solvent or its mixture with water and fixed in electrolyzers [81], comprise a cation-exchange membrane prepared by polymerizing a carboxy diene in the presence of a SO_3H-containing polyfluorocarbon polymer [82], ion-exchange membrane comprising a perfluoroalkyl [83] or tetrafluoroethylene copolymer [84], or a aminostyrene-tetrafluoroethylene-vinylsulfonyl chloride copolymer [85], improved current efficiency of cation-exchange membranes [86], improved fluoropolymer resin cation exchange membrane [87], cation exchange membrane and its use for aqueous alkali chloride electrolysis [88], membrane having layers of fluoropolymers with hydroxy fluoroalkyl side groups [89], double-layered fluorocarbon cation exchange membranes [90] and membranes with juxtapositioned parallel ionic bands [91].

The durability of fluoropolymer cation exchangers was improved by coating them with a compact surface layer of fluorooligomers [92] and with an appropriate treatment [93], inhomogeneous cation exchange membrane [94]. To avoid expansion the ion exchange membrane was moistened with an aqueous alkali metal acetate solution [95]. Improvement of fluorohydrocarbon cation exchange membrane is claimed by an appropriate treatment [96]. Preparation and properties of other fluoropolymer cation exchange membranes are also claimed [97 to 100]. Regeneration of anion-exchange membranes was made by treating with an oxidizing agent [101].

Other patents on ion exchange membranes include a high-efficiency dipolar membrane made from a single film [102], ion-exchange membranes with high physicomechanical indexes [103], ion exchange membranes selectively permeable to ions with low charge [104], reinforcement of fluoropolymer cation-exchange membranes [105] ,laminated cation- and anion-exchange membrane [106] and treatment of cation exchange membrane for NaCl electrolysis [107].

An ion- and water-impermeable polymer membrane was made ion permeable, suitable for electrodialysis, by a layer of a porous material on one or both surfaces of the membrane [108].

ION SELECTIVITY

The exchange rates in the conversion of four different chloride-loaded anion exchangers into the sulfate form appear to be controlled by a pseudo first-order homogeneous chemical reaction. The mathematical model of such a reaction was analyzed and experimental evidence supported the proposal model [110]. Ions of different valencies can be separated thanks to the fixed charges in the membrane. For membranes with low hydrodynamic permeability the separation process can be

described by the Nernst-Planck equations, for which exact sòlutions exist in
parametric form. The analysis shows that at relatively high fields the coions
are more effectively separated than the counter-ions, although the separation of
the latter is more economical [111].

The exchange capacities of the halogens increased with temperature for F and
were independent of temperature for Cl, Br and J. Retention energies increase in
the order F < J < Cl < Br [112]. In the absorption of NaCl and CsCl electrolytes
by anion exchange membranes, the presence of cation exchange functional groups
capture a marked concentration of cations from the outer solution [113]. Published
results and new data for selectivity of various membranes for Na^+-Cs^+ and the Cu^{2+}
during dialysis and electrodialysis were analyzed. Selectivities were determined
by radio-tracer and ESR methods. Theoretical dependences of selectivity on various
system parameters were established. Possibilities and difficulties in quantitative
evaluation of the influence of various factors on selectivity were described [189].

Patents. A N-containing compound was plasma-polymerized on the surface of a
cation-exchange membrane to form a N-containing polymer film and render the membrane monovalent cation-selective [114].

ION AND WATER TRANSPORT

To elucidate the mechanism of cation permselectivity, electrodialysis experiments and interdiffusion studies were carried out. The carrier concentration
profile thus obtained agreed with theoretical predictions. The ion concentration
levels in the membrane indicate the presence of additional anionic species. Further experiments suggest that they originate from the aqueous system and offer
an explanation for the cation permselectivity observed, as well as for the electric
characteristics of neutral carrier membranes at extreme voltages [115]. An expression was derived describing the transport of ions and solution from the desalting
chambers to the concentrating chambers in a seawater electrodialyzer [116].

An electrochemical study with a cation exchange membrane-aqueous $CaCl_2$ system
is presented as an experimental verification of a theory for ion transport through
membranes based on nonequilibrium thermodynamics [117]. In studying the change of
transport properties by various polyelectrolytes, the properties studied were the
relative transport number of calcium ion to sodium ion, the current efficiency of
cations and the electric resistance of the membrane. If the polyelectrolyte could
be made to adhere to the surface of the membrane to form a compact coiled structure,
any cationic polyelectrolyte was effective in producing a remarkable decrease in
the Ca/Na relative transport number of any cation-exchange membrane. [118]. Changes
of membrane properties, when immersed in a cationic polyelectrolyte solution, are:
permselectivity between cations with different electric charges, current efficiency
and electric resistance of the membrane. An electrodeposition method was adopted
and a change in the permselectivity of the resultant cation exchange membrane was
observed. The permselectivity value dropped to about 0.3 from about 2.5 of the
untreated membrane during electrodialysis of the sodium chloride-calcium chloride
system and an increase in the electric resistance of the membrane, i,e., organic
fouling, due to a cationic surface-active agent could be prevented [119]. An electrochemical study with a cation exchange membrane-aqueous $CaCl_2$ system, was undertaken for the experimental verification of a theory for ion transport through membranes, in which the cross terms for the phenomenological coefficients are explicit-
ly accounted for on the basis of nonequilibrium thermodynamics [120].

An electrodialysis method for the determination of transport numbers of cations
and anions through a bipolar ion-exchange membrane was presented. The decrease in
current efficiency of H^+ and OH^- with decreasing current density is related to an
increase in diffusion transport of the electrolyte through the bipolar membrane [121].

The electric transport phenomena in mixed electrolyte solutions were analysed
from the phenomenological point of view. The expressions for the equivalent conductivity and the stoichiometric transport numbers were derived using the equivalent
fractions, the total equivalent concentration and the conventional ionic conducti-

vities of the ion constituents. General expressions for the ionic conductivities and the excess conductivity of ternary electrolyte solutions are given [122]. Experimental data for several systems were presented. Within the experimental error and with the few experimental data available for four alkali-metal chloride mixtures, the limiting law of the transport number is confirmed. Limiting ionic conductivities of the trace ion constituents were calculated and compared with their ionic conductivities in the binary solutions and with the values predicted by Van Rysselbergh's approximation. The latter are consistent with the experimental results [123].

The effect of H^+ ions on the membrane potential and electrical resistance of synthetic membranes has been investigated theoretically and experimentally for salt solutions of typical pH values. Employing the equations of Schlögl, membrane potentials and resistances have been calculated for binary and ternary electrolyte solutions. The ratios of the mobilities of the cations and the anions have been estimated at relatively large external salt concentrations from membrane potential curves. From measurements of the electrical resistance of homogeneous cellulose acetate membranes at different salt concentrations, individual ion diffusion coefficients have been estimated. Further, the concentration dependence of the ion diffusion coefficients in the membrane has been demonstrated. The counterion diffusion coefficients exhibit a strong change of their diffusion coefficients, whereas the coion diffusion coefficients and the H^+-ion diffusion coefficient vary slightly with the external electrolyte concentration [124].

Experiments on mass transport across non-isothermal porous partitions were reported. Solvent and solute fluxes were separately assessed and simple methods for their evaluation from experimental data were given. An interesting conclusion is reached that often the observed fluxes move against chemical potential and hydrostatic pressure gradients [125].

Other work includes mass transfer in ion exchange membranes [126], mechanism of ion transfer through membranes and effect of flux conjugation [127], rapid mechanism of ion transfer in ion exchange membranes [128], mechanism of zinc ion transport in cation-exchange membranes [129], precision characteristics of membrane transport phenomena [130], diffusion of fluoride ions through a homogeneous anion exchange membrane [131], anisotropic behavior of ion-exchange membranes [132], mobility of cations in ion-exchange membrane [133], and study of water transfer through cation-exchange membranes during electrolysis [134].

ELECTRICAL PROPERTIES

Membrane resistivity was determined for two well-characterized cation-exchange membranes in ambient solutions of $MgCl_2$, $CaCl_2$, $SrCl_2$, and $BaCl_2$. An attempt was made to compare the membrane resistivity to the membrane selectivity of these ions. For an ACI membrane, resistivity decreases in the order Mg > Ca > Sr > Ba and increases with hydrated radius, indicating that hydrated ions interact with the membrane phase. On the contrary, the AMF membrane resistivity follows the order of bare ion radius and increases in the order Mg < Ca < Sr < Ba. This indicates that dehydration in the membrane phase must take place [135]. On the basis of measured electrical conductivity and diffusion coefficients the conditions were examined of load and mass transfer by cation-exchange and anion-exchange membranes on contacting with aqueous acetone solutions. The electrical conductivity of membranes decreased with increased molar participation of acetone in solution [136].

The dependence of the streaming potential during streaming through ion-exchange membranes is nonlinear, depending on pressure differences. It changes sign when the pressure exceeds 50 cm of liquid column [137]. An attempt was made to explain the results in terms of changes in the structure of the electric double layer at the membrane-permeant interface under the action of steaming pressure [138].

The chronopotentiometric method was used to study the current-voltage (I-E) characteristics of a bipolar membrane under conditions of acid-base generation. An equation was presented describing the I-E characteristics by taking into account

heating of the desalination layer. From the I-E characteristics were calculated
the activation energy, electrical conductivity and heat transfer coefficient of
the membrane. The magnitude of the membrane potential of a bipolar membrane
under current does not depend on the temperature of the desalination layer [139].

The dielectric behavior of heterogeneous membranes is characterized by three
dispersion regions, the frequency of which depends on the water content of the
sample [140].

Other work includes a study of the electric conductivity and physicochemical
properties of ion exchange membranes of different structure [141], electrical conductivity and current-voltage characteristics of membranes [142] and bi-ionic potentials in anion exchange membranes [143].

Patents. To improve efficiency carbon particles are added and electricity conductive metal nets are lined with the carbon particles containing ion exchange
resins to obtain electricity conductive ion exchange resin nets useful in seawater
desalination apparatus in order to improve efficiency [144].

CONCENTRATION POLARIZATION

The limit current densities for cation exchange membranes are generally lower
than those of anion exchangers and are independent of flow rate at low salt concentrations. The limit current densities increase greatly with increasing salt concentration. The quotient of limit current density and specific conductivity is
nearly constant over a broad concentration range [145]. Two regimes were found for
the variation of the limiting current with the external salt concentration. At low
NaCl concentration, the limiting current did not follow the standard pattern. The
relation between the concentration polarization and electro-osmotic flow as a
function of current density is discussed for different convection-diffusion types
at the membrane solution interfaces [146].

An extension of the Nernst model of the current against voltage relation for
ion-selective membranes is proposed. It assumes that the salt concentration at the
surface of the membrane on the side of the depleted layer reaches a lower limit
when the current approaches the Nernst limit of high current density. Beyond that
limit the depleted layer is assumed to separate into two sub-layers, one containing
that lower limit concentration and the other a concentration gradient towards the
bulk concentration. The physical basis for this modification of the Nernst model
is indicated to be the existence of a laminar convection at high current densities,
which injects salt into the depleted layer. Beyond the Nernst limiting current
density the convection turns turbulent, eventually, as deduced from the analysis
of the current noise. Turbulence enhances the injection of salt into the depleted
layer, thus increasing the conductance with increasing current density [147]. The
voltage-current curves of cation exchange membranes have a characteristic shape
with a plateau region of slow current variation followed by a region of accelerated
current growth as the voltage is increased. Currents greater than the plateau
value are conducted across the membrane mainly by cations with little contribution
from water splitting. Suppression of convection is not responsible for the accel-
eration of current growth above the plateau region. The passage of current greater
than the plateau current was explained on the basis of a simple theoretical model
in terms of the formation of a region of volume charge near the membrane reflected
in the model by replacing the electroneutrality condition by the full Poisson
equation [148].

The calculation of the limiting current may be used for the prediction of the
limiting current on ionic membranes in electrodialysis with continuous doses of
the solutions under laminar hydrodynamic conditions. The use of generalized variables made it possible to extend the required dependences on apparatus of arbitrary
construction [149].

Other work includes the effect of the diffusion layer on the ionic current from
a solution into an ion exchange membrane [150], polarization characteristics of
bipolar membrane in hydrochloric acid and sodium hydroxide solutions [151], quan-

titative treatment of oscillatory phenomena in coarse-grained ion exchange membranes [152], comparison of theoretical and experimental determinations of the limiting current of electrodialyzers [153] and evaluation of industrial electrodialysis plant performance by overpotential measurements [154].

ELECTROOSMOSIS

A method was developed for the electroosmotic concentration of ionic forms of elements from extremely diluted aqueous solutions. During the process of electroosmotic transport of water the premeation of the ion, which has the same charge as the membrane, is determined by the radius of the membrane pores and by the concentration of the ion in the solution. Membrane thickness and the rate of the electroosmotic water transport do not affect practically the ion permeation [155].

The electrochemical transfer of hydrated protons across cation exchange membranes produces a net displacement of water. Thus, using hydrogen as a carrier water can be removed from aqueous solutions. Hydrogen is electrochemically oxidized to protons needed to transfer water, recombined under pressure at the cathode and recycled, while liquid water is recovered. The process is applicable to desalination and concentration of aqueous solutions [156]. The direction and magnitude of electroosmotic flow in asymmetric cellulose acetate membranes in contact with sodium chloride solutions depends on the solution concentrations contacting the membrane. If the solution concentrations are below 0.2 M NaCl, the membranes are cation selective and above 0.1 M NaCl, they are anion selective [109].

Other work refers to the effect of electroosmosis on the regeneration of pulp-production processing liquors by electrodialysis [157], point of inversion in electroosmosis and phase transformation of Onsager coefficients [158], study of electroosmosis through electrochemically inactive diaphragms with variable cross sections [159], effect of highly dispersed solid phase concentration on the manifestation of electroosmosis and electrophoresis [160], electroosmotic permeability of composite clay membranes [161], electroosmotic transmission of glyoxal through ion-exchange membranes [162], electroosmotic transport of alcohol-water mixtures through ion-exchange membranes [163] and electroosmosis of water through liquid membranes [164].

Patents. An apparatus used in electroosmotic and electrophoretic processes for the desalination of seawater etc was described [165]. The electrolytic stability of graphite electrodes is maintained by eleminating electrolysis of water and the reaction of the graphite anode with oxygen [166]. Composite electrodes contain two conductive nonmetallic foils between which in parallel connected conductive wires are embedded [167]. Electrodes are arranged in parallel or in series to generate a definite number of mutually superimposed individual fields forming a general electrical field [168].

An electrolytic cell apparatus for adjusting the alkali ion concentration of potable water was described [169], which can be used for electrolysis and electroosmosis. The product potable water, containing alkali ions, is therapeutic [170]. A control valve is provided in either of the two water outlets [171]. Means for heating the cell with an a.c. current and a heat sensor are provided [172]. The same process is also described in another patent [173].

SCALING AND FOULING

The efficiency of electrodialysis for the purification of effluents is impaired by the adsorption of surfactants by the membrane. In the absence of an applied electrical current, the sorption of the surfactant by the membrane increased with the time of contact and the surfactant concentration. Inorganic electrolytes decreased the sorption. Langmuir-type sorption isotherms were obtained only for the sorption by cationic resin membranes. Discontinuities in the isotherms for anionic resin membranes indicated a change in the sorption mechanism during sorption. The application of an electrical current, as in electrodialysis, accelerated the sorption process and, in the absence of an inorganic electrolyte, led to a high level of adsorption, even at low current densities. The presence of electrolytes decreased the rate of sorption, although the final level of sorption was not affected [174]. The kinetics of the formation of deposits on ion exchange membranes were investigated [175].

Patents. The concentration chambers in electrodialysis were classified into two types for the respective concentrated solutions, which are mixed outside the electrodialysis to deposit difficultly soluble salts. The mother liquor is recycled free from those salts. The apparatus was used for desalination of machinery wastewater and underground brine [176]. Ion exchange membranes are washed with aqueous solutions containing anionic surfactant, N_2H_4 or its hydrate or salt, and H_2O_2 or H_2O_2 generating agent. The membranes are regenerated without deterioration [177]. The area between the electricity conducting section and the ducts in a membrane is impregnated with polysulfide rubber to prevent the formation of alkali scale [178]. To decrease deposition of scale the electrolytic cell contains a rotating nonconductive core. The cathode exposed to the deposition of the scale consists of an elastically deformable metal mesh [179]. An acid or a low pH solution is dialyzed in an electrodialyzer to remove scale and fouling from the membranes without dismantiling the apparatus [180]. An exhausted anion exchange membrane, contaminated with a surfactant or waste water, is regenerated by applying reverse polyrity [181]. A free chlorine containing alkali metal chloride solution is circulated through the electrodialyzer to clean the membranes [182]. $CaSO_4$ or gypsum scale deposits are removed from surfaces, in electrodialysis stacks or reverse osmosis membrane modules, by an aqueous solution containing a polyamine carboxylic acid derivative., e.g., EDTA tetra-Na salt [183].

INORGANIC ION-EXCHANGE MEMBRANES

Diffusion rates of various electrolytes through polystyrene supported aluminium ferrocyanide membranes increased with temperature. The activation energy values for different electrolytes through the membrane were higher than those for free diffusion. The free energy and entropy values increased regularly with the size of the ions [184]. Thermodynamically effective fixed charge density, which is an important parameter governing the membrane phenomena, has been evaluated by recently developed theories, based on the principles of the irreversible thermodynamics, to predict the bi-ionic potentials developed across the membranes. Theoretical predictions were borne out quite satisfactorily by experimental results [185].

LITERATURE TO 3C.

1. Southern Research Institute (NTIS Rept. PB-288391/6GA [1978] 53p)
2. S. Tabaddor, F. Fazilate, R. Gouloubandi (J. Radiat. Curing 5, No 4 [1978] 17/22.- C.A. 90 [1979] 187754)
3. M.P. Blinova, E.E. Ergozhin, E. Zh. Menligaziev (Izv. Akad. Nauk Kaz. SSR, Ser. Khim. 28, No.2 [1978] 82/84.- 29, No 1 [1979] 19/23.- C.A. 89 [1978] 90516, 90 [1979] 152894)
4. E.E. Ergozhin, T. Chukenova, E.Zh. Menligaziev (Izv, Akad. Nauk Kaz. SSR. Ser. Khim. 29, No 1 [1979] 69/71.- C.A. 90 [1979] 169379)
5. T. Saegusa, A. Yamada, S. Kobayashi, S. Yamashita (J. Appl. Pol. Sci. 23 [1979] 2343/2353)
6. N.D. Rozenblyum, L.L. Kocherginskaya, V.P. Baranov, L.G. Zhitkova, I.V. Smirnova, E.V. Burmakina, K.V. Bazenkova (Plast. Massy No. 4 [1979] 13/15.- C.A. 90 [1979] 204939)

7. A.S. Tevlina, T.P. Akulova, A.M. Ivankin (Plast. Massy No 9 [1978] 31/32.-
C.A. 89 [1978] 216231)

8. J. Ceynowa (Polymer 19, No 1 [1978] 73/76)

9. T. Sata, R. Izuo (Colloid Polym. Sei. 256 [1978] 757/769.- C.A. 89 [1978]
164582)

10. C.C. Gryte, J. Chen, V. Kevorkian, H.P. Gregor (J. Appl. Pol. Sci. 23 [1979]
2611/2625)

11. M.L. Srivastava, S.N. Lal (Colloid Polym. Sci. 257 [1979] 427/430)

12. K. Inenaga, H. Kimizuka (Mem. Fac. Sci. Kyushu Univ., Ser C, 11 [1979] 261/264.-
C.A. 90 [1979] 157483)

13. C. Heitner-Wirguin, D. Hall (J. Mem. Sci. 5 [1979] 1/14)

14. V.D. Volgin, F. Ya. Shmidel, Yu. E. Sinyak, S.V. Chizhor (Teor. Osn. Khim.
Tekhnol. 13, No 1 [1979] 114/116.- C.A. 90 [1979] 210591)

15. G.Z. Nefedova, Z.V. Klimova, A.B. Pashkov, K.P. Brande, L.I. Skakal'skaya,
M.M. Zhatkina, N.A. Titova, G.D. Bazikova, Yu. G. Freidlin, M.A. Zhukov
(Nauch. Tr. Kuban. Un-t 232 No 2 [1977] 3/15)

16. H. Ihara (Kogyo Yosui 239 [1978] 18/23)

17. N.I. Men'shakova, I.I. Krishtalik, V.L. Kubasov (Gov. Electrochem. 14 [1978]
293/297)

18. V.P. Greben, N. Ya. Pivovarov, N.Ya. Kovarskii, G.Z. Nefedova (Zh. Fiz. Khim.
52 [1978] 2641/2645.- C.A. 90 [1979] 92978)

19. N.I. Nikolaev, A.S. Tevlina, V.V. Korshak, G.G. Chuvileva, V.P. Demin, I.A.
Semenova (Zh. Fiz. Khim. 52 [1979] 2700.- C.A. 90 [1979] 72719)

20. V.D. Grebenyak, T.D. Gudrit, S.V. Grechko, R.D. Chebotareva, V.I. Pisaruk,
N.M. Gukova, T.V. Kuznetsova (Teoriya i Praktika Sorbtsion Protsessov No 12
[1978] 96/103)

21. I. Kreja, W. Grochowski (Inz. Chem. 8 [1978] 561/566)

22. S.R. Seharay, A.S. Basu (Indian J. Technol. 16, No 3 [1978] 102/104.- C.A. 89
[1978] 221458)

23. H.L. Yeager, B. Kipling (J. Phys. Chem. 83 [1979] 1836/1839.- C.A. 91 [1979] 57899)

24. F.G. Will (J. Electrochem. Soc. 126, No 1 [1979] 36/43)

25. T.C. Huang, P.H. Lian (I.E.C. Fundam. 18 [1979] 221/226.- C.A. 91 [1979] 59426)

26. M. Nagasawa, T. Fujimoto, S. Imai, H. Yokoyama (Jap. 77.88.595, 25 Jul 1977.-
C.A. 89 [1978] 7022)

27. H. Ukihashi, T. Asawa, M. Yamabe, H. Miyake (Jap. 77.116.790, 30 Sep.1977.-
C.A. 89 [1978] 111658)

28. M. Seko, Y. Yamagoe, K. Miyauchi, K. Kimoto, T. Hane, M. Fukumoto (Jap. 77.120.
983, 11 Oct 1977.- C.A. 89 [1978] 44783)

29. T. Gunjima, M. Yamabe, M. Motoki (Jap. 77.144.388, 1 Dec 1977.-C.A. 88 [1978]
137619)

30. K. Yamaguchi, H. Harada (Jap. 77.149.287, 12 Dec 1977.- C.A. 88 [1978] 137385)

31. H. Ukihashi, T. Asawa, M. Yamabe, H. Miyake (Jap. 77.150.792, 14 Dec 1977.-
C.A. 89 [1978] 111711)

32. I. Kiyota, K. Takahashi, S. Asami, A. Shimizu (Jap. 78.04.786, 17 Jan 1978.-
C.A. 89 [1978] 111668)

33. K. Kimoto, M. Seko, Y. Yamakoshi, H. Miyauchi, M. Fukumoto, S. Yokoyama,
S. Tsushima, I. Watanabe (Jap. 78.15.282, 10 Feb 1978.- C.A. 89 [1978] 44789)

34. T. Sata, A. Nakahara, Y. Murata, S. Murakami, J. Ito (Jap 78.22.580, 2 Mar 1978.-
C.A. 89 [1978] 25638)

35. T. Sata, A. Nakahara, Y. Murata, S. Murakami, J. Ito (Jap 78.26.285, 10 Mar
1978.- C.A. 89 [1978] 25654)

36. T. Kiyota, K. Takahashi, S. Asami, A. Shimizu (Jap. 78.29.290, 18 Mar 1978.-
C.A. 89 [1978] 7338)

37. K. Takahashi, T. Kiyota, S. Asami, A. Shimizu (Jap. 78.29.291, 18 Mar 1978.-
C.A. 89 [1978] 44807)

38. N. Murayama, T. Sakagami, M. Fukuda, S. Suzuki (Jap. 78.46.489, 26 Apr 1978.-
C.A. 89 [1978] 90835)

39. N. Murayama, T. Sakagami, M. Fukuda, S. Suzuki (Jap. 78.61.579, 2 Jun 1978.-
C.A. 89 [1978] 147868)

40. T. Seida K. Takahashi, S. Asami, A. Shimizu (Jap. 78.70.090, 22 Jun 1978.-
C.A. 89 [1978] 164409)

41. M. Hamada, M. Seko, Y. Yamagoe, H. Miyauchi, F. Yamamoto (Jap. 78.73.484, 29 Jun 1978.- C.A. 89 [1978] 147602)
42. H. Harada (Jap. 78.79.781, 14 Jul 1978.- C.A. 89 [1978] 164377)
43. T. Hane, Y. Yamagoshi, K. Miyauchi, K. Kimoto (Jap. 78.82.684, 21 Jul 1978.- C.A. 89 [1978] 147829)
44. M. Seko, Y. Yamagoshi, H. Miyauchi, Y. Kimoto (Jap. 78.94.289, 18 Aug 1978.- C.A. 90 [1979] 24414)
45. K. Takahashi, T. Seida, S. Asami, A. Shimizu (Jap. 78.97.988, 26 Aug 1978.- C.A. 90 [1979] 104857)
46. M. Tamaru (Jap. 78.103.989, 9 Sep 1978.- C.A. 89 [1978] 216296)
47. N. Kuramoto, H. Matsui, Y. Kakihara, S. Matsuura (Jap.78.110.976, 28 Sep 1978.- C.A. 90 [1979] 39847)
48. M. Seko, Y. Yamagoshi, H. Miyauchi, M. Fukumoto, K. Kimoto, T. Hane, M. Hamada (Jap. 78.116.287, 11 Oct 1978.- C.A. 90 [1979] 104873)
49. M. Seko, Y. Yamagoshi, H. Miyauchi, Y. Kimoto, T. Hane (Jap. 78.131.292, 15 Nov 1978.- C.A. 90 [1979] 104900)
50. S. Imai, S. Fujii, M. Asada (Jap. 78.131.293, 15 Nov 1978.- C.A. 90 [1979] 105218)
51. H. Kojima, T. Uchino (Jap. 78.134.088, 22 Nov 1978.- C.A. 90 [1979] 104997)
52. T. Sata, A. Nakahara, J. Ito (Jap. 78.137.888, 1 Dec 1978.- C.A. 90 [1979] 122492)
53. T. Sata, A. Nakahara (Jap. 78.141.187, 8 Dec 1978.- C.A. 90 [1979] 138590)
54. M. Hamada, M. Seko, Y. Yamagoshi, H. Miyauchi, Y. Kimoto (Jap. 78.141.188, 8 Dec 1978.- C.A. 90 [1979] 122497)
55. M. Seko, Y. Yamagoshi, H. Miyauchi, Y. Kimoto, M. Ebizawa (Jap. 78.149.188, 26 Dec 1978.- C.A. 90 [1979] 153214)
56. K. Takahashi, T. Seita, S. Asami, A. Shimizu (U.S. 4.136.237, 23 Jan 1979, Jap. appl. 76.109.346.- C.A. 90 [1979] 153197)
57. M. Hamada, M. Seko, Y. Yamagoshi, H. Miyauchi, F. Yamamoto (U.S. 4.126.589, 21 Nov 1978, Jap. appl. 76.136.244.- C.A. 90 [1979] 882259)
58. T. Gunjima, I. Takeshita (W. Ger. 2.822.824, 7 Dec 1978, Jap. appl. 77.59.345.- C.A. 90 [1979] 138767)
59. T. Sano, T. Shimomura (Jap. 77.91.789, 2 Aug 1977.- C.A. 89 [1978] 7023)
60. S. Imai, S. Fujii (Jap. 78.14.684, 9 Feb 1978.- C.A. 89 [1978] 25614)
61. S. Imai, S. Fujii, M. Asada (Jap. 79.05.889, 17 Jan 1979.- C.A. 90 [1979] 169703)
62. I. Yoshimura, T. Nomura, S. Kimura, H. Nakano, E. Ohno, N. Shiradori (Jap. 77.156.779, 27 Dec 1977.- C.A. 89 [1978] 148730)
63. S. Machi, T. Sugo, A. Sugishita, S. Kanai, H. Fujiwara (Jap. 78.08.692, 26 Jan 1978.- C.A. 89 [1978] 25600)
64. T. Kiyota, S. Asami, A. Shimizu (Jap. 77.143.988, 30 Nov 1977.- C.A. 88 [1978] 122144.- Jap. 78.60.388, 30 May 1978.- C.A. 89 [1978] 164391)
65. T. Kiyota, S. Asami, A. Shimizu (Jap. 77.151.683 and 684, 16 Dec 1977.- C.A.88 [1978] 137382, 137383)
66. T. Kiyota, A. Shimizu (Jap. 79.04.290, 79. 04.291, 12 Jan 1979.- C.A. 90 [1979] 205142, 169592)
67. T. Kiyota, K. Takahashi, S. Asami, A. Shimizu (Jap. 79.06.885, 79.06.886, 19 Jan 1979,- C.A. 90 [1979] 169701, 169702)
68. S. Imai, N. Nagata, T. Ikeuchi, S. Fujii (Jap. 77.91.788, 2 Aug 1977.- C.A. 88 [1978] 106409)
69. A. Udagawa, T. Sasuga, Y. Kusama, N. Morishita, M. Takehisa (Jap. 78.37.789, 7 Apr 1978.- C.A. 89 [1978] 130660)
70. S. Asami, T. Seida, A. Shimizu (Jap. 78.106.678, 16 Sep 1978.- C.A. 90 [1979] 24383)
71. S. Machi, H. Taniguchi, A. Sugishita, K. Tawara, H. Fujiwara (Jap. 78.08.690, 26 Jan 1978.- C.A. 89 [1978] 25310)
72. S. Machi, T. Sugo, H. Taniguchi, A. Sugishita, H. Fujiwara (Jap. 78.08.691, 26 Jan 1978,- C.A. 89 [1978] 25311)
73. K. Takada (Jap. 78.22.887, 2 Mar 1978.- C.A. 90 [1979] 138761)
74. Y. Mizutani, K. Kusumoto, Y. Mizumoto, K. Takada (Jap. 78.59.782, 29 May 1978.- C.A. 89 [1978] 90853)

75. K. Kihara, R. Tanaka, M. Yokomizo (Jap. 78.88.678, 4 Aug 1978.- C.A. 90 [1979] 24405)
76. M. Takesute, Y. Miyamatsu (Jap. 78.106.395, 18 Sep 1978.- C.A. 90 [1979] 24180)
77. A. Suzuki, S. Tokuda, Y. Kanaya, T. Shimizu (Jap. 78.140.289, 7 Dec 1978.- C.A. 90 [1979] 122493)
78. A. Suzuki, S. Tokuda, Y. Kanaya, K. Shimizu (Jap. 79.26.287, 27 Feb 1979.- C.A. 91 [1979] 21748)
79. R.B. Hodgdon (U.S. 4.110.265, 29 Aug 1978.- C.A. 90 [1979] 105188)
80. A. Rembaum, C.J. Wallace (U.S. 4.119.581, 10 Oct 1978.- C.A. 90 [1979] 39908)
81. N. Murayama, M. Fukuda, T. Sakagami, S. Susuki (Jap. 77.72.398, 16 Jun 1977.- C.A. 87 [1977] 203743)
82. T. Seida, K. Takahashi, S. Asami, A. Shimizu (Jap. 78.28.589, 16 Mar 1978.- C.A. 89 [1978] 130608)
83. K. Takahashi, T. Kiyota, S. Asami, A. Shimizu (Jap. 78.34.690, 31 Mar 1978.- C.A. 89 [1978] 111198)
84. S. Asami, T. Seida, A. Shimizu (Jap. 78.70.984, 23 Jun 1978.- C.A. 89 [1978] 147809)
85. T. Seida, K. Takahashi, S. Asami, A. Shimizu (Jap. 78.71.692, 26 Jun 1978.- C.A. 89 [1978] 180975)
86. T. Seida, A. Shimizu (Jap. 79.06.869, 79.06.870, 19 Jan 1979.- C.A. 90 [1979] 188197, 188198)
87. K. Takahashi, Y. Kanaya, K. Shimizu (Jap. 79.26.976, 28 Feb 1979.- C.A. 90 [1979] 205444)
88. K. Takahashi, T. Kiyota, A. Shimizu (Jap. 79.32.189, 9 Mar 1979,- C.A. 91 [1979] 21751)
89. M. Seko, Y. Yamagoshi, H. Miyauchi, K. Kimoto, T. Hane (Jap. 78.125.283, 1 Nov 1978.- C.A. 90 [1979] 88271)
90. M. Seko, Y. Yamagoshi, H. Miyauchi, M. Fukumoto, Y. Kimoto, T. Hane, M. Hamada (Jap. 79.06.887, 19 Jan 1979.- C.A. 91 [1979] 6080)
91. J.P. Quentin (French 2.353.322, 30 Dec 1977.- C.A. 89 [1978] 111642)
92. T. Sata, A. Nakahara, Y. Murata, S. Murakami, J. Ito (Jap. 78.26.284, 10 Mar 1978.- C.A. 89 [1978] 25655)
93. R. Tanaka, Y. Yamagoshi, K. Miyauchi, Y. Masuda (Jap. 78.99.087, 30 Aug 1978.- C.A. 89 [1978] 216502)
94. R. Tanaka, S. Yasukawa (Jap. 79.05.888, 17 Jan 1979.- C.A. 90 [1979] 169593)
95. T. Oku, N. Kuramoto (Jap. 78.56.172, 22 May 1978.- C.A. 89 [1978] 137669)
96. T. Sata, A. Nakahara, Y. Murata, J. Ito, M. Shiromizu (Jap. 78.58.493, 26 May 1978.- C.A. 89 [1978] 130707)
97. T. Sata, A. Nakahara, Y. Murata, J. Ito, M. Shiromizu (Jap. 78.108.888, 22 Sep 1978.- C.A. 90 [1979] 55939)
98. T. Sata, Y. Murata, A. Nakahara, J. Ito (Jap. 78.123.390, 27 Oct 1978.- C.A. 90 [1979] 105217)
99. T. Sata, A. Nakahara, J. Ito, M. Shiromizu (Jap. 79.20.981, 16 Feb 1979.- C.A. 90 [1979] 205169)
100. T. Sata, A. Nakahara, J. Ito, M. Shiromizu (Jap. 79.21.478, 17 Feb 1979.- C.A. 91 [1979] 5885)
101. A. Yamaguchi, M. Kamaya, K. Tsuchida (Jap. 78.65.281, 10 Jun 1978.- C.A. 89 [1978] 203967)
102. Allied Chemical Corp. (Neth. 77.08.679, 7 Feb 1979.- C.A. 91 [1979] 21780)
103. N. Ya. Lyubman, F.T. Shostak, G.K. Imangazieva, D.M. Romanova, S.M. Serikbaeva (USSR 633.875, 25 Nov 1978.- C.A. 90 [1979] 55707)
104. K. Motani, T. Oku, T. Sata, A. Nakahara (Jap. 79.24.284, 23 Feb 1979.- C.A. 90 [1979] 205170)
105. H. Ukihashi, T. Asawa, T. Gunjima (Jap. 79.01.283, 8 Jan 1979.- C.A. 90 [1979] 169829)
106. T. Eguchi, S. Mori, T. Shimokawa (Jap. 79.14.389, 2 Feb 1979.- C.A. 91 [1979] 22059)
107. H. Yamamoto, Y. Kakihava (Jap. 79.26.286, 27 Feb 1979.- C.A. 91 [1979] 40318)
108. K. Moeglich (U.S. 4.124.458, 7 Nov 1978.- C.A. 90 [1979] 73927)
109. J.H. MacLeish, K.S. Spiegler (AIchE 71st Ann. Meet., Miami Beach, [1978] Paper 109f)

110. L. Liberti, G. Schmuckler (Desalination 27 [1978] 253/260)
111. R. Schloegl (Symp. Membr. Desalin. Waste Water Treatm., Jerusalem, Jan. 1978)
112. J.D. Lopez Gonzales, C. Valenzuela Calahosro, A. Garcia Rodriguez (An. Quim. 73 [1977] 485/490)
113. V.V. Sitnikova, N.N. Nikolaev, G.G. Chuvileva (Zh. Fiz. Khim. 52 [1978] 1704/1708.- C.A. 89 [1978] 186464)
114. K. Yomo, S. Matsuoka, T. Shigemune (Jap. 78.116.286, 11 Oct 1978.- C.A. 90 [1979] 153191)
115. A.P. Thoma, A. Viviani-Naver, S. Arvanitis, W.E. Morf, D.W. Simon (Anal. Chem. 49 [1977] 1567/1572)
116. Y. Tanaka (Denki Kagaku Oyobi Kogyo Butsuri Kagaku 45 [1977] 630/635.- C.A. 89 [1978] 152423)
117. H. Kimikuza, K. Kaibara, E. Kumamoto, M. Shirozu (J. Membrane Sci 4 [1978] 81/98)
118. T. Sata (J.Polym. Sci., Polym. Chem. Ed. 16 [1978] 1063/1080.- C.A. 89 [1978] 111597)
119. T. Sata, Y. Mizutani (J. Polym. Sci., Polym. Chem. Ed. 17 [1979] 1199/1213.- C.A. 90 [1979] 205092)
120. M. Shirozu, K. Kaibara, N. Yoshida, H. Kimizuka (Mem. Fac. Sci. Kyushu Univ., Ser. C 10, No 3 [1978] 165/178.- C.A. 89 [1978] 31324)
121. V.P. Greben, V.P. Nechuanev (Zh. Prikl. Khim. 51 [1978] 1986/1989.-J. Appl. Chem. USSR 51 [1978] 1879/1882)
122. E.O. Timmermann (Ber. Bunsenges.Phys. Chem. 83 [1979] 257/263)
123. E.O. Timmermann (Ber. Bunsenges.Phys. Chem. 83 [1979] 263/270)
124. H.U. Demisch, W. Pusch (J. Colloid Interf. Sci.69 [1979] 247/270)
125. F. Bellucci, E. Drioli, F.G. Summa, F.S. Gaeta, D.G. Mita, N. Pagliaca (J. Chem. Soc., Faraday Trans. 2, 75 [1979] 247/259)
126. N.P. Gnusin, N.P. Berezina, V.P. Beketova, A.A. Turro (Deposited Doc VINITI 959-77 [1977] 14p.- C.A. 90 [1979] 170742)
127. M.M. Shul'ts, O.K. Stefanova (Tezisy Dokl.- Vses. Konf. Ekstr. 3 [1977] 60/64.- C.A. 89 [1978] 186473)
128. N.I. Nikolaev, G.G. Chuvileva (Nauchn. Tr. Kuban. Unit 232, No 2 [1977] 55/57.- Ref. Zh. Khim. [1978] 13 B 1782)
129. V. Svetlicic, Z. Konrad (J. Colloid Interface Sci. 66 [1978] 207/212.- C.A. 89 [1978] 153260)
130. P. Sullo, A. Zelman (AlChE Ann. Meet. Miami Beach [1978] Paper 91f)
131. M.E. Kornelli, Z.I. Ivasyuk (Zh. Fiz. Khim. 52 [1978] 2947/2948.- C.A. 90 [1979] 44280)
132. R. Kumar (J. Coll.Polym. Sci. 257 [1979] 550/553.- C.A. 91 [1979] 57952)
133. V.V. Sitnikova, N.I. Nikolaev, G.G. Chuvileva, A.S. Tevlina, T.P. Akulova, A.N. Ivankin (Zh. Fiz. Khim. 53 [1979] 1295/1296.- C.A. 91 [1979] 63184)
134. O.P. Romashin, M.M. Fioshin, R.G. Erenburg, E.F. Ryabov, V.L. Kubasov, L.I. Krishtalik (Elektrokhimiya 15 [1979] 653/659.- C.A. 91 [1979] 46352)
135. V. Svetlicic, Z. Konrad (J. Membrane Sci. 5 [1979] 129/134)
136. S.A. Mechkovskii (Zh. Fiz. Khim. 52 [1978] 1084/1085.- Russ. J. Phys. Chem. 52 [1978] 620/621
137. R.P. Rastogi, K. Singh, R. Kumar, S.A. Khan (J. Phys. Chem. 81 [1977] 2114/2118.-C.A. 88 [1978] 12371)
138. P. Kumar (Colloid Polym. Sci. 257 [1979] 206/209.- C.A. 90 [1979] 142547)
139. V.P. Greben, N. Ya. Kovarskii (Zh. Fiz.Khim. 52 [1978] 2304/2307.- C.A. 89 [1978] 206491)
140. T. Z. Sotskova, A.I. Derevyanko (Zh. Fiz. Khim. 53 [1979] 1058.- C.A. 90 [1979] 205053)
141. T.N. Toroptseva, S.M. Kanaeva, I.V. Sporykhina, R.G. Sokolova (Nanch. Tr. Kuban. Un-t 232, No 2 [1977] 16/23.- Ref. Zh., Khim. [1978] 13 B 1783)
142. V.N. Golubev, B. Purina, T.A. Filatova (Latv. PSR Zinat. Akad. Vestis, Kim. Ser. No 5 [1978] 546/548, 549/552.-C.A. 90 [1979] 29459, 29460)
143. C.H. Lee (Inst. Eng. Austr. Electr. Eng. Trans. 21 [1977] 851/853)
144. O. Mihara (Jap. 78.22.067, 6 Jul 1978.- C.A. 89 [1978] 185875)
145. K. Hattenbach, K. Kneifel (GKSS Rept. 77/E/42 [1977] 23 p)

146. G. Khedr. A. Schmitt, R. Varoqui (J. Colloid Interface Sci. 66 [1978] 516/530.- C.A. 89 [1978] 221486)
147. B. Gavish, S. Lifson (J. Chem. Soc., Favaday Trans. 1, 75 [1979] 463/472)
148. I.Rubinstein, L. Shtilman (J. Chem. Soc., Favaday Trans 2, 75 [1979] 231/246)
149. V.A. Shaposhnik, I.V. Drobysheva (Elektrokhimiya 15 [1979] 252/254)
150. V.K. Indushekar, P. Meares (Pap. Conf. Physicochem. Hydrodyn. 2 [1977] 1031/1043.- C.A. 89 [1978] 131623)
151. V.P. Greben, N. Ya. Kovarskii (Zh. Fiz. Khim. 52 [1978] 3160/3165.- C.A. 90 [1979]92993)
152. U.F. Franck (Electrochim. Acta 23 [1978] 1081/1091)
153. V.L. Sigal, V.V. Yagodkin (Elektrokhimiya 14 [1978] 1309.- C.A. 89 [1978] 186477)
154. R. Passino, C. Marra, A. Rozzi, G. Tiravanti (AlChE 71st Ann. Meet. Miami Beach [1978] Paper 91c)
155. L.N. Moskvin, N.N. Kalinin, L.A. Godon (Zh. Anal. Khim. 32 [1979] 1899/1903.- C.A. 88 [1978] 110298)
156. H.J.R. Maget (AIChE 71st Ann. Meet., Miami Beach, [1978] Paper 91b)
157. Yu. N. Nepenin, B.N. Filatov, V.V. Sokolova, E.F. Shtreis (Sb. Tr. NSNll Bumagi No 14 [1977] 86/91.- Ref. Zh., Khim. [1978] 22 B 1799)
158. G. Dickel, R. Kretner (J. Chem. Soc., Favaday Trans. 2, 74 [1978] 2225/2234.- C.A. 90 [1979] 61743)
159. L.A. Shishkanova, K.P. Tikhomolova (Vestn. Leningr. Univ., Fiz. Khim. No 3 [1978]94/97.- C.A. 90 [1979] 61718)
160. F.D. Ovcharenko, Yu. P. Boiko, O.L. Alekseev, L.A. Chubirka (Dokl. Akad. Nauk SSSR 244 [1979] 1415/1417.- C.A. 90 [1979] 157502)
161. R.C. Srivastava, M.G. Abraham (J. Non-Equilib. Thermodyn. 4, No 2 [1979] 107/118.- C.A. 90 [1979] 192984)
162. B.M. Fel'dman, L.L. Blyakhman, E.M. Balavadze (Proizvod. Prerab. Plastmass Sint. Smol No 10 [1978] 38/40.- C.A. 90 [1979] 210597)
163. R. Kumar (J. Membrane Sci. 5 [1979] 51/61)
164. R.C. Srivastava, S. Yadav (J. Colloid Interface Sci. 69 [1979] 280/286.- C.A. 90 [1979] 192927)
165. H.W. Tenge (W. Ger. 2.705.813, 17 Aug 1978.- C.A. 90 [1979] 12135)
166. H.W. Tenge (W. Ger. 2.705.814, 17 Aug 1978.- C.A. 90 [1979] 28854)
167. H.W. Tenge (W. Ger. 2,706.172, 17 Aug 1978.- C.A. 90 [1979] 28852)
168. H.W. Tenge (W. Ger. 2.706.193, 17 Aug 1978.- C.A. 90]1979] 28853.- Ref.165 to 168 additions to W. Ger 2.503.670)
169. T. Okazaki (Jap. 77.96.983, 15 Aug 1977.- C.A. 89 [1978] 30537)
170. T. Okazaki (Jap. 77.96.984, 77.96.985, 15 Aug 1977.- C.A. 89 [1978] 30538, 30539)
171. T. Okazaki (Jap. 77.138.067, 17 Nov 1977.- C.A. 89 [1978] 48771)
172. T. Okazaki (Jap. 78.30.987, 23 Mar 1978.- C.A. 89 [1978] 203978)
173. T. Okazaki (Braz. 76.01.598, 13 Sep 1977.- C.A. 89 [1978] 30551)
174. L.D. Kireeva, V.V. Kotov, R.I. Zolotareva (Izv. Vyssh. Uchebn. Zaved., Khim. Khim. Tekhnol. 21 [1978] 1788/1790.-C.A. 90 [1979] 192009)
175. O.V. Bobreshova, V.M. Logacheva, A. Ya. Shatalov (Nauch. Tr. Kuban. Un-t 232 No 2 [1977] 44/47.- Ref. Zh., Khim. [1978] 13 B 1781)
176. O. Kuroda, S. Yoshikawa (Jap. 77.124.482, 19 Oct 1977.- C.A. 89 [1978] 48769)
177. K. Ueno, N. Ozawa, H. Oki, T. Ishida, K. Nakajima, N. Sudo (Jap. 78.73.483, 29 Jun 1978.- C.A. 90 [1979] 76355)
178. M. Torii (Jap. 78.144.473, 15 Dec 1978.- C.A. 90 [1979] 122743)
179. F. Stummer, J. Mueller (W. Ger. 2.649.649, 3 May 1978.- C.A. 90 [1979] 28858)
180. K. Tsuchiya, Y. Hiyama (Jap. 78.23.878, 4 Mar 1978.-C.A. 89 [1978] 135187)
181. O. Kuroda, S. Takahashi, H. Matsuzaki, S. Koike (Jap. 78.88.671, 78.88.672, 4 Aug 1978.- C.A. 89 [1978] 220517, 220531)
182. T. Uehara, Y. Terada, H. Saeki, K. Doi (Jap. 78.112.289, 30 Sep 1978.- C.A. 90 [1979] 109773)
183. J. Block (U.S. 4.144.185, 13 Mar 1979.- C.A. 91 [1979] 9337)
184. S.P. Aroza, D.K. Rastogi (J. Electrochem. Soc. India 27, No 1 [1978] 60/61.- C.A. 90 [1979] 128077)
185. M. Nasim Beg, F.A. Siddiqi, S.P. Singh, P. Prakash, V. Gupta (Electrochimica Acta 24 [1979] 85/88)

186. T. Eguchi, S. Mori, M. Shimokawa (Maku 3 [1978] 289/294.- C.A. 89 [1978]
     216439)
187. A.P. Platonov, V.S. Soldatov, A.F. Pestrak (Kolloidn. Zh. 40 [1978] 793/796.-
     C.A. 89 [1978] 111055)
188. R.D. Chebotareva, V.D. Grebenyuk, I.I. Shamolina (Zh. Prikl. Khim. 51 [1978]
     2197/2201.- J. Appl. Chem. USSR 51 [1979] 2093/2096)
189. N.I. Nikolaev, V.I. Volkov, G.A. Grigor'eva, Yu. M. Popkov, V.V. Sitnikova,
     G.G. Chuvileva (Elekrokhimiya 15 [1979] 451/461)

3D. Electrodialysis

A bibliography of citations from the U.S. National Technical Information
Service data base with 145 abstracts on electrodialysis desalination, its theory,
membrane preparation and performance, as well as test and pilot plant operations
was published [1]. A similar bibliography with 91 abstracts of citations from
the Engineering Index data base, covering world wide engineering research, was
also presented [2].

Kinetic equations describing desalination processes in a potentiostatic electrodialysis device were derived. Calculated current-time and concentration-time
curves were in good agreement with corresponding experimental data [3]. Optimum
operating parameters, e.g., current, pH, salinity, and dialysis time were determined for an apparatus producing 1 m^3/day of drinking water by electrodialysis
desalination of seawater [4]. The energy consumption required for decreasing the
salinity of 1 m^3 of seawater from 26 to 1 g/l was 29.5 kWh. The capacity of the
device working 8 hours a day for 3 months without a substantial change of parameters was 30 l/h [5].

The performance of an electrodialysis unit depends greatly upon the structure of
spacers which are inserted between ion-exchange membranes. Out of various net-
type spacers as turbulence promoters studied, a honeycomb-type spacer and laminated
thin nets exhibited higher limiting current density, less current shadowing and
less fouling than a single net, though some rise in pressure drop was observed [6].
An electrodialyzer with membrane spacing of 0.3 mm and string-type inserts was
used in desalination of NaCl solution. The power consumption coefficients favored
the use of these electrodialyzers for the desalination of highly-mineralized
solutions [7]. Properties of a new type electrodialyzer with string-type gaskets
characterized by intermembrane distance less than 0.1 mm, high coefficient of
utilization of the membrane area (0.9) and low coefficient of the electrical resistivity increase because of nohconductive gaskets (1.12) indicated its use for
demineralization of water with wide salt concentration range (514 g/l) with low
energy losses [8].

A method of calculation for predicting the number of units required in a series
of electrodialysis stacks used for desalination requires knowledge of the transport numbers within the membranes, the substitution resistance of one membrane
pair and the potential drop at the electrodes. Experimental operating conditions
are such that the same potential difference is applied to each electrodialysis
stack and the current passing through each is less than the critical current.
This may be determined by two different methods, which do not require intensity
potential curves, provided the current is lower than the limiting current.
Experimental and calculated results agree for seven stacks in series [9].

Electrodialysis may be used as a pretreatment before ion exchange in order
to achieve a high degree of desalting [10]. Electrodialysis was selected for
the first stage of deionization in the manufacture of plated-through-hole printed
wiring boards and ion exchange for the second step [11].

A standardized brackish water desalination plant, using the electrodialysis
process, can treat up to 500000 gpd water containing 7000 ppm total dissolved
solids. By the use of modules in parallel, it can produce up to 1.5 Mgd of treated
water [12]. The Japanese electrodialysis process is designed to separate

dissolved dissociated salts from aqueous solutions by virtue of the permselectivity
of the ion exchange membrane. The ion exchange membrane is a thin film made of
styrene copolymer [13]. Results of some basic studies of the electrodialysis process were reported. Basic equations expressing the batch-type process were
developed and factors controlling a newly developed electrodialysis system were
determined. A simulation program for the multi-stage continuous process was also
presented [14].

A solar photoelectric electrodialysis desalination plant was proposed. The
energy and technological characteristics include a photogenerator with a capacity
of 200 W for an output of 9 l/h desalted water. The specific electricity consumption is 14 kWh/m^3 drinking water or 0.9 kWh/kg removed soluble salts [15].

In designing a continuous, single pass, electrodialysis plant to desalt seawater from 35000 ppm to less than 4 ppm, or approximately 10,000 fold reduction in
concentration, the number and mix of hydraulic and electrical stages are key
design variables to maximize the cut achieved by each electrodialysis stack and to
minimize the number of stacks in the plant. The total number of stages in the
plant and the flow rate per stage determine the plant's specific production, while
the pressure drop per stage sets the requirement for interstage pumping. The
number of electrical stages limits the total current density applied and also
determines the specific energy consumption of the plant. The design analysis is
applied, using experimental data, to a plant application requiring minimum investment and simplified plant operation [16].

Other work includes the significance of electrodialysis for the desalination of
brackish and seawater [17], cooling tower effluent reduction by electrodialysis [18],
theorotical estimation of the boundary layer thickness in an electrodialysis desalination plant and its effect on plant performance and design [19], brackish
water desalting by electrodialysis [20], EDEA-1, a mobile electrodialysis desalter
[21], electrodialysis desalting plants in industries [22], the marine desalination
unit by electrodialysis for ships [23], tanks for electrodialysis [24], a review
on electrodialysis of aqueous solutions [25], prospects for using an electrochemical method to prepare water for industrial boilers [26], effect of electric
field on the treatment providing low permeability for bivalent ions to cation exchange membrane [27], change of the efficiency of low permeability treatment with
time for bivalent ions of cation exchange menbrane [28], effect of leakage of solutions on material transport in an electrodialyzer for the concentration of seawater [29], a review of fundamentals on electrodialysis [30] and an historical
review on the first twenty-five years of electrodialysis [31].

Patents. Arrangements of membranes and chambers of an electrodialysis apparatus are described [32]. For the performance of selective electrodialysis, the
membrane is made of a material capable to induce selective migration of a selected
group of charged ions [33].

The limit current density is increased by increasing the pressure of the
liquid [34]. The dialysis is carried out by setting the gap between a pair of
cation and anion exchange membrane surfaces at up to 3 mm and greater than the
thickness of the deposited material [35]. Air bubbles are introduced into the
desalination chambers in a greater amount than into the concentration chambers to
decrease the resistance between the electrodes and to control membrane vibration
[36]. Magnitude of pressures in the desalination chamber and the concentration
chamber are reversed temporarily by varying amounts of gas bubbles or flow rates.
The resistance between the electrodes is decreased and the membrane vibration is
minimized [37].

The design is given of a multichamber filter press type bipolar diaphragm cell,
capable of furnishing a large current at a relatively low voltage [38]. An electrodialyzer consists of 2 to 4 stages, which are connected in series with one pump.
Construction details are given [39]. An insulating film is provided along the
edge of an anion exchange membrane facing the ducts of the apparatus. The service
life is prolonged because the discoloration, embrittlement and deterioration of
the membrane are avoided [40]. In a part of the electrodialysis apparatus an
anion exchange membrane is used as the diaphragm for the anode side of the compartment and a cation exchange membrane as the diaphragm for the cathode side. In a

second part of the apparatus, anion exchange membranes are used as the diaphragm for both the anode as well as cathode side of the compartment through which the diluted solution is being passed. This makes it possible to select the appropriate anion/cation concentration ratio [41]. The diluted portion is electrodialyzed (a.c.) to remove heavy metal ions, polychlorinated biphenyl, organomercury compounds and oils. Residual NaCl is removed by ion exchange [42]. The feed solution is treated in a series of electrodialyzer units by circulating the concentrated solution from a previous unit to the dilute solution side of the next unit for further concentration. Thus a low concentration solution can be concentrated for reclaiming or desalination and a high efficiency is achieved [43].

Sodium ions are removed by a cation exchange membrane having up to 30 μ thick hydrogel film containing up to 30% water [44]. An oxidizing solution (NaClO) is added to the electrolyte. The method is useful for seawater desalination and the apparatus is easily cleaned [45]. Sensors for electric current and fluid flows, control circuits and appropriate valves are provided for automatic control in electrodialyzers [46].

A MnO_2-coated carbon was used as an anode in electrodialysis of water. The anode was stable for over 10 h without any oxidative damage, whereas a carbon electrode showed considerable damage after 2 h [47]. A purification process and apparatus include provision for subjecting the concentrate stream to electrolysis after the electrodialysis step [48]. Acidified mine waters were treated by electrodialysis to prevent poisoning of the ion-exchange membranes by polyvalent ions and to reduce the amount of highly mineralized effluents [49].

ELECTRODIALYSIS AND ION EXCHANGE

The spacers, serving as turbulence devices, are usually made of dielectric materials and exert no active influence on the electrodialyzer. The use of turbulence spacers having ion-exchange characteristics lowers membrane polarization and raises the conductivity of the electrodialyzer compartments. Conductivity was measured in a cell assembled from a cation-exchange and an anion-exchange membrane with a spacer of the usual or ion-exchange type between them. The effect due to the ion-exchange spacers was then found from the difference between the measured conductivities [50]. A new electrodialysis stack configuration, recently proposed by Kedem, assumes the availability of woven spacers made of ion-exchange fibers, both cation- and anion-exchangers. Ion-exchange fibers are characterized by large exchange surface, low flow resistance and ease of handling. A new brand of ion-exchangers from low-density melt-spun polyethylene fibers utilizes the special geometry of the fibers and is tailored to achieve high mechanical strength and optimal radial distribution of ion-exchange sites. The fiber thus obtained has average ion-exchange capacity of 1.6 meq/g dry weight, electrical conductivity of the order of 100 Ω.cm and fairly high mechanical strength [51]. A model was proposed describing electrodialysis in a cell with ion-exchange fibers. The actual heterogeneous medium is replaced, with the aid of averaging, by a homogeneous one with three interacting phases, corresponding to anion- and cation-exchange fibers and to the bulk solution [52].

Some investigations of preliminary softening and reduction of bicarbonate alkalinity of drinking-type water in two-chamber electrodialysis apparatus with cation exchanger charging of the anode chamber were carried out. The relations presented in graph form permit evaluation of the interrelations between technological values and specific energy consumption for the appropriate treatment. They also permit calculation of the main design parameters of a two-chamber electrodialyzer [53].

Other work includes the effect of an external constant electric field on mass transfer in a diffusion layer in the ion exchanger two-component solution system [54] and determination of optimum conditions of combined desalination by electrodialysis and ion exchange [55].

Patents. The space between the membranes of the electrodialyzer is partly filled with conductive elements, which have greater conductivity than the dialysate, namely an anion exchange material is in contact with the anion selective membrane and a cation exchange material is in contact with the cation selective membrane [56].

ELECTRODIALYSIS REVERSAL

In an electrodialysis process, in which the current is reversed cyclically, there is only one set of flow channels, the usual second set being replaced by a set of storage compartments. The design and operating parameters indicated that the process is potentially applicable to the desalting of brackish water [57]. A review was presented on the electrodialysis reversal process, which does not require any chemicals for scale control [58].

Patents. The formation of deposits on the cathodes is prevented by periodically changing the polarity of part of the cathodes, so as to clean such cathodes, with continuing normal operation of the electrolysis cell with remaining cathodes [59]. In a polarity reversal method for controlling scale build up in electrodialysis, the direct current is reversed periodically and, on reversal, anolyte is withdrawn from the chamber that was previously the cathode chamber and fed into the anolyte storage tank, and catholyte is withdrawn from the chamber that was previously the anode chamber [60].

HIGH TEMPERATURE ELECTRODIALYSIS

In an investigation into the suitability of commercial ion-exchange membranes for electrodialysis at high temperature, it was found that the temperature dependence is the same for nearly all of the membranes, the resistance at 75° is about 30% of that at 25°, the permselectivity decreases, as expected, with increasing concentration and has only a little dependence on the temperature, and the limiting current density is a function of the concentrations, the flow velocity and, above all, the temperature [61]. Experiments showed that the specific membrane requirement for desalination is 60% smaller at 75°C than at 25°C and that the energy requirement is very favorable compared to other processes. Capital costs are 35 to 40% lower than those for conventional electrodialysis plants and are practically independent of the salt content of the raw feed water [62].

Phase VI of previous work has been dedicated to field testing the stacks. To interpret the data a computer model of the unit was developed and used to simulate the performance under a wide variety of conditions. These simulations with appropriate economic factors were used to calculate the cost of desalting seawater. Under reasonable conditions, the cost of product water ranged from $1.00 to $1.75/ 1000 gal. The increase over previous estimates is due primarily to increased costs of power and increased costs of components [63].

HYGIENIC EVALUATION

The less the initial microbiological contamination, the higher the current density and the longer the desalination cycle, the less is the contamination in the final effluent. Additional decontamination is necessary and with high initial pollution, preliminary decontamination is also necessary. It is possible for the membranes to become coated with a habitat for bacteria requiring periodic cleaning. Some memrane types actually favor bacterial growth [64].

OPERATING EXPERIENCE

Greece. In a tender for the supply of a 15000 m^3/day plant of potable water to meet the needs of the town of Corfu, eleven bids were received from companies in seven countries. The proposals covered diverse methods of supply. The electro-dialysis reversal process was given the preference [65]. In the initial 18 month period of operation the plant produced potable water (550 ppm) from a blend of brackish sources of salinities up to 2000 ppm. The plant and integration of the plant into the municipal system, as well as the unique system employed to segregate the treatment of the different brackish waters, employing only the highest salinity water for blowdown, are described. Operating cost data for this period are given [66].

Japan. An electrodialysis desalting plant at Noshima with a capacity of 120 m^3/day has been in operation for 4 years. Electricity consumption was 16,21 kWh/m^3 and acid consumption 0.3 kg/m^3 [67]. An electrodialysis plant with a capacity of 200 m3/day was constructed at Hatsushima island [68]. A report was given on the design, construction and operation of the desalination process set-up in Ohshima island, including the damage caused by an earthquake with magnitude 7.0 on the Richter scale which occurred on January 14, 1978 [69]. A discussion was given on electrodialysis desalting plants for boiler water of power plants in the Kashima industrial complex [70].

Libya. Many electrodialysis plants have been installed in the last seven years in Libya, including Benina, Dahra-Benina and Dahra. Other electrodialysis plants are also considered and in each case the effects of the deteriorating input of brackish water, the pretreatment, the design gaps and sometimes faults are indi-cated together with the solutions developed on-site [71]. One of the major problems in a plant in the Libyan desert was the very high calcium sulfate content of the existing 36.000 ppm well. The design and operation of the plant were described [72].

U.S.A. Three seawater pretreatment systems, three reverse osmosis membranes and two electrodialysis pilot plants were tested at Wrightsville Beach Test Facility [73]. The electrodialysis process has been used commercially in the U.S. to desalt brack-ish waters of over 20 years. The first municipal installation in Coalinga, Calif-ornia, was commissioned in 1959. The first plant continuously used to produce the total water supply for a community was installed in Buckeye, Arizona, in 1962. Development of the electrodialysis reversal process in the early 1970's resulted in a system featuring reduced operation and maintenance due to the elimination of the need to continuously feed acids and chemicals. New membrane systems, membrane stack design and applications are under study. High temperature membranes and spacers have been developed. The combination of solar electric power and electro-dialysis is also under study for use in purifying water in remote locations [74].

TREATMENT OF WASTE WATERS

Electrodialysis is widely used in the treatment of various waste waters, as well as in separation techniques. Data were given on the treatment of secondary treated water at Kawasaki municipal waste water treatment plant [75].

Separation of inorganics. Relévent work includes a study of electrodialysis treatment of an atmospheric moisture condensate [76], electrodialysis of trialkyl-amine containing hydriodic acid [77] or containing rhenium [78], graphical calcu-lations of salt crystals deposited by cooling mother liquors in salt making by electrodialysis of seawater [79], electric resistivity of ion-exchange membranes in waste waters from titanium and magnesium production [80], determination of the current-voltage characteristics of ion-exchange membranes in waste waters from titanium and magnesium production [81], electrodialytic treatment of waste water

in the production of ammonium nitrate [82], advances in the commercial use of
electrodialysis for the concentration, control and modification of inorganic
solutions [83], use of neutral-membrane electrodialysis for the concentration of
potassium salt solutions [84], use of ion exchange and electrodialysis for decon-
taminating liquid radioactive wastes [85], removal of radioactive ions from nuclear
waste solutions by electrodialysis [86], desalination of lime-coagulated water of
neutral mineralization by electrodialysis [87], properties of ion-exchange membranes in strongly acidic media containing copper ions [88], electrodialysis of
osmium (IV) sulfate solutions [89] and electrodialytic recovery of waste water in
the nickel electroplating process [116].

Patents. Cyanide is removed from waste water by anodic oxidation in an electrodialyzer [90]. Waste solutions from first and second rinsing stations of a pickled
copper wire are electrodialyzed for recycling of the rinse water [91]. An electrodialyzer useful for treating metal pickling rinse waters is described [92]. Regeneration of sulfuric acid from metal pickling is effected by electrodialysis [93].
Spent solution from waste gas scrubbing is regenerated for recycling by means of
electrodialysis [94,95]. Spent photographic bleach-fix solutions are treated by
electrodialysis to efficiently remove silver [96].
The aqueous NaCl solution, after removing Ca^{2+} and SO_4^{2-} by strongly ion exchange
resins, is subjected to electrodialysis. The cation-exchange resin regeneration
waste solution, containing a high concentration of Ca^{2+}, and the anion-exchange
resin containing a high SO_4^{2-} concentration are mixed to deposit $CaSO_4.2H_2o$ crystals.
The mother liquor is used as the regeneration solution of the ion-exchange resins
[97]. Blowdown waste water from waste gases desulfurization with lime or gypsum
is electrodialyzed to remove dithionate, which is oxidized to SO_4^{2-} [98]. A waste
solution containing acids and metal salts is treated by electrodialysis to remove
most of the acids and then the solution is led into the cathodic chamber of an
electrolysis cell, having a cation exchange membrane to deposit most of the metals.
The overflow is then separated into a concentrated and a diluted solution by
electrodialysis and the two solutions are recycled [99].
Heavy metal ions in wastewater are removed by electrodialysis. Thus a Na_2SO_4-
$Al_2(SO_4)_3$ solution is electrodialyzed and an electrolyte (Na_2SO_4 solution) is
circulated. The cathode chamber was purified after 10h of operation [100]. Industrial wastewater is first desalinated by reverse osmosis, then by electrodialysis.
The desalinated brine is returned for further reverse osmosis treatment [101]. An
electrodialysis method was used to produce sols of such elements as Al, Sb, Cr, Mn
etc [102]. An electrodialysis apparatus is equipped with mechanical means for
removing precipitated particulate material from the electrodes for avoiding fre-
quent disassembly of the cell [103].

Separation of organics. Further applications of electrodialysis include the
purification of intermediate products of sugar production [104], the possibility
for dealkalinization of phenolic alcohols by electrodialysis [105], the application of electrodialysis to food industry [106] and apparatus for the electrochemical purification of waste waters from polymer production [107].

Patents. Dodecylbenzenesulfonate containing waters are electrolyzed to prepare
purified water and a concentrated solution [108]. Molasses are purified by pretreatment with weakly basic anion exchangers and electrodialysis [109]. An electrodialysis method is described to remove lactic acid from whey and citric acid
from orange juice [110]. An apparatus for ultrafiltration and electrodialysis is
described, which is suitable for separating proteins from whey [111]. Inorganic
electrolyte containing organic anions is electrodialyzed in the presence of an
oxidant to prevent organic scale formation on the anion exchange membrane [112].

OTHER ELECTROCHEMICAL PROCESSES

Electrosorption. The solution to be desalted is passed through chambers filled
with activated charcoal which is electrically charged via electrodes. The elec-

trode polarity is reversed in the subsequent desorption step. In a complete operation absorption and desorption steps alternate [113].

Donnan dialysis. Donnan dialysis is a relatively new ion-exchange process using
ion exchange membranes to permit continuous operation. Potential applications
exist in many areas where traditional columnar ion exchange, requiring a regeneration cycle, is used [114]. Separation of the samples from concentrated receiver
electrolytes by ion exchange membranes, which results in Donnan dialysis, was
shown to provide rapid, precise and accurate transfers of test ions from the
samples into the electrolytes. Optimum receiver electrolytes were described for
cations, anions of weak acids and weakly basic anions such as nitrate and chloride. The methods were shown to be superior in several respects, including simplicity, to solvent extraction, columnar ion exchange and chemical digestions in
comparative tests on surface water and sewage samples [115].

Electrochemical pumping. An electrochemical column, consisting of two high
surface porous carbon electrodes and a thin separator, was operated in 4-action
electrochemical parametric pumping cycles at the double layer potential range.
A significant concentration profile was built up along the column indicating that
several theoretical plates were attained. The separative properties of a single
theoretical plate were derived. The method seems promising for industrial water
processing [117].

LITERATURE TO 3D.

1. D.M. Cavagnaro (NTIS Rpt. PS-78/0243 [1978] 150 p)
2. D.M. Cavagnaro (NTIS Rept. PS-78/0244 [1978] 98 p)
3. N.N. Zubets, Z.D. Lavrova, E.S. Vorontsov, A. Ya. Shatalov, G.A. Kuznetsov,
   N.A. Boshok, A.K. Korolev (Nek. Vopr. Sovrem. Elektrokhim. Kinet. 1 [1976]
   120/126.- C.A. 89 [1978] 48704)
4. A.K. Korolev, G.A. Kuznetsov, N.I. Boshok, V.B. Bogdanovich, A.A. Levchenko,
   V.I. Tyrin (Nek. Vopr. Sovrem. Elektrokhim. Kinet. 1 [1976] 127/132.- C.A. 89
   [1978] 48705)
5. N.N. Zubets, G.A. Bedyukh, I.A. Anishchenko, V.V. Cherevkov, G.A. Kuznetsov,
   A.A. Levchenko, A.K. Kozolev (Nek. Vopr. Sovrem. Elektrokhim. Kinet 1 [1976]
   116/120.- C.A. 89 [1978] 48703)
6. T. Ichiki, T. Asawa (Asahi Garasu Kenkyu Hokoku 27, No 2 [1977] 115/121.-
   C.A. 89 [1978] 217173)
7. V.D. Grebenyuk, A.A. Vinnichenko, L.V. Lysenko (Zh. Prikl. Khim. 51 [1978]
   793/797.- C.A. 89 [1978] 152424)
8. V.D. Grebenyuk, L.V. Lysenko, S.I. Vdovenko (Zh. Prikl. Khim. 51 [1978]
   2261/2265.- J. Appl. Chem. USSR 51 [1979] 2152/2157)
9. R. Audinos, E. Casademont, V. Sanchez (Electrochim. Acta 23 [1978] 271/278)
10. A.V. Dvoretskii, I.S. Lavrov, O.V. Smirnov (Zh. Prikl. Khim. 50 [1977]
    2108/2110.- J. Appl. Chem. USSR 50 [1978] 2009/2011)
11. G.A. Lordi, R. Vankirk (NWSIA 6th Ann. Conf.,Sarasota, Fla., [1978] 21 p)
12. E.P. Geishecker (NWSIA 5th Ann. Conf., San Diego, Cal. [1977] Paper No 13)
13. S. Itoi, I. Nakamura (Chem. Econ. Eng. Rev. 10, No 1 [1978] 26/31.- C.A. 89
    [1978] 64750)
14. M. Kishi, T. Fujiwara, S. Serizawa, M. Negoro, H. Miyamoto (Tech. Rev.
    Mitsubishi Heavy Ind. 15, No 1 [1978] 29/36.- C.A. 89 [1978] 168876)
15. I.G. Savchenko, B.V. Tarnizhevskii, T.M. Kotina, B.I. Lemasov (Geliotekhnica
    14, No 3 [1978] 45/50.- Appl. Solar Energy 14 No 3 [1978] 34/38)
16. J.E. Lundstrom (Desalination 32 [1980] 259/277)
17. H. Strathmann (Fortschr. Ber. VDI-Z, R3, No 47 [1977] 185/189)
18. D. R. Jordan, W. F. McIlhenny, G. T. Westbrook (Proc. Amer. Power Conf. 38
    [1976] 980/987)

19. M. A. Chandry (Nucleus, Karachi, 14, No 2 [1977] 17/24.- C. A. 90 [1979] 12071)
20. A. Maurel (Rev. Gen. Electr. 86 [1977] 480/486)
21. K. Kneifel, K. Hattenbach, U. Martens (GKSS Rept. 78/E/4 [1978] 30 p)
22. T. Saito, S. Yoshida (Kogyo Yosui 239 [1978] 54/61)
23. Y. Tani, K. Doi, Y. Terada, M. Yokota, M. Wakayama (Kogyo Yosui 239 [1978] 86/89)
24. M. Urabe, K. Doi (Kogyo Yosui 239 [1978] 24/28)
25. I. F. Miller (Tech. Electrochem. 3 [1978] 437/487)
26. A. N. Bukhantsev, A. P. Voinov, A. A. Avrashkova (Prom. Energ. No 10 [1978] 22/25.- C. A. 90 [1979] 109739)
27. Y. Tanaka, N. Kanai (Nippon Kaisui Gakkaishi 32, No 2 [1978] 95/99.- C. A. 90 [1979] 209859)
28. Y. Tanaka (Nippon Kaisui Gakkaishi 32, No 2 [1978] 100/103.- C. A. 91 [1979] 9294)
29. Y. Tanaka (Denki Kagaku Oyobi Kogyo Butsuri Kagaku 46 [1978] 208/211.- C. A. 89 [1978] 65865)
30. K. Urano (Kogyo Yosui 239 [1978] 4/8)
31. W. E. Katz (NWSIA 5th Ann. Conf., San Diego, Cal. [1977] 3 pp)
32. A. J. Giuffrida (U.S. 4.057.483, 8 Nov 1977)
33. M. Perry, M. Rubinstein, O. Kedem (Isr. 49.675, 31 Aug 1978)
34. H. Saeki, H. Shinozuka, M. Urabe, K. Toi (Jap. 77.78.680, 2 Jul 1977.- C. A. 88 [1978] 123243)
35. H. Matsuzaki, K. Ebara, S. Takahashi (Jap. 77.124.481, 19 Oct 1977.- C. A. 89 [1978] 48770)
36. H. Matsuzaki, O. Kuroda, S. Koike, T. Takahashi (Jap. 77. 151.677, 16 Dec 1977.- C. A. 89 [1978] 168914)
37. H. Matsuzaki, O. Kuroda, S. Koike (Jap. 77.151.678, 16 Dec 1977.- C. A. 89 [1978] 131720)
38. K. Sato, T. Hamano, Y. Sajima (Jap. 78.13.427, 10 May 1978.- C. A. 89 [1978] 137668)
39. A. Yamaguchi, K. Miyaso, M. Kamaya (Jap. 78.16.374, 15 Feb 1978.- C. A. 89 [1978] 135625)
40. Y. Hagari, T. Maki, K. Asada (Jap. 78.22.165, 1 Mar 1978.- C. A. 89 [1978] 131568)
41. H. Kagechika, H. Yamagishi, M. Tanaka (Jap. 78.28.574.- C. A. 89 [1978] 203959)
42. K. Morita (Jap. 78.70.085, 22 Jun 1978.- C. A. 90 [1979] 28849)
43. T. Kuwabara, Y. Egashira (Jap. 78.132.481, 18 Nov 1978.- C. A. 90 [1979] 174467)
44. T. Gunjima, K. Arai (Jap. 78.137.083, 30 Nov 1978.- C. A. 90 [1979] 192354)
45. T. Uehara, K. Doi, K. Oshima (Jap. 78.137.879, 1 Dec 1978.- C. A. 90 [1979] 192343)
46. E. M. Balaradze, K. M. Salmadze, V. G. Stepanov, I. M. Tseitlin, R. G. Milovidov, B. A. Plicin (French 2.381.549, 27 Oct 1978)
47. I. Shimokawabe, S. Takahashi (Jap. 79.56.248, 7 May 1979.- C. A. 91 [1979] 46446)
48. Zan-Al Ltd (Isr. 49.839, 31 Jan 1979)
49. V. I. Pisaruk, I. I. Penkalo, S. I. Mukha, V. D. Grebenyuk (USSR 655.653, 5 Apr 1979.- C. A. 91 [1979] 62213)
50. A. V. Bezprozvannykh, B. M. Vrevskii, Yu. A. Kononov, A. M. Famintsyn, L. A. Vol'f (Zh. Prikl. Khim. 50 [1977] 1972/1975.- J. Appl. Chem. USSR 50 [1977] 1880/1882)
51. R. Messalem, C. Forgacs, I. Michael, O. Kedem (J. Appl. Polymer. Sci., Appl. Polymer Symp. 31 [1977] 383/388)
52. O. Kedem, I. Rubinstein, L. A. Segel (Desalination 27 [1978] 143/156)
53. I. V. Pasechnik, V. A. Ruban (Izv. Vyssh. Uchebn. Zaved. Energ. No 4 [1977] 140/143.- E. I. 16 [1978] 14785)
54. N. P. Gnusin, V. I. Zabolotskii, V. V. Nikonenko (Elektrokhimiya 14 [1978] 660/666.- C. A. 89 [1978] 136288)
55. N. N. Zubets, A. A. Mazo, A. E. Serebryakov (Nauch. Tr. Kuban. Un-t 232, No 2 [1977] 111/113.- Ref. Zh., Khim. [1978] 12 I 393)
56. Yeda R & D Co. (Isr. 48.473, 15 Jun 1978)

57. M.E. Abu-Goukh (Ph.D. Diss. Univ. Of Brit. Columbia.- Diss. Abst. B 38 [1977] 2279 B)
58. K. Tomiie (Kogyo Yosui 239 [1978] 74/79)
59. I. Malkin (U.S. 4.088.550, 9 May 1978.- C.A. 89 [1978] 135636)
60. E.J. Parsi (U.S. 4.115.225, 19 Sep 1978.- C.A. 90 [1979]106460)
61. H. Behret, H. Binder, A. Koehling (Chem. Ing. Tech. 50 [1978] 397)
62. R. Eggersdorfer, H.J. Hampel, A. Koehling, K.H. Scherer (Chem. Ing. Tech. 50 [1978] 395)
63. F.B. Leitz, H.I. Viklund, A.D. Jha (NTIS Rept PB–288002/9GA ,[1979] 93 p)
64. G.I. Sidorenko, Yu. A. Rakhmanin, G.I. Rozhnov, A.I. Mel'nikova, Yu. N. Nikitina (Gig. Sanit. No 11 [1978] 14/20.- C.A. 90 [1979] 127300)
65. G. Andreadis, J.W. Arnold (6th Ann. Conf. NWSIA, Sarasota, Florida, [1978] 13 p)
66. J.W. Arnold (Desalination 30 [1979] 145/153)
67. K. Miyaso (Kogyo Yosui 239 [1978] 80/85.- C.A. 90 [1979] 127306)
68. N. Matsumoto (Kogyo Yosui 239 [1978] 48/53.- C.A. 90 [1979] 127324)
69. S. Koga, Y. Mitsugami (Kogyo Yosui 239 [1978] 41/47.- C.A. 90 [1979] 127304)
70. I. Nakamura (Kogyo Yosui 239 [1978] 62/69.- C.A. 90 [1979] 127307)
71. M. Aswed, H. El Hares (Desalination 27 [1978] 51/57)
72. H.C. Valcour (NWSIA 5th Ann. Conf., San Diego, Cal. [1977], 7 pp)
73. K. Patel, F. Harris (NTIS Rept PB - 287987/2GA [1976] 138 p. - C.A. 91 [1979] 27061)
74. A.L. Goldstein (Desalination 30 [1979] 49/58)
75. T. Fukuji (Kogyo Yosui 239 [1978] 94/97.- C.A. 90 [1979] 91834)
76. V.N. Smagin, I.N. Medvedev, V.I. Kharchuk, V.A. Chukhin (Nanch. Tr. Kuban. Un-t 232, No 2 [1977] 135/148.- Ref Zh., Khim [1978] 12 I 347)
77. J. Licis (Vses. Konf. Ekstr. 3 [1977] 50/53.- C.A. 89 [1978] 186472)
78. J. Licis (Vses. Konf. Ekstr. 3 [1977] 39/43.- C.A. 89 [1978] 150114)
79. S. Oka (Nippon Kaisui Gakkaishi 31 [1977] 164/178.- C.A. 90 [1979] 123934)
80. R.A. Karvatskaya, L.A. Kostromina, T.P. Pisotskaya (Zh. Prikl. Khim. 50 [1977] 1965/1969.- J. Appl. Chem. USSR 50 [1978] 1874/1877)
81. R.A. Karvatskaya, N.A. Akimova, L.A. Kostromina (Zh. Prikl. Khim. 50 [1977] 1962/1965.- J. Appl. Chem. USSR 50 [1978] 1871/1874)
82. N.P. Gnusin, V.I. Zabolotskii, V.F. Pismenskii (Izv. Sev.- Kavk. Nauchn. Tsentra Vyssh. Shk., Ser. Tekh. Nauk 6. No 1[1978] 103/105.- C.A. 89 [1978] 230395)
83. R.E. Horn (AIChE 71st Ann. Meet., Miami Beach [1978] Paper 91a)
84. M. Szetela, I. Trzepierczynska, M. Manczak, A. Szaynok (Pr. Nauk. Inst. Inz. Ochr. Srodowiska Politech. Wroclaw 40 [1978] 79/87.- C.A. 90 [1979] 206624)
85. F.V. Rauzen, N.F. Kuleshov, N.P. Trushkov, S.N. Dudnik (At. Energ. 45, No 1 [1978]49/53.- C.A. 89 [1978] 203598)
86. S. Sugimoto (J. Nucl. Sci. Technol. 15 [1978] 753/759.- C.A. 90 [1979] 141847)
87. S.P. Vysotskii, O.M. Kopylova (Teploenergetica No 2 [1979] 48/52.- C.A. 91 [1979] 27057)
88. G.A. Grishaeva, N.Ya. Shulepova, G.I. Novikov, B.A. Butylin, T.I. Savel'eva, N.A. Dreiman (Elektrokhimiya 15 [1979] 607.- C.A. 90 [1979] 210630)
89. V.N. Golubev, B. Purins, T.A. Filatova (Elektrokhimiya 15 [1979] 595/597.- C.A. 90 [1979] 210629)
90. J.P. Bernat (W. Ger. 2.749.208, 18 May 1978.- C.A. 89 [1978] 117295)
91. K. Oka, H. Takatama (Jap. 78.23.879, 4 Mar 1978.- C.A. 89 [1978] 135188)
92. K. Oka, H. Takatama, K. Ogasawara (Jap. 78.23.880.- C.A. 89 [1978] 135189)
93. K. Oka, H. Takatama (Jap. 78.41.058, 14 Apr 1978.- C.A. 89 [1978] 168555)
94. M. Shiraishi, M. Okamura, S. Ono, K. Ninomiya (Jap. 78.34.675, 31 Mar 1978.- C.A. 89 [1978] 168272)
95. M. Shiraishi, M. Okamura, S. Ono, K. Ninomiya (Jap. 78.34.676, 31 Mar 1978.- C.A. 89 [1978] 168273)
96. T. Ono, S. Iribe, M. Watanabe (Jap. 78.60.371, 30 May 1978.- C.A. 89 [1978] 155557)
97. O. Kuroda, S. Takahashi, A. Nakaoka, T. Hayashida (Jap. 77.89.577, 27 Jul 1977.- C.A. 89 [1978] 91715)
98. T. Sawa, S. Kikkawa, I. Shimokobe, K. Otani (Jap. 78.14.672, 9 Feb 1978.- C.A. 89 [1978] 94662)

99. Y. Nomiyama, T. Kawahara, H. Shibata, K. Asada (Jap. 78.19.171, 22 Feb 1978.-
    C.A. 89 [1978] 117270)
100. T. Fukutsuka, T. Matsumura, S. Suda (Jap. 78.149.182, 26 Dec 1978.- C.A. 90
    [1979] 192176)
101. T. Kuwahara, Y. Egashira (Jap. 79.08.180, 22 Jan 1979.- C.A. 90 [1979] 192345)
102. B.A. Schenker, T.T. Sugano, N.W. Stillman, K.J. O'Leary (U.S. 4.147.605,
    3 Apr 1979.- C.A. 90 [1979] 193033)
103. J.C. Beatty (U.S. 4.105.534, 8 Aug 1978.- C.A. 90 [1979] 174232)
104. M.P. Kupchik, M.I. Ponomarev, I.G. Bazhal (Nauch. Tr. Kuban Un-t 232 [1977]
    122/127.- Ref. Zh., Khim. [1978] 12 R 434)
105. A.M. Egorov, I.I. Tezikov, R.P. Pozdeeva, T.A. Krasnova (Tezisy Dokl.-Vses.
    Konf. Ekstr. 3 [1977] 81/84.- C.A. 90 [1979] 40671)
106. P. Pierrard (Rev. Gen. Electr. 86 [1977] 491/495)
107. V.S. Zhurkov (Plast. Massy No 2 [1978] 58/60.- C.A. 89 [1978] 135018)
108. O. Kuroda, S. Takahashi, H. Matsuzaki, S. Koike (Jap. 78.88.673, 4 Aug 1978.-
    C.A. 89 [1978] 220530)
109. T. Matsushita, H. Hayashi, T. Ebashi, T. Hiramoto, T. Kaga (Jap. 78.69.841,
    21 Jun 1978.- C.A. 89 [1978] 131467)
110. R.M. Ahlgren, B.M. Schneider (U.S. 4.110.175, 29 Aug 1978.- C.A. 90 [1979]
    73930)
111. Coca Cola Co. (Neth. 77.09.134, 20 Feb 1979.- C.A. 91 [1979] 59303)
112. A. Yamaguchi, T. Yamane, Y. Kageura (Jap. 79.21.971, 19 Feb 1979.-
    C.A. 91 [1979] 27148)
113. V.V. Krokhv (Zh. Prikl. Khim 50 [1977] 2211/2216.- J. Appl. Chem. USSR 50
    [1978] 2118/2123)
114. M.A. Lake, S.S. Melsheimer (AIChE J. 24, No 1 [1978] 130/137.- C.A. 89 [1978]
    8220)
115. J.A. Cox (Illinois Univ., Water Resour. Cent. Res. Rep. No 138 [1978] 72 p.-
    E.I. 17 [1979] 21538)
116. S. Itoi, I.Nakamura, T. Kawahara (Desalination 32 [1980] 383/389)
117. Y. Oren, A. Soffer (J. Electrochem. Soc. 115 [1978] 869/875)

3E. Reverse Osmosis and Ultrafiltration

Economic comparisons indicate reverse osmosis to be more cost effective than
distillation for large operations [1]. Important progress has been made in synthetic semipermeable membranes, leading to a wider scope of reverse osmosis applications, including the purification of used water and saline water for drinking,
the supply of ultra-pure water for the pharmaceutical and electronic industries
and for feeding boilers [2]. Along with the mass transfer within the membrane
surface is important. Reverse osmosis and ultrafiltration processes can be described by heat-transfer equations, but in contrast to the design of a heat exchanger, the equations for calculating flows and concentrations at membranes can
only be solved iteratively. For the most important membrane configurations (tube,
hollow-fiber, and spiral-wound modules) equations are given for describing their
transfer characteristics [3].

An optimization of flow conditions was presented to minimize permeate product
cost in ultrafiltration in laminar channel flow. The methodology for the pressure-
driven membrane system is based on a hydrodynamic optimization scheme developed by
Sonin and Isaacson for electrochemical systems. Experimental data are presented
for permeate flux as a function of transmembrane pressure for the ultrafiltration
of bovine serum albumin solution in laminar channel flow with and without detached
strip-type turbulence promoters. Empirical expressions are established for the
limiting permeate fluxes as a function of the flow rate and interpromoter spacing.
Applying the economic optimization scheme, the performance characteristics of the
convection-promoted systems are assessed. The optimum values for the interpromoter
spacing, the permeate flux, and the Reynolds number are also determined [36].

A historical review of the first decade of commercial reverse osmosis desalting 1968 to 1978 was presented by Ferguson [4].

A large number of papers was published reviewing reverse osmosis and ultrafiltration. A selection includes reverse osmosis techniques and economics in process water purification [5], water treatment by reverse osmosis methods [6], design and application of apparatus for reverse osmosis in desalination of seawater [7], a modern water treatment system for bakers [8], water quality improvement by reverse osmosis [9], reverse osmosis for preconcentration [10], ultrafiltration and reverse osmosis equipment [11], separation by reverse osmosis [12], reverse osmosis moves forward [13], drinking water purification by new methods [14], a review on ultrafiltration giving costs of equipment and operation [37], reviews on developments from ultrafiltration to microfiltration [38], on progress and development of ultrafiltration [39], and on reverse osmosis and ultrafiltration processes [40].

A monograph on the treatment of water by reverse osmosis and ultrafiltration was published [15].

MECHANISM OF REVERSE OSMOSIS

The problem of application of boundary-layer theory to ultrafiltration requires solution of the equations of continuity, momentum and diffusion. The equations of motion and diffusion are coupled. A numerical solution has been developed based on the implicit finite-difference method. Solutions are presented for the permeation velocity, solute concentration at the wall, filtration rate, and wall shear stress for the range $1 \leq Sc < \infty$. For large Schmidt numbers, the effect of concentration polarization dominates. The results show that as the membrane permeability is enlarged, the total filtration rate approaches a limiting value [16]. The phenomenon of the increase of selectivity of solution separation on semipermeable membranes is accompanied by an increase of flow rate. The phenomenon is observed at the separation of a low molecular-weight compound from solvent (reverse osmosis), as well as at the separation of a high molecular-weight compound from solvent (ultrafiltration) [17].

Other work includes mechanisms, membranes, transport and applications of reverse osmosis [18] and mechanism for the separation of solutes by reverse osmosis [19].

OSMOSIS

The osmotic extraction of a solvent is a prospective industrial method for obtaining fresh drinking water, regenerating valuable substances with simultaneous production of pure water or solutions and concentrating alimentary products and solutions. The method involves the passage of a solvent into an extraction solution through a semipermeable membrane, where the extraction solution has a higher osmotic pressure than the starting solution [20].

The forward osmosis extractor design considered is one in which counterflowing solutions are separated by a semipermeable membrane. Fresh water is extracted osmotically from sea or brackish water into a concentrated solution of human nutrients or of fertilizer. A nutrient powder, for instance, could extract six times its own weight of potable water from seawater, thus reducing sevenfold the load of food and water stored in a lifeboat. Other examples involving the extraction of water from brackish water into concentrated fertilizer solution for agriculture, extraction from seawater into ethanol and extraction from seawater into sulphur dioxide solution with recycling of the sulphur dioxide are given [21].

Patents. A liquid having a high osmotic pressure is contacted to a liquid having a low osmotic pressure with a semipermeable membrane. The first liquid is also contacted with another liquid having a low osmotic pressure with a movable sealing, e.g. piston. Good quality fresh water can be produced [22]. A flowing

medium is contacted with an osmosis membrane to prevent the accumulation of suspended, colloidal, precipitated or deposited materials on the membrane. The flow
motion is applied by the feed solution. Membrane contamination is prevented [23].
A floculating medium, for depositing organic and/or inorganic materials, is added
to the feed water to prevent concentration polarization, crystallization and
flocculation deposition of the materials being concentrated near the membrane
surface [24]. Seawater or brine is passed through permeation membranes at 0.2 to
100 kg/cm^2 and the filtrate is then passed through reverse osmosis membranes at
10 to 100 kg/cm^2 to give concentrated liquid and desalted water [25]. Osmotic
separation of solutions into solvent and concentrate uses a two-parts vessel
separated by a membrane [26].

THERMOOSMOSIS

The pseudothermoosmotic pressure of a composite membrane system, constituted
by two cellulose membranes and an aqueous polyethylene glycol approached about
6.0 X 10^5 dyn/cm^2 which was higher than that measured with simple membranes. The
thermodynamics of irreversible processes were used to derive the composite membrane phenomenological equations containing both the classical thermoosmosis and
the thermal diffusion. A value of 3.91 kcal/mol was calculated for the apparent
activation energy, indicating the diffusive nature of the solvent flow through the
composite membrane [27].

In studying thermoosmosis of pure water through cellulose acetate membranes, two
kinds of experiments have been carried out. In the first, the temperature gradient
was the only thermodynamic driving force acting on the water molecules. In the
second, a hydrostatic pressure gradient was added simultaneously. The results of
these experiments disagree with other published studies which report that thermoosmosis is only possible with charged membranes and electrolyte solutions. In
this study, the numerical values are lower than, or similar to, those published by
other authors. The measured thermoosmotic permeabilities are of the order of
10^{-16} mol/cm-s-K and they increase both with temperature gradient and mean temperature. The heats of transport are of the order of 10^{-1} cal/mol and increase with
the temperature gradient [28]. Thermoosmotic mass flow of one component liquids
in capillary systems such as membranes, films, or soil is due to formation of
pressure drops at the boundaries and may be brought to a stationary state by
external counterforces. The incluence of temperature gradients on the formation
of pressure drops is analyzed considering quasithermodynamic equilibrium. The
effect in electrolyte solutions is combined with electrophoretic transport [29].

A review was presented on measurement and temperature dependence of thermal
membrane potential, effect of ions on thermal membrane potential and thermoosmosis [30].

OSMOTIC ENERGY PRODUCTION

A review was presented on water salination as a source of energy [31]. A
large source of energy exists at the interface between water bodies of different
salinities. The pressure-retarded osmosis and dialytic batteries or reverse
electrodialysis are the two main techniques with the most immediate potential for
this energy extraction. Although the present cost of membranes suitable to these
methods is too high, this salinity gradient might become energy competitive with
other energy sources [32].

PRESSURE RETARDED OSMOSIS

Experimental and analytical results demonstrate that internal polarization may
have a profound adverse effect on the water permeation rate of an asymmetric semi-

permeable membrane under pressure-retarded osmosis [33]. The effect of high
osmotic pressures on the performance of membranes was that while the water permeation coefficient decreased at high osmotic pressures, no permanent adverse change
occurred in the osmosis characteristics of the B-9 flat sheet membrane. The
performance of B-9 sheet improved in reverse osmosis after soaking in concentrated
brine solutions. By contrast,an improvement in salt rejection together with a
modest decline in water permeation rate was observed for the B-10 fiber. In
pressure retarded osmosis, however, the water permeation coefficient decreased as
a result of permanent change in the porous substructure of the membrane. For these
membranes, which must always be kept wet, the change in the porous substructure may
be as a result of osmotic dehydration [34].

In pressure-retarded osmosis (PRO), the osmotic pressure gradient exceeds the
hydraulic pressure gradient. Hence water permeation flux should be uphill with
regard to the latter gradient. While this usually occurred in PRO tests with the
Permasep B-10 polyamide fiber, it frequently did not occur with the Fiber Research
Laboratory (FRL) composite fiber, i.e., flux was negative. Results with both of
these asymmetric fibers could be correlated by using an equation for the water
flux, containing two permeation coefficients. The two-coefficient equation can be
understood physically as an extension to PRO of the patchwork model frequently
applied to the membrane skin to explain reverse osmosis results. The extension
requires consideration of the porous substructure and the semipermeable patches
of the membrane skin as two resistances in series impeding osmotic pressure-driven
transport through the semipermeable region. The extended model enabled correlation
of attempted PRO tests with skin and porous substructure properties in both the
FRL and Permasep fibers. The two coefficient equation appears to be a useful
method for categorizing membranes in PRO operation [35].

LITERATURE TO 3E.

1. C.S. Crowe, F.F. Zdenek (Johnson Drillers J.49, No 6 [1977]6/8)
2. E. Chiriea (Engergetica, Bucharest, 25 [1977] 367/375)
3. R. Rautenbach, K. Rauch (Intern. Chem. Eng. 18 [1978] 417/425. - Chem.Ing. Tech. 49 [1977] 223/231. - Verfahrenstechnik 12 [1978] 309/315)
4. P.V. Ferguson (Desalination 32 [1980] 5/12)
5. J. Goodall (Filtr. Separ. 14 [1977] 649/654)
6. E. Tatsuata (Hyomen 15 [1977] 483/495)
7. K. Marquardt (in Taschenb. Abwasserbehandl. Metallverarb. Ind., L. Hartinger, Editor, Munich, Hauser, 2 [1977] 149/164)
8. R. Jarrett (Baker's Dig. 52, No 1 [1978] 44/50)
9. I. Nusbanm, A.B. Riedinger (Water Treat. Plant Des. Pract. Eng [1978] 623/652)
10. D. Pepper (Chem. Eng., London, No 339 [1978] 916/918)
11. P.C. Freschi (Chim. e Ind., Milan, 60 [1978] 138/141)
12. K.Y. Kim, Y.M. Lee (Pollimo 2 [1978] 228/241)
13. Anonymous (Chem. Week 123, No 13 [1978] 67/68)
14. K. Haberer (Umsch. Wiss. Tech. 79, No 3 [1979] 80/86)
15. A.A. Yasminov, A.K. Orlov, F.N. Karelin, Ya D. Rapoport (Stroiizdat, Moscow [1978] 121 p)
16. H.D. Papenfuss, J.F. Gross, F. Sanchez-Ruiz (AIChE Symp. Ser. 74, No 172 [1978] 218/225)
17. A.P. Bogdanov, K.M. Saldadze (Dokl. Akad. Nauk SSSR 242 [1978] 860/863)
18. S. Sourirajan (Pure Appl. Chem. 50 [1978] 593/615)
19. M. Igawa (Hyomen 16 [1978] 399/412)
20. L.A. Kul'skii, T.V. Knyaz'kova, E.G. Zaritskii (khim. Prom-st. No 3 [1978] 188/191)
21. C.D. Moody (NTIS Rept. PB-289251/1GA [1977] 150 p)
22. T. Tsukamoto (Jap. 77.78.678, 2 Jul 1977.- C.A. 88 [1978] 176969)
23. T. Tsukamoto (Jap. 77.94.877, 9 Aug 1977.- C.A. 89 [1978] 8316)
24. T. Tsukamoto (Jap. 77.104.471, 1 Sep 1977.- C.A. 88 [1978] 176970)

25. T. Tsukamoto (Jap. 77.104.472, 1 Sep 1977.- C.A. 88 [1978] 176967)
26. M. Theoni (Brit. 1.526.836, 4 Oct 1978)
27. L. D'Ilario, M. Canella (Polym. J. 9 [1977] 253/260)
28. J.I. Mengual, J. Aguilar (J. Membrane Sci. 4 [1978] 209/219)
29. B.M. Mogilevskii, V.N. Sokolov (Zh. Tekh. Fiz. 48 [1978] 1297/1299)
30. M. Tasaka (Maku 2 [1977] 119/129)
31. A. Emren (Kem. Tidskr. 89, No 11 [1977] 24/27)
32. G.L. Wick (Energy, Oxford, 3, No 1 [1978] 95/100)
33. G.D. Mehta, S. Loeb (J. Membrane Sci. 4 [1978] 261/265)
34. G.D. Mehta, S. Loeb (J. Membrane Sci. 4 [1979] 335/349)
35. S. Loeb, G.D. Mehta (J. Membrane Sci. 4 [1979] 351/362)
36. J.J.S. Shen, R.F. Probstein (I.E.C., Process Des. Dev. 18 [1979] 547/554)
37. P.R. Klinkowski (Chem. Eng., New York, 85, No 11 [1978] 164/173)
38. P. Aptel (Inf. Chim. 170 [1977] 135/139)
39. X. Marze (Actual. Chim. No 7 [1978] 37/42)
40. D. Pepper (Inst. Chem. Eng. Symp. Ser. 54 [1978] 247/252)

3F. Reverse Osmosis Membranes

A bibliography of citations from the U.S. National Technical Information Service data base with 183 abstracts on membranes for reverse osmosis desalination, electrodialysis desalination and other osmotic desalting processes was published, covering Federally funded research for the period 1973 to February 1978 [1]. A newer edition contains 198 abstracts, 15 of which are new entries, covers the period up to February 1979 and superseds all previous editions [2]. A similar bibliography of citations from the Engineering Index data base citations covers world wide engineering research and contains 255 abstracts for the period 1974 to February 1978 [3]. A newer edition contains no new entries [4]. A supplement edition contains 51 abstracts, all of which are new entries and cover the period 1978 to February 1979 [5].

Proportionality exists between the ion fluxes and the electrochemical potentials but not between the water flux and the membrane patential, because the latter does not vanish in the point of inversion. Using the Nernst-Planck equation, frictional coefficients are obtained for which the ratios do not differ essentially from the values found in solutions [6]. A mathematical model of membrane filtration was developed, in which the number and height of transfer units are used [7]. Stability of the membranes to high and low pH and to oxidation by Cl was examined for various types of membrane material. Membrane flux rates decreased as NaCl rejection increased and pore size decreased [8].

Theoretical and experimental results for free-convection-governed mass transfer to membranes proved that free convection governed mass transfer can be described by the analogous heat-transfer laws for isothermal walls. The analogy holds in the whole flow-regime despite sometimes different boundary conditions [9]. The time variation of filtration-system variables, e.g. concentration, volume and permeation rate was simulated. Rigorous mathematical solutions, are illustrated for a physical model with stepwise approximation [10].

Examining the next-generation reverse osmosis membranes, the following topics are covered: the development of reverse osmosis membranes, the development of asymmetric cellulose acetate-type membranes, reverse osmosis membranes in actual use, problems with membranes in current use, solution of existing problems and the Teijin poly(benzimidazolone) reverse osmosis membrane [11].

Other review papers include water treatment by reverse osmosis methods [12], developments in functional membranes [13], developments in membranes, module and system [14], semipermeable and permselective membranes [15], interaction of membranes in water 258, reverse permeation of ions across a cellulose membrane [259] and structure and mechanism of membranes [260].

Patents. An electrical method and apparatus are described to detect the presence of undesirably large holes in membranes [16]. The ranges from two compounds are selected and the processing of a mixture of the two compounds to form semi-permeable membranes are described [17].

Preparation of porous supports for reverse osmosis membranes having high dry bursting strength is described [18]. A high strength support tube is cast from a mixture of epoxy resin and a particulate filler [19]. A flat woven material suitable as support for semipermeable membranes is formed from a fibrous cloth support on which is rolled a covering layer of a thermoplastic [20]. A supporting-plate structure and fittings are presented for the water-tight fitting of semiper-meable membranes [21].

WATER TRANSPORT AND SALT REJECTION

The transport equations, based on capillary flow diffusion mechanism, were com-bined together to represent the membrane phenomenon, in the absence of concentra-tion polarization. The model was tested at various operating pressures with two different membranes and is applicable within small range of operating pressures and at low concentration polarization levels [22]. A brief description of the existing mathematical models of membrane transfer processes was given and two diffusional mass transfer models dependent nonlinearly on the concentration gradient of the transferred substance were proposed. For low concentration gradients, they coincide with the Fick's first law. Also, a system of different-ial equations of the unsteady state mass transfer in a membrane by diffusional and filtrational mechanisms involving chemical reactions was derived [23].

Equations for transport through membranes were derived from basic principles of statistical mechanics, with the classical-mechanical Liouville equation as the starting point. The membrane is taken as one component of the mixture held fixed in space, and the structure of the membrane is assumed to lead only to a geomet-rical space-filling role. The assumption of nonseparative viscous flow allows the transport equations to be decoupled from the details of membrane structure. The results have the same final form as those obtained earlier by heuristic gen-eralization of gas transport equations. Comparison is made with previously pro-posed phenomenological equations for membrane transport [24].

A method was developed for characterizing the transport properties of ultra-filtration membranes. A solution of a polydisperse polymer was used and a steady-state sample of the retentate and ultrafiltrate was taken. Analysis of the amount of polymer in each stream and determination of its molecular weight distribution by gel permeation chromatography gave a complete characterization of the ultra-filtration membrane [25].

A flat blend membrane of cellulose diacetate and cellulose triacetate was evaluated for desalination of saline water by reverse osmosis. The water vapor sorption by the dense film occurs in the order cellulose diacetate film, blend film, cellulose triacetate film, while the asymmetric blend membrane has higher total water content than cellulose diacetate [26]. In reverse osmosis experiments, at the beginning both bulk phases of sucrose and water have equal composition. During this process there exists a certain range of compositions of the bulk phases in which concentration of that component for which the membrane is less permeable (sucrose) increases simultaneously in both bulk phases. The existence of the phenomenon (obligatory deviation) was predicted by R. Schloegl. The change of composition of both bulk phases during hyperfiltration is calculated from the concentration of solute in the transported mixture [27].

The heat capacities of homogeneous and asymmetric cellulose acetate membranes have been measured at different water contents within the temperature range of -40 to +20°C. The results for the partial heat capacity of water within the membranes, as well as for the heat of fusion, were interpreted by assuming two different states of water: unfreezing bound water due to a sorption process and unbound water due to capillary phenomena, which freezes with a freezing point depression and a reduced heat of fusion [28].

Liquid-solid chromatography data on retention volumes of selected reference solutes offer a means of characterizing membrane materials for reverse osmosis. Four cellulosic and four noncellulosic polymer materials have been characterized by a parameter called the β-parameter. The values of β exhibit unique correlations with other parameters governing solute separations in reverse osmosis systems, where water is preferentially sorbed at the membrane-solution interface. Using data on β-parameter for the polymer, and only one set of reverse osmosis data for a reference $NaCl-H_2O$ feed solution for any membrane made from the above polymer material, reverse osmosis separations obtainable with the membrane for a number of other solutes can be predicted. This is illustrated with respect to reverse osmosis systems involving 8 polymer membrane materials, 15 membranes of different surface porosities and 22 organic and inorganic solutes in single-solute dilute aqueous feed solutions [29].

In an investigation of transport phenomena with cellulose acetate membranes, the rejections of metal complexes involving organic sequestering agents such as EDTA or citric acid were much higher than those of the corresponding metal ions. In the case of metal complexes involving small inorganic ligands, such as NH_3 or $SCN-$, their rejections did not necessarily increase with the increase in the coordination numbers of the metal ions. To more precisely understand such transport behaviors, the distribution and the diffusion coefficients of metal complexes were obtained by desorption-rate measurements with dense cellulose acetate membranes. The results revealed that the distribution of a metal ion to the membrane was largely depended on the coexisting ligands. Attempts were also made to explain the distribution coefficient from the microscopic point of view by using Glueckauf's equation [30].

The transport properties of inorganic solutes through cellulose acetate membranes were studied to clarify the effects of both pH and other solutes on the rejection of a particular solute. The rejection of a metal ion is influenced by the pH of a feed solution. The rejection of a particular solute in a mixed-salt solution system is different from that in a single-salt solution system. The degree of change in rejection depends on the species, concentration and permeation of solutes present. The pH dependence of rejection is attributable to the problem of coexisting solute in mixed-salt solution systems [31, 32].

The analytical technique was extended for predicting the reverse osmosis performance of cellulose acetate membranes of different surface porosities with different aqueous feed solutions. This technique needs only a single set of experimental data on membrane specifications given in terms of the pure-water permeability constant and the solute transport parameter and the applicable mass transfer coefficient for the feed solution. The validity of the prediction technique was verified experimentally [33].

Transport properties were determined for various Loeb-type membranes which exhibit strong structure differences from one to another. Using linear relations of thermodynamics of irreversible processes for the homogeneous membranes, correlations were found between the specific transport coefficients and the states of water in the membrane medium. In the case of heterogeneous membranes a schematic multilayer model, using in an appropriate way Darcy's law, was worked out in accordance with electron micrographs of membrane cross sections. It enables to correlate the membrane hydraulic permeability with the relative extent of four typical structures. In addition, this model provides both the hydraulic specific permeabilities of these types of structure and their hydration characteristics [34]. The volume flux and the salt rejection of various cellulose acetate membranes, annealed at different temperatures, were determined from reverse osmosis experiments. The hydraulic permeability decreases sharply in the temperature range 70 to 80°C, while the membrane selectivity increases just as sharply, as function of the annealing temperature. Both can be qualitatively explained by a decrease in the free water content of the membrane and an increase in the ratio of polymer-polymer to polymer-water hydrogen bonding. Practical efficiency of the heat-treatment depends on the formation conditions of the as cast membranes, owing to different sensitivity to pressure effects of each type of membrane structure [35].

Other work includes transport models in osmotic membranes [36], mechanism of solution separation on semipermeable membranes [37], mass transfer in and through

membranes [38], material transport through membranes [39]and permeability of solutes and salt rejection by grafted membranes [40].

CONCENTRATION POLARIZATION

A variety of techniques were discussed for controlling concentration polarization, including prefiltration, backwashing, pulsed flow, ultrasonics and cross
flow [41]. The effects of concentration polarization in water purification and
desalination by membrane processes were discussed [42].

DEGRADATION AND FOULING

The deleterious effect of fouling and compaction was reduced with increased
membrane curing temperature and reduced pressure. The asymmetric cellulose acetate
membranes behaved according to the solution-diffusion model and exhibited increased
resistance to permeation with curing temperature but appeared to be independent of
feed type when analyzed by the modified filtration theory. Increased fluxes (78%)
were obtained with colloidal SiO_2 and polystyrene latex spheres covering the unpro-
tected membrane [43].

Laboratory wastewater was used, after treatment as a feed for desalination by
reverse osmosis. After 1111 h of continuous operation, 0.9 g SiO_2 per m^2 of membrane was deposited. Electrolysis with aluminum electrodes or flocculation with
NaOH and $FeCl_3$ was effective for removal of SiO_2 as the pretreatment for reverse
osmosis [44, 45].

Insoluble yellow iron oxide fouling of reverse osmosis membranes was removed
by acid cleaning and prevented by a different technique of acid cleaning the water
wells [46]. A technique using osmosis and/or electro-osmosis was developed to
clean and possibly decompact contaminated modified cellulose acetate membranes.
The rejuvenation technique developed is called molecular backwash. When contaminated with ferric hydroxide, the membranes were observed to have reduced flux and
salt rejection. After molecular backwashing, both the hyperfiltration flux and
salt rejection improved. Flux loss recovered by molecular backwashing varied from
30% to over 100%, i.e., the flux of the compacted membrane was greater than before
contamination but not greater than the flux of an entirely new membrane [47, 48].
A number of in-situ physical cleaning methods were tested as alternatives to
using chemicals to clean reverse osmosis and ultrafiltration membranes. Flow
surging, air surging, continuous air addition and ultrasonic cavitation were used
alone and in combination with each other as cleaning methods. Ultrasonic cavitation alone and in conjunction with flow surging were the most effective of the
methods tested [49].

In developing of low cost membrane cleaning agents in a first laboratory phase,
scales were deposited on cellulose acetate reverse osmosis membranes. Cleaning
tests were performed with various chemicals and combinations of chemicals including several commercially available scale inhibitors. The field-test portion of
the program showed that the predominant scale deposit obtained was a clay-like
material composed of Si, Al, Ca, Mg and Fe. These deposits are extremely difficult to remove and were only partially removed by some of the better performing
cleaners tested in the laboratory [50].

An organo phosphonate compound was proposed as pretreatment agent to control
calcium carbonate and calcium sulfate scale [51]. Small celloidal particles,
intermediate in size between suspended solids and true dissolved solids, become
concentrated at the surface of the membrane during the reverse osmosis process,
plugging the membrane. To avoid this, it was suggested that the reverse osmosis
pretreatment system be designed properly [52]. The modified fouling index was
recently developed to determine the fouling characteristics of reverse osmosis
feed water [53].

Experiments were conducted with secondary effluent in order to determine the potential of the sponge ball cleaning system as an advanced reverse osmosis cleaning technique. As a result, the product water flux was maintained at 0.65 to 0.75 m^3/m^2.day at 25°C and membrane rejection was more stable. No damage to the membrane was recognized. Tight membranes were more suitable, because it was easier to remove membrane fouling, the product flux was nearly equal and the product water was of better quality [54].

The feasibility of removing scale forming ions from solutions of complex salt mixtures by reverse osmosis with selective cellulose acetate membranes was investi- gated. The rejection of Al^{3+}, Fe^{3+}, Mn^{2+}, Sr^{2+}, Ca^{2+} and Mg^{2+} from their binary mixture with NaCl solutions, as well as the effect of temperature of heat treatment of membranes on the ionic rejection are given in graphs [261].

Patents. Reverse osmosis membranes are contacted with a treating agent poly (Me vinyl ether) at high concentration and treated again by the same agent at a lower concentration. The desalination efficiency decreased after 120 days operation, versus 70 days without the treatment [55].

Scaled or slimed membranes are cleaned by an alkaline solution and an O_3 solu- tion [56]. Ozone and up to one transition metal ions or alkaline metal ions are added to the feed solution to improve membrane performance [57]. Ozone is passed through a slimed or scaled ultrafilter or reverse osmosis unit and the resulting solution recirculated to remove scale and slime [58].

CO_2 is blown into water containing Ca^{2+} and Mg^{2+} and the water is passed through reverse osmosis membranes to remove Ca^{2+} and Mg^{2+}. The CO_2 dissolution in water prevents Ca or Mg based scale deposition [59]. Solutions of citrus juices are used to remove scale from membranes used for the treatment of liquids containing inorganic substances [60].

A liquid containing solids and capable of bubble generation is filtered with a tubular membrane module. A concentrated liquid from the module is recycled and gas bubbles generated in the recycle conduit are discharged. Plugging of the permeable membrane is prevented [61]. To prevent clogging of semipermeable mem- brane during reverse osmosis, hollow fiber fabric is wound on a perforated tube. The particle size of the deposit is greater than the space between the fibers [62]. In the membrane separation process, a pressurized gas is passed through the mem- brane for filtration together with the solution. The materials deposited can be removed readily without damaging the membrane [63]. A cleaning liquid is passed through the apparatus and its flow is pulsated to improve the cleaning efficiency [64]. Air bubbles are passed through the apparatus to remove organic and inorg- anic deposits from the membranes [65]. To remove material deposited on the mem- brane, an object having a spiral passage and external cushions is inserted in a tubular reverse osmosis apparatus and is moved forward a backward by the feed solution [66], or a spring with elastic objects placed in the gaps between the spirals [67], or a spring with external cushions [68], or an object having a spiral passage [69]. Contaminants deposited on the semipermeable membrane are removed by washing with a liquid introduced into a wash line different from the treating line [70]. Cleaning compounds were prepared from surfactants, hydrazine and H_2O_2 [71].

Membranes, fouled during the ultrafiltration of a cataphoretic paint containing Pb pigments, are cleaned by an aqueous solution of an organic solvent and an acid that forms a soluble lead salt [72]. An aqueous solution of an acid, a fatty amine and a water soluble organic solvent are used to clean ultrafiltration mem- branes fouled with cationic electrophoretic paints [73]. Reverse osmosis mem- branes, such as cellulose acetate and saponified poly(vinyl acetate), were regen- erated with acidic solutions of aldehydes. The NaCl-removing capacity, decreased from 96.7% to 86.3% after the membrane has operated for 20 h , was regenerated to NaCl-removing capacity of 97.1% [74]. Desalination of wastewaters, seawater and groundwaters is done in the presence of ultrasound to prevent fouling of the mem- branes [75]. A used tubular membrane was cleaned with sponge-balls, then rinsed with a solution containing 0.2% EDTA and 0.4% $NaBO_3$. The performance of the cleaned membrane was about 97% versus 88% by cleaning with a solution of pH = 2 [76].

DIFFUSION

A continuous-diffusion model, based on capillary-porous structure of hydrophilic membrane, was developed to predict selectivity over a wide range of flow rates through the membrane or applied pressures. The flux and degree of filtration of salt calculated by using this model are in reasonable agreement with experimental values [77].

A previously developed semi-empirical thermodynamic-diffusive model is generalized. Increasing the downstream pressure can either increase or decrease the concentration of the more volatile component in the permeate depending upon the membrane selectivity. It is suggested that for systems in which pervaporation is superior to partial vaporization, including azeotropic mixtures, the best separations will be achieved with high vacuum downstream. The processes of reverse osmosis and pervaporation are compared [78]. A general procedure was presented for indirect determination of diffusivity and gel concentration of solutions of large molecules by comparing measured ultrafiltration limiting fluxes in plane, laminar and turbulent channel flow with theoretical fluxes obtained from analytical mass transfer solutions. A method is also given for determination of diffusivity of macromolecular solutions, as a function of concentration by comparison ultrafiltrate flux-pressure curves in laminar channel flow with a theoretical closed-form mass transfer solution for the flux-pressure behavior. [79].

CELLULOSE ACETATE MEMBRANES

The relationship between ionic mobilities and ionic radii in cellulose acetate membranes having salt rejections lower than 80% is almost the same as that in aqueous solutions. This implies that the ions in these membranes behave as if they exist in bulk water. However, the ionic mobilities in the membranes having salt rejections higher than 86% differ significantly depending on the ionic radii. It seems probable that the bound water influences the ionic mobilities in these membranes [80]. Cellulose acetate semipermeable membranes developed with low air-exposure periods by Sirkar et al. from novel Manjikian-type casting solutions having high cellulose acetate content, partial ·replacement of acetone by dioxane, and small amounts of $ZnCl_2$, have been tested for various combinations of operating pressure and feed brine concentration. In the pressure range of 250 to 600 lb/in.^{2} the pure water permeability constants of membranes for a given salt transport parameter are greater than those of similar membranes tested earlier at 600 psig with a 5000 ppm brine feed. The pure water permeability constants decrease faster with pressure as the casting solution CA concentration is reduced and the dioxane content of the solvent is raised. The intrinsic performances of membranes from a 32% cellulose acetate casting solution with 40% dioxane improve as the pressure is increased from 250 to 600 psig. The solution structure-rate of evaporation concept along with the phenomenon of fingerlike cavities and protrusions in membranes from high dioxane casting solutions explain most of the observations of this work. The superior low-pressure performance of these cellulose acetate membranes provide a basis for systematically lowering the operating pressure in reverse osmosis desalination [81].

Preparation and properties. In preparing porous cellulose acetate membranes, with a pore diameter of 0.01 to 10 μ and a porosity of 30 to 85%, $CaCl_2$ as additive increases the pore density [82]. From the same starting material, it is possible to prepare a variety of membranes which differ markedly in their selective and permeation properties. Ultrafiltration membranes may be prepared from low molecular weight forms of the acetate cast from nonvolatile solvents, while reverse osmosis, or high pressure operating membranes, may be prepared from high molecular weight forms of the acetate cast from very volatile solvents such as acetone or tetrahydrofuran [83]. Ultrafilter membranes were manufactured by coating paper stock with a solution of cellulose acetate. The elasticity of the cellulose acetate coating was improved by plasticization with rubber, without affecting the water permeability [84].

The water permeability rates of cellulose acetate membranes were remarkably influenced by the evaporation period of Me_2CO and formamide as solvents and the casting solution composition. For short solvent evaporation periods, the water permeability rate increased with decreasing content of Me_2CO-cellulose acetate and increasing content of formamide-Me_2CO and formamide-cellulose acetate ratios in the casting solution. Permeation experiments showed that the changes of solvent evaporation period and casting solution composition were related to the change of microporous structure of the resulting cellulose acetate membranes [85].

In studying the effect of ethanol-water mixture as gelation medium during formation of cellulose acetate reverse osmosis membranes two different casting solution compositions were used. With composition I, with pure water gelation medium at 0^oC, the resulting membrane gave a solute separation of 5% and product rate of 220 g/h, whereas with 95% alcohol as gelation medium, the resulting membrane gave a solute separation of approximately 1% and product rate of 1240 g/h under otherwise identical experimental conditions. With composition II membranes, the maximum product rate of 360 g/h with the corresponding minimum solute separation of approximately 1% was obtained with 71.2% alcohol-water gelation medium at 0^oC. These results offer a basis for the development of cellulose acetate ultrafiltration membranes [86]. Experimental data on membrane performance in reverse osmosis fell in four or five distinct regions with progressive increase in ethanol concentration in the gelation medium. Starting with pure water as the gelation medium, in the initial region, the rate of membrane permeated product decreased and the corresponding solute separation increased with increase in ethanol concentration in the gelation medium. In the second region the product rate increased and solute separation decreased with further increase in ethanol concentration in the gelation medium. In the subsequent regions 3 to 5, with still further increase in ethanol concentration in the gelation medium, the product rates obtained were relatively high, and they passed through successive maxima and minima while the corresponding solute separations were relatively low or practically zero. The magnitude of the above changes also depended on the composition of the film casting solution, solvent evaporation period prior to gelation and the temperature of the gelation medium [87].

Reverse osmosis membranes with fairly high salt rejection and productivity characteristics have been prepared by blending cellulose di- and triacetates available in India. A notable feature of the method developed is the elimination of the post-cast heat treatment step necessary in the conventional method. Such unheated membranes have shown satisfactory performance for brackish water desalination. Comparative studies on flat and tubular membranes prepared from the blended dope under similar casting conditions revealed non-uniformity in the tubular membranes. This has been attributed to the acetonic atmosphere prevailing in the casting tube [88]. In the casting of cellulose acetate reverse osmosis membranes, the control of Me_2CO evaporation time from the surface of the membrane and the annealing temperature at the prevailing relative humidity require extreme care to obtain productive membranes under atmospheric conditions in India [89].

The permeability of the volume flux driven by pressure difference decreased whereas the selectivity and reflection coefficient increased with increasing temperature of heat treatment of the Loeb type cellulose acetate membrane. The volume fraction of cellulose acetate in a swollen membrane also increased with increasing temperature. Heat treatment at 74° induced a large compaction of the dense layer of the membrane as determined by transmission electron microscopy. The heat of fusion of freeze-dried membrane was not changed with increasing temperature suggesting that the crystallinity of the cellulose acetate membrane does not increase by heat treatment [90].

Investigating the mechanism of formation of asymmetric membranes used in reverse osmosis, a distinction between segregation of compositions due to spinodal decomposition and separation of phases due to nucleation and growth has been drawn by Smolders et al. The mechanism of formation of the asymmetric membrane used in reverse osmosis has been considered by Strathmann and Scheible in terms only of nucleation and growth. An attempt was made to describe the membrane formation in terms of both mechanisms [91]. Different types of phase separation are responsible for the build-up of the dense skin layer and the porous supporting layer in asymmetric membranes of several materials: cellulose acetate, poly-sulfone polyacrylonitrile and polydimethylphenylene oxide. The formation of the porous sublayer is explained in terms of liquid-liquid phase separation, coales-cence and gelation. A possible explanation is given on thermodynamic and kinetic grounds for the formation of nodular structure in the skin of the membranes [92].

Light scattering from concentrated solutions of cellulose acetate in acetone is discussed in terms of exponential correlation functions for refractive-index fluc-tuations. These are related to intermolecular aggregation. Light scattering, tensile creep and recovery, and hyperfiltration studies with films of cellulose acetate cast from acetone are discussed in terms of the state of aggregation in the films. It is concluded that intermolecular aggregates are probably small and imperfectly formed regions of microcrystallinity stable up to about 210°C, with the overall state one of near amorphous disorder [93]. Physicochemical properties of cellulose acetate membranes used for purification of mineralized water by re-verse osmosis were reviewed [94].

A cellulose acetate membrane was deacetylated with alkalies or alkali-alkoxides to give a product which exhibited increased resistance to organic solvents [95].

The development of Dry-RO membrane program treated two aspects of ionogenic cellulosic reverse osmosis membranes: The feasibility of incorporating anionic (sulfonate groups) charges into cellulose acetate and the feasibility of the cova-lent attachment of quaternary ammonium groups via carbamate, rather than ester linkages. The second part of the program resulted in the simplified approach of in situ formation of cellulose acetate carbamates. This approach consisted of pre-paring blocked isocyanate monomers bearing quaternary ammonium groups and the inc-lusion of these monomers in ordinary cellulose acetate casting solutions such that they wound up incorporated within the final membrane [96].

Structure. The structure of cellulose acetate solutions became more homogen-eous with increasing content of formamide in the casting solution, reaching greatest homogeneity at 30% formamide. The membranes obtained from such solutions were the most productive and the best for water desalting [97]. Investigating the effects of the structure of cellulose acetate membranes, transport coefficients under dialysis conditions were determined and the flow rates and salt rejection measured for seven membranes that had been formed under different evaporation tem-peratures, coagulation conditions and heat treatments, showing that the applied pressure modifies the membrane, affecting its properties [98]. A review was pub-lished discussing the chemical and physical structures of asymmetric cellulose acetate membranes and their application in reverse osmosis [99]. Using experimen-tal sorption data and corresponding experimental findings of calorimetric investi-gations, the state of water in cellulose acetate membranes was discussed by apply-ing a theoretical treatment of sorption. The sorption of water can be attributed to a gain in surface energy at the polymer/vapor interface. Using differential thermodynamic potentials of sorbed water together with experimentally determined heat capacities of sorbed water, the thermodynamic potentials G, H and S of sorbed water are estimated for the temperature interval -40°C to +40°C. At constant tem-perature, each thermodynamic potential depends on the water content. The resulting distribution function of G indicates that the sorbed water exists in different states. Comparing the Gibbs free energy of sorbed water with that of ice or liquid water at the same temperature leads to the conclusion that none of the sorbed water freezes to ice within the temperature interval used. Based on the Gibbs free energy of water in electrolyte solutions and the distribution function of G for sorbed water, partition coefficients of salts within cellulose acetate-membranes may be estimated. The results are in good agreement with experimentally determined par-tition coefficients which are available from the literature. As the partition coefficient of a salt is directly related to the salt rejection of the membrane this provides a method of estimating the desalination performance of a membrane from its water sorption isotherm [100].

Sorption of water in cellulose acetate membranes of two acetyl contents has been studied by infrared spectroscopy. A striking resemblance between the spectra of

water in these membranes and those of water dissolved in ethyl acetate has been
noted. It is concluded from the infrared spectra that, even near saturation,
water sorbed in the membranes has a low degree of association and that bonds between
water and cellulose acetate are considerably weaker than those in liquid water.
Similar conclusions can be derived from the form of the sorption isotherms of water
by cellulose acetate. The weak bonding of water to the membrane is consistent with
its high mobility, while its low degree of association explains the low solubility
of ions and hence the low permeability of the membrane to salts [101].

Patents. A cellulose acetate sheet is treated with an aqueous solution of at
least dihydric alcohol, especially glycerol, and dried at above 50°C. It regains
desalting effectiveness after wetting with water [102]. A filter element is manufactured from cellulose acetate semipermeable membranes by soaking the membrane with
glycerin and joining its edges by heating [103].

Preparation of cellulose acetate membranes having good mechanical strength
(240 kg/cm^2) is described [104]. Permeability and salt rejection of semipermeable
cellulose acetate membranes are controlled by treatment with water-miscible solvents, e.g. 50% aqueous EtOH [105]. A solution of 3.5% NaCl was treated with a
cellulose acetate based membrane and gave a water recovery of 68.2% at 60 kg/cm^2
[106]. A semipermeable cellulose acetate membrane is formed on a porous core, surr-
ounded by a porous membrane support, to prepare a reverse osmosis element[107].
Cellulose acetate membranes were prepared, which exhibited 86% polyethylene glycol
retention and 92% selectivity [108]. Casting cellulose acetate with molecular mass
over 30000 Dalton from solutions gave asymmetric membranes for use in separation of
substances [109]. Continuous applying cellulose acetate solution on organic solv-
ent-impregnated paper, compact fleece or water permeable film gave asymmetric membrane with high mechanical strength and substance diffusion [110].

Casting cellulose acetate solutions and treating the formed film in a gelation
bath, consisting of EtOH, gave ultrafiltration membranes with high water flux rates
[111]. A microporous plastic film or gauze or taffeta is unsymmetrically incorporated into a membrane, which becomes stronger and more water-permeable than that
without the reinforcement [112]. Semipermeable membranes having good permeability
were prepared by casting acetone solutions containing 18 to 25% cellulose acetate
and 10 to 40 parts lactic acid or citric acid per 100 parts of cellulose acetate
[113]. Reverse osmosis membranes were made from solutions containing cellulose
derivatives, organic solvent and $R(CO_2H)_4$, where R is a C_{2-10} aliphatic or alicy-
clic organic group [114].

MODIFIED CELLULOSE ACETATE MEMBRANES

The effects of annealing temperature on the mechanical properties and the desalination performance of modified cellulose acetate membranes have been investigated.
A critical temperature exists at which all membrane properties change abruptly. In
addition, the effects of low concentrations of phenol in the solution on the mechanical properties and on the water content of the membranes have been established.
At the same concentration at which the phenol uptake sharply increases, a significant decline of water content and mechanical properties have been found. The results confirm close relationships between the physical properties and the desalination performance [115].

Cellulose acetate butyrate membranes were cast from five different formulations.
The pure water flux through the membrane increased with evaporation period. The
separation of 4000 ppm NaCl aqueous solution remained unchanged until it reached a
critical flux. At that point, separation decreased inversely proportional to the
flux. Scanning electron microscope photography of the membranes corresponding to
each evaporation period is reported [116]. The data on free-energy parameters for
alkali metal and halide ions, and on structural group contributions to free energy
of hydration, and correlations allowed calculation of the values of solute transport parameter of cellulose acetate propionate membranes of different surface porosities and aqueous feed solutions involving 12 alkali metal halides and 24 nonion-

ized polar organic solutes. From data on membrane specifications given in terms of water permeability constant and solute transport parameter for NaCl at any single feed concentration,reverse osmosis separations of all the other solutes could be predicted for cellulose acetate propionate membranes [117].

Ultrafiltration membranes were prepared by casting nitrocellulose from ether-alcohol solutions and were characterized by using 1% aqueous dextran solution containing 1500 ppm NaCl. The variation of the ether-alcohol ratio in the solvent mixture, evaporation time of the cast solution, heat treatment of the film, solvent addition in the casting solution, as well as feed concentration and flow rate, operation temperature and pressure and molecular weight of dextran affected swelling degree, compaction, water permeability, pole diameter and ultrafiltration rate of the resulting membranes [118].

Casting of cellulose nitrate from mixtures of $HCONH_2$ with low-boiling solvents gave membrane with highest permeability rate for water [119]. The water permeation rate of cellulose nitrate and cellulose membranes, obtained by regeneration of cellulose nitrate, in the presence of polyethylene glycol decreased with increasing molecular weight of polyethylene glycol and depended on the viscosity of its aqueous solutions. The permeation rate of cellulose membranes was greater than that of cellulose nitrate membranes [120]. The ultrafiltration and adsorption characteristics of anisotropic polymer membranes obtained from a system of cellulose nitrate/methyl alcohol/formamide/activated charcoal were investigated for possible use of such membranes in artificial kidneys. The ultrafiltration rate and the extent of solute adsorption were influenced markedly by the length of the solvent evaporation period during membrane formation, the amount of activated charcoal added, the heat treatment temperature, the feed concentration, the nature of the solute, and the operating temperature and pressure [262]. The ultrafiltration rate, the hydrolysis and adsorption characteristics for urea and BSP of solvent-cast semipermeable nitrocellulose membrane containing urease, partially esterified maleic anhydride-styrene copolymer and activated charcoal were influenced significantly by the solvent composition evaporation period, additional amount of charcoal and operating temperature and pressure [263].

An 80:20 cellulose acetate-poly(Me methacrylate) blend membrane was superior to that of cellulose acetate for desalination by reverse osmosis and had about 40% better tensile strength [122]. Solvent flux values of these blend membranes, determined experimentally and calculated on the basis of theory, agreed fairly well. The solute concentration at the membrane wall was nearly double the feed concentration, so that the true salt rejection was a little greater than observed [123]. The blend membranes produced 97 to 98% NaCl rejection at a product flux rate 500 l/m^2 day at outlet pressure about 42 kg/cm^2 [124].

Patents. Cellulose acetate membranes with improved salt rejection performance for water purification are prepared by treating with a 3% solution of exo-cis-3,6-endoxo-4-tetrahydrophthalic acid - vinyl acetate copolymer [125] or with a copolymer of a vinyl ester with an unsaturated imide [126] or an aldenyde to acetylate the polymer [127]. A method for manufacturing phosphorylated cellulose ester membranes is described [128].

POLYMER FILM MEMBRANES

In the transport from the liquid phase through homogeneous polymeric membranes, the concentration of permeant in the polymer is usually high and has a large influence on diffusion coefficients. The driving force for transport, the gradient of chemical potential of the liquid, may be controlled by the application of a pressure. Frequently the chemical potential gradients in liquid transport are small. The driving force may also be controlled by varying the composition of the liquid phases, i.e. by setting up an osmotic pressure difference across the membrane. This can only be done with systems of more than one component and the study of transport from the liquid phase therefore leads readily into the study of several simultaneous fluxes and the possibilities of membrane selectivity and separation. The mechanisms of steady flow of, principally, organic liquids through polymer membranes under pressure from the liquid phase into either another liquid phase, as in hyperfiltration, or into a vapor phase as in pervaporation are reviewed [129].

The rejection coefficient for rigid and flexible polyelectrolytes during ultrafiltration was exerimentally determined by a transient material balance technique. Parameters such as supporting electrolyte (NaCl) concentration, pore radius and pore flow velocity were varied to study their effects on the degree of rejection. The membranes were fabricated by the track-etch process and offered the advantage of a well-defined pore structure, in which constrictions and tortuosity effects were minimal. At a given pore radius, the NaCl concentration strongly affected the rejection coefficient by altering the long-range electrostatic forces between pore wall and macromolecule for the rigid polyelectrolyte, and by substantially altering the molecular conformation of the flexible polyelectrolyte. In all cases, the rejection coefficient increased as the ionic strength decreased. With the flexible macromolecules, the shear rate in the pore also appeared to be a significant factor. The most surprising result is that the flexible polyelectrolyte was able to pass through pores of radius less than one-half the hydrodynamic radius of the macromolecule, indicating that flexible macromolecules cannot be modeled as rigid spheres in predicting the rejection coefficient [130].

In evaluating and selecting polymeric materials for reverse osmosis membranes, three classes of polymers were studied: cellulose acetate, aromatic polyhydrazides and aromatic polyamides. The overall superior performance of aromatic polyamides, combined with the economics of large scale manufacture, made them the choice as the best polymer type for use in commerical permeators [131].

Other work includes reviews on grafting of reverse osmosis membranes with polymers [132] and polymeric membranes based on heterochain aromatic polymers [133].

Preparation and properties of polymer film membranes are summarized in the following according to the alphabetical order of the main material.

Acrylic acid derivatives. Experimental results were presented dealing with the water permeance and diffusion rate in two-phase membranes consisting of hydrophilic domains of polyacrylic acid chemically bound by grafting to hydrophobic matrices, which are either rigid crystalline or elastomeric amorphous. Permeance is higher in rigid matrices, which is attributed to the wide net of microfractures originated from swelling of the hydrophilic domains whose formation is only possible in the case of a rigid matrix. A calculation method was proposed, which allows the evaluation of the apparent diffusion coefficient of both water and salt, starting from dilatometric measurements of membranes contacted with water and salt water. The apparent diffusion coefficient of water in rigid membranes is of the order of 10^{-8} cm^2/s, whereas in elastomeric membranes it appeared about 100 times lower. The apparent diffusion coefficient of salt is of the order of 10^{-12} cm^2/s for both rigid-matrix and elastomeric membranes when previously swollen [134].

A study of the ionic crosslinking of poly(acrylic acid) membranes for possible applications in dialysis and reverse osmosis consists of casting a film of poly (acrylic acid) neutralized with sodium hydroxide, followed by immersion in appropriate metal salts (aluminum, zinc, and chromium salts). A qualitative rate model has been developed to guide this synthesis. Since both metal cations and protons in solution compete for the carboxylic acid sites, acid-base properties of the metal and polycarboxylic acid appear to be important for successful membrane formation. The use of a nonsolvent for the polymer in the crosslinking solution was tested and found to give improved membranes under some conditions, dimethylformamide being the most successful [135]. The reverse osmosis properties of ionically crosslinked polyacrylic acid membranes were investigated in terms of the salt separation of a 0.1% NaCl solution and water flux. The effect of such variables as the polymer concentration in the casting solution, the ratios of solvents used, the ratio of monomer to the crosslinking agent, the evaporation time and temperature, and the nonsolvent nature and treatment times were studied in some detail. The most important variable was found to be the length and nature of the treatment in the nonsolvents acetone and methanol. In the best series of the membranes that were synthesized, fluxes of more than 3.0 gfd, with salt separations at the 80% to 85% level, were obtained [136].

The permeability of H^+, Na^+, insulin, and hemoglobin through porous membranes made of crosslinked poly(methacrylic acid) was investigated at different pH values on both sides of the membrane [137]. Acrylonitrile-styrene copolymer membranes did not have sufficient reproducibility and permeation stability, but these were improved by the addition of ethylene glycol or glycerol. These membranes show better separation and concentrating properties than hydrophobic membranes [138]. Preparation and properties of hydrophilic grafted polymer membranes were reviewed [139].

Aromatic polyamide polymers. Polypiperazinamide reverse osmosis membranes were prepared, as a special class of polymers which have osmotic characteristics similar to those of cellulose acetate, but are superior with regard to chemical resistance and also resistance to bacterial attacks. This was due to lack of amidic hydrogen, which is present in the traditional aromatic polyamides. Flat asymmetric membranes with variable osmotic characteristics, which can replace the cellulose acetate membranes, were reproducibly prepared [140].

Poly(xyleneadipamide) membranes may be feasible for ultrafiltration rather than reverse osmosis under conditions of high temperature and high or low pH [141]. Hydrophilic membranes based on N-substituted (meth)acrylamides showed a pronounced deswelling in salt solutions [142]. A novel hydrophilic polyamide membrane exhibited an excellent fractional solute rejection behavior from aqueous solutions, having a comparatively sharp boundary at the molecular weight range of about 200 to 1400, together with an extremely high water permeability [143].

Permeability of polyamide membranes can be controlled by variations in the technological parameters of the formation process. The temperature of the substrate affects the structure and properties of membranes. The number of pores formed at the polymer-substrate interface was extremely dependent on substrate temperature. The total porosity of the polyamide membranes decreased with increasing temperature [144]. Membranes prepared from PAK 60/40 and P548 have a high porosity which allows their use in ultrafiltration of aggressive media. The film characteristics of the polyamides depended on the preparation method. The mechanical strength of the membranes increased and the porosity and the productivity decreased with a decrease in the phase separation break of the system. Films prepared from melts and by the precipitation method had low and high porosity, respectively. A correlation existed between the porosity and the degree of ordering of the polymer, which suggested a proportionality between the degree of amorphousness and the porosity of the film [145].

Development of chlorine-resistant reverse osmosis membranes for use in desalination of sea water was carried out using in situ formation of a thin, dense coating on a microporous support. Three types of polymer systems were studied: polypiperazineamide membranes formed by interfacial deposition, membranes formed by crosslinking, water-soluble polysulfone-sulfonic acids in situ, and linear vinyl or acrylic polymer membranes crosslinked in situ by chemical reactions or ultraviolet radiation. Most promising results were obtained with polypiperazineamide membranes which were prepared with fluxes of 20-28 gfd at a salt rejection of about 98%, tested at 1500 psig ub 3.5% synthetic sea water. One variation of this method produced a membrane with salt rejection of 99.3% [146].

Aromatic polyamide membranes had over 99% NaCl rejection during pressure filtration at 100 bars. Poly(amidehydrazides) had quantitative retention for $MgSO_4$ but only 90 to 98% for NaCl [147]. A study of the structure of ultrafiltration membranes based on aromatic poly(amide imide) was made [148]. The characteristics of aromatic polyamide membranes and of the ultrapure water they produce were given [149]. Preparation and characterization of aliphatic/aromatic copolyoxamide membranes were reported [150].

Low molecular weight salts contained in aromatic polyamide reverse osmosis membranes greatly increase membrane fluxes without a detrimental effect on rejection. Highly dissociated salts, e.g. $LiClO_4$ or $Mg(ClO_4)_2$, exert a stronger influence than the commonly used LiCl. Mixtures of different salts may give stronger effects than those obtained with a single additive. The salt effect is caused by a general effect on solvent activity and thus on the kinetics and equilibrium associated with evaporation and coagulation processes [151]. The tensile strength of poly (m-phenylene-isophthalamide) membranes depends linearly on temperature. The permeability is directly proportional to the pressure and increases linearly with temperature [152].

In preparing Nylon-6 and Nylon-12 polyamide films, irradiation was compared to chemical grafting, either with or without crosslinking by means of diacryloxylhexamethylenediamine from acrylic acid solutions in the presence of ferrous ammonium sulfate or cerous ammonium sulfate catalysts. Grafted and crosslinked membranes were tested in a reverse osmosis system for desalting water solutions containing 10000 ppm NaCl. Daily flow was only 20 to 25 l/m^2 and the salt rejection varied from 75 to 95% [153]. Polyamide membranes grafted with 4-vinylpyridine were useful for desalting water. The water permeability and salt rejection varied with the composition of the membrane [154]. The permeation characteristics and the burst strength of nylon 12 membranes treated with heat in various solutions were significantly influenced by the treatment solution, temperature and time and the concentration of acid and alkali in the treatment solution. In particular, nylon 12 membranes treated with formic acid/formalin remarkably improved the permeation characteristics and the burst strength [264]. The permeability characteristics of nylon-cellulose acetate polymer blend membranes in the separation of poly(vinyl alcohol) from aqueous solution were markedly influenced by the blend composition, the feed concentration and the operating temperature and pressure. These effects were related to the microporous structure of the membranes [265].

Ethylene derivatives. The effect of polyethylne glycol on the kinetics of some nucleophilic reactions and the selective and active transport of alkali metal ions with acyclic ethylene oxide homologs was investigated and the ability of 14 α-carboxy-ω-hydroxy polyethers containing ethylene, 1,2-phenylene and tetrahydrofuranyl units as model compounds for ionophores to transport alkali metal ions selectively and actively through membranes was examined in various solvents [155].

Poly(methyl-methacrylate) polymers. Membranes, formed by making use of the stereocomplex phenomena which occurs in solution when isotactic and syndiotactic poly(Me methacrylate) are dissolved, were highly permeable to water when compared with commercial cellulosic membranes, while their NaCl permeabilities were the same. The permeabilities of the membranes were explained by a simple capillary model. The finest membrane is formed when the isotactic content is about 30%. This shows the close relation between complex formation and membrane permeability [156].

Poly(vinyl alcohol) polymers. Changes in permeability of membranes grafted with various vinyl monomers relate to changes in physical structure, and chemical structure. By comparing the permeation characteristics for several feed solutes of the grafted membranes, Me methacrylate-vinyl alcohol graft copolymer and Me methacrylate poly-ethylene glycol graft copolymer, the importance of the interaction between solute molecules and the membrane was established [157]. The permeation stability of alkali-bridged poly(vinyl alcohol) treated with m-cresol is dependent on structural changes based on an increase of H-bond content and on crystallinity [158].

When acrylonitrile-vinyl alcohol graft copolymer membranes are treated in mixtures of DMF and Me_2SO and immersed in water, the permeabilities of uric acid, vitamin B_{12} and water through the membranes increased substantially without decreasing the mechanical strength of the membrane. Values obtained were superior to those of the cuprophane membranes used in artificial kidneys [159]. Two polyanions and two polycations were prepared by the esterification of poly(vinyl alcohol). Properties of their neutral polyelectrolyte complex membranes were compared with those of polyelectrolyte membranes. The charged state on the membranes was related closely to the mechanochemical reaction, the salt rejection and the solution permeability. Both the salt rejection and the water flux of neutral polyelectrolyte complex membranes were lower than those of the corresponding acidic or basic complex membranes [160].

Polyvinylpyrrolidone polymers. The stability of solutions of bisphenol A polycarbonate in N-methylpyrrolidinone was studied and ultrafiltration membranes of various pore sizes were prepared from the solutions 161 . The pervaporation performances of membranes obtained by radiation grafting of N-vinylpyrrolidone onto

poly(tetrafluoroethylene), polyethylene, polybutene and poly(dimethylsiloxane)
films were studied using a water-dioxane azeotropic mixture. An attempt has been
made to interpret the results in terms of variations in the number, size and density of the grafted domains which form the diffusion paths for the permeating molecules. Based on this interpretation, a general rule is proposed for selecting
adapted methods to prepare efficient pervaporation membranes [162].

Other polymeric membrane materials. Membranes cast from PVC solutions are
stable and might be used for the ultrafiltration of acids, bases and organic liquids [163]. The permeability of Ftorlon 2M membranes increases and their selectivity decreases with increasing Ftorlon content or decreasing temperature of solutions used for their preparation [164].

An electron-microscopic method was described for determination of the structure
of asymmetric membranes prepared from poly(tri-methylvinylsilane). The membranes
consisted of nonporous layers and a porous substrate. The porous substrate contained pores differing in size from 0.1μm to several μm. The dense surface layers
had no pores [165].

Three novel polymer membranes for water desalination by reverse osmosis were
originated and characterized. Two of these, alloy membranes, were composed of
cellulose acetate/poly(bromophenylene oxide phosphonate) and cellulose acetate/
poly(4-vinyl pyridine). For the first alloy, two different structures have been
identified for the asymmetric membrane: the well known dense skin resting on an
open-celled foam and skin resting on a porous layer which displays a two-phase
morphology. In the latter, dense spheres appear to grow out of a continuous
polymer network. The second alloy has the same morphology as cellulose acetate
when cast as an asymmetric membrane. Both membranes have high water permeabilities
and high salt rejections. The third membrane was an asymmetric PPOBrP membrane.
This polymer was cast into an asymmetric membrane which displays a sharp boundary
between the skin and the porous structure, resembling a composite membrane. The
three novel membranes were tested under hydraulic pressures up to 1200 psi, producing fluxes of 5 tb 40 gfd and salt rejections greater than 90% [166].

Membranes of synthetic poly(α-amino acids), were prepared and their permeabilities of oxygen dissolved in water were measured in the 8 to 50°C temperature range
using an oxygen electrode [167].

Patents. The ranges of component composition of nitrogen-containing aromatic
polymers are given [168]. The range of composition of added supplemental polymer,
incorporated in the membrane to improve rejection performance, is sepcified [169].
The subsurface region of the membrane has at least one polymer or copolymer contained in at least some of the pores [170].

A polytriazine, a polyquinoxaline, a polyquinoline and similar polymers were
useful as ultrafiltration membranes [171]. Desalination membranes prepared from
3-aminobenzhydrazide-4-aminobenzhydrazide-isophthaloyl chloride-terephthaloyl
chloride copolymer containing $LiNO_3$ were extracted with aqueous $Me_3N+CH(C_{14}H_{29})$ CO_2
and gave water permeability 1222 l/m^2/day and NaCl rejection 97.8%, compared with
978 and 98.4% for a membrane extracted with water [172]. Membranes prepared from
tert-Bu-crotonate-tert Bu isocyanate block copolymer gave a filter having MeOH
permeation 3.1 ml/cm^2 at pressure difference 650 mm [173]. A dope containing
poly(p-phenyleneterephthalamide was cast to a film with a liquid transmission of
1.1 l/m^2. day and salt rejection 84% [174]. A blend containing poly(2.5-p-phenylene-
1.3.4-oxadiazole and poly(ethyleneterephthalamide) was cast and treated to give a
membrane with tensile strength 6.5 kg/cm^2 and water transmission 15 ml/cm^2.h [175].
A 1.4-phenylenediamine-terephthaloyl dichloride copolymer solution was extruded at
80° and treated to give a membrane having average pore size 0.02 μ [176].

Reverse osmosis asymmetric membranes, having high salt rejection and flux, comprised a porous polysulfone support film coated with polyethyleneimine or epichlorohydrin ethylenediamine - propylene oxide copolymer [177]. A polyethylenimine
film, deposited on a porous polysulfone substrate, gave a membrane with 16.3 gal/
ft^2.day flux and 99.8% salt rejection [178]. A membrane obtained from isotactic
polypropylne and low density polyethylene gave 99.8% salt rejection and 514 l/m^2.day
water permeability [179]. Manufacture of membranes based on polytetrafluoroethylene

is described [180]. Membranes prepared from polyethylene glycol monostearate gave a water permeability 668 ml/m^2/day at 0.07 kg/cm^2 and salt rejection 94.8% [181]. Triethylene tetramine and triglycidyl isocyanurate were treated with an isophthaloyl chloride - trimesoyl chloride mixture to prepare a membrane with a flow rate 151.9 l/m^2/h and salt rejection 98.2% [182]. Tydex-12 was grafted with acronitrile to give a graft polymer, which after treatment gave an oxidation-resistant membrane with salt rejection about 99.6% [183]. Membranes of ethylene vinyl alcohol copolymer were treated with an adduct of ethylene oxide with glycerol to improve softness and permeation stability [184]. Membranes comprising ethylene - vinyl alcohol copolymer were found useful for hemodialysis, ultrafiltration, separation of proteins and microorganisms [185]. A mixture of high-density polyethylene, SiO$_2$ powder and diatomaceous earth was used to prepare a 500 µ membrane, which gave a water permeability 35 ml/cm^2.min at 3 kg/cm^2 pressure [186].

A doping solution containing cellulose acetate was cast on a poly(ethylene tetraphthalate) to prepare a membrane having salt rejection 99.8% and water permeation 2.1 ton/m^2.day [187]. A similar membrane exhibited 99.4% salt rejection and 1.97 ton/m^2.day water permeation [188]. Poly[oxy(2,6-diphenyl-1,4-phenylene)] is sulfonated to prepare reverse osmosis membranes by casting or coating on porous matrixes [189].

A (4-HOC$_6$H$_4$)$_2$ SO$_2$-epichlorohydrin copolymer was treated to give a polymer membrane, which had slightly lower salt rejection, but higher water permeability than cellulose acetate membranes [190].

Preparation of poly(vinylidene)fluoride Millipore membranes, suitable for chemical and pharmaceutical uses, is described [191]. Poly(vinylidene) fluoride was sulfonated and treated to give a membrane that showed separating power over 99.9% for mineral oil and over 97% for ABS surfactant at permeability of 20000 l/m^2 [192]. A mixture of 3.3′-methylenebis(anthranilic) acid, m-phenylenediamine, N-methyl-2-pyrrolidone and propylene oxide was stirred with isophthaloyl chloride and terephthaloyl and treated to give a membrane with 1.11 m^3/m^2.day flux and 93.8% NaCl rejection [193]. Semipermeable membranes were prepared from sulfonate hydantoin polymer, which showed 82.7% salt rejection and 4.9 x 10^{-4} g/cm^2.s water permeability at 80 kg/cm^2 [194]. Vinyl-pyrrolidone-vinyl acetate copolymer was mixed with acronitrile-vinyl acetate copolymer and treated to obtain a 0.2 mm thick membrane. The total of an aqueous paint was removed with a water permeation of 45 l/m^2.h [195]. A benzimidazolone polymer desalination membrane showed a water permeation of 17.5 l/m^2.h at 40 kg/cm^2 and 97.2% salt rejection [196]. Pendant sulfo group containing polybenzimidazolone membranes exhibited 29 l/m^2.h water permeability and 35.2% salt rejection [197].

Semipermeable membranes prepared from aromatic polyether-polysulfones containing phenylene units had 4.3 m^3/m^2.day water permeability per kg/cm^2 [198], from poly(vinylimidazoline) bisulfate gave a flux rate of 38 gal/ft^2.day and a salt rejection of 83% [199], and from 4.4′-diphenylmethanediisocyanate, HCN and N-methyl-pyrrolidinone with which 98.2% of the NaCl was removed in the dialysis of a 0.2% NaCl solution [200].

A polyester fiber net is incorporated in a N-methyl-2-pyrrolidinone membrane [201] or in cellulose acetate based membrane [202] to increase strength and water permeability. Membranes were also formed from a poly(acrylethersulfone) [203], an acrylonitrile polymer [204], sulfonated polybenzoxazine-1,3-dione-2,4 [205] and of aromatic polyamides containing specified disulfimide groups [206]. A membrane prepared from Teflon powder had a water permeability 0.13 ml/cm^2.min at 0.76 kg/cm^2 filtration pressure [207].

Preparation of semipermeable membranes from isophorone diisocyanate, polyethylene glycol and 2-hydroxyethyl acrylate [208], from acrylonitrile-diethylene glycol dimethacrylate-2-hydroxyethyl methacrylate copolymer [209], and from acetyl cellulose and polyester film [211] was reported. Semipermeable membranes were cast from polyacrylonitrile or copolymers [212]. A semipermeable membrane exhibited water permeability 9.4 x 10^{-10} cm^3/s.atm and salt rejection 95.5% at 80 kg/cm^2 [213]. Reverse osmosis membranes were prepared from polyamides with pendant CONH$_2$ groups [214], by coating a substrate with a polymer liquor [215], from a γ ray irradiated polypropylene film [216], from poly(vinylidene)fluoride

[217], from cellulose acetate and 1,4-butanedicarboxylic acid ethylene glycol copolymer [218], from acrylonitrile-vinyl acetate copolymer [219], from poly-ethylenimine crosslinked with a polyfunctional agent [220], from crosslinked ethylene-vinyl alcohol copolymer [221], from poly(p-phenylene-1,3,4-oxadiazole) with appropriate treatment [222] and from polyethylene glycol [223].

Permeable membranes were also prepared from acrylonitrile-Me acrylate copolymer [224] and from the same material mixed with acrylonitrile-vinylidene chloride copolymer [225]. An interpolymer membrane is based on a polyelectrolyte and an inert copolymer film, for example acrylonitrile-vinyl chloride copolymer [226]. An anisotropic polyamide membrane was prepared by spreading the casting solution onto a suitable support [227]. Semipermeable membranes were prepared from poly(vinyl-imidazoline bisulfate [228] and from acrylonitrile copolymers [229]. Polysulfone ultrafiltration membranes, with uniform pore size, were prepared by dissolving a polysulfone in a C_{2-4} saturated polyhydric alcohol and a volatile solvent and casting the mixture [121].

ULTRATHIN AND COMPOSITE MEMBRANES

Reverse osmosis composite membranes prepared by plasma polymerization of allyl-amine-nitrogen mixtures over a porous substate were evaluated for the rejection of components present in washwater. The membranes exhibited high rejections for sodium chloride, potassium chloride, detergent and dextrose. High rejections of urea could be achieved, but only at a sacrifice of water flux. Lactic acid could also be rejected but caused a degradation of the plasma-deposited layer [230]. A new thin composite reverse osmosis membrane, designated FT-30, that shows excellent potential for single-pass seawater desalination, was formed by depositing a proprietary thin polymer coating on a microporous polysulfone support layer. Tested at an operating pressure of 1000 psi with synthetic seawater the membrane exhibited fluxes of about 30 g/ft^2.day and salt rejections as high as 99.6%. At 800 psi the flux dropped to 23 g/ft^2.day with little decrease in salt rejection. The membranes provided also excellent results with brackish waters at 200 to 600 psi [231].

Patents. A semipermeable membrane was manufactured from a polysulfone ultra-filtration membrane and a acronitrile copolymer film supported by a polyester fiber web. A flow rate of 400 l/m^2.day was observed with 5 g/l aqueous NaCl and 30 bar differential pressure. Rejection was 91% [232]. A composite membrane was prepared from polyester cloth and acetyl cellulose [233]. Crosslinked polyethyl-enimine membranes are formed on porous substrates to give semipermeable composite membranes with water permeability 0.82 m^3/m^2.day and salt rejection 97.7%, when treating a water sample containing 5000 ppm NaCl at 40 kg/cm^2 [234]. A porous polymer membrane was coated with a mixture of butadiene-styrene copolymer latex and treated to give a sucrose rejection 81% [235].

A film-forming polymer is reacted with a monomer under specified conditions so that a crosslinked polymeric ultrathin film is formed [236]. An ultrathin semi-permeable film is deposited on the surface of a microporous substrate to form a semipermeable membrane [237]. Semipermeable composite membranes were prepared from acylonitrile copolymer. The membranes have water permeability more than 10,5 m^3/m^2.day kg/cm^2 [238].

OTHER TYPES OF MEMBRANES

Dynamically formed membranes. Porous graphite tubes were used for the manufacture of dynamic semipermeable membranes for wastewater purification. The membranes impregnated with polyacrylic acid are used successfully for the clarification of wastewaters from the Baikal cellulose plant by ultrafiltration at 10 atm. [239]. Membranes were dynamically formed from poly(acrylic acid), poly(1-methyl-2-vinyl-pyridinium iodide) and methylcellulose on a millipore filter. The primary factors that affect membrane performance were ascertained to be the kind and concentration of salts and the pH of the preparing solutions, the concentration and the polymerization degree of the membrane materials, the pore size of filters and the pressure and circulation velocity during membrane formation [240].

The properties of the dynamic layer depend on the base pore radius, solution circulation and filtration rate. The highest selectivity was observed in MGA-100 membranes with pore size less than 0.01 µ [241]. Dynamic formation of polyacrylamide membranes on a Millipore filter was optimized as to the rejection of a standard NaCl aqueous solution. Exclusion of salts with alkali cations depend on the kind of anion and is in the order

$$SO_4^{2-} > F^- > HCO_3^- > Cl^- > NO_3^- > J^-$$

Rejection of salts with alkaline earth cations is generally lower than corresponding salts with alkalies. The membranes are not useful for seawater desalination, but they exhibit a high rejection of Na_2SO_4, which is 93.6% for a 5 g/l aqueous solution at a flux of 820 $1/m^2$.day [242].

The formation of dynamic membranes was examined by using polyethylenimine, carboxymethylcellulose, hydrous Zr oxide-polyacrylic acid and gelatin. In the case of reverse osmosis membranes with the three first membrane types, the rejection of NaCl was high when they were formed in the pH range in which the viscosity of the solutions was low. The rejection of NaCl increased with increase in the concentration of NaCl in the pH range in which the viscosity decreased by the addition of NaCl. The relation between the rejection of NaCl and the water content showed that membranes with high rejection formed in the pH range of low viscosity were compact and had small water content and high electric charge density. In the case of filtration type membranes with gelatin, the self-rejection was high, when the membrane was formed in the range of pH less than 7 and was low in the alkaline solution. The rejection decreased by the addition of NaCl [243].

Hydrous Zr(IV)-oxide-polyacrylate dynamically formed membranes have been studied on microporous supports and feed crossflow velocities of 10, 15 and 25 ft/s. Only minor effects of support pore size are evident from the dynamic membrane properties, which are most significantly influenced by the crossflow velocity. At a pressure of 1000 psi, hydrous Zr(IV)-oxide membranes vary in their water flux from 300 to 1000 gfd, while the salt rejection for a 0.05M NaCl feed is 40 to 50%. Zr(IV)-polyacrylate dual treatment membranes give 70 to 100 gfd and 85-96% salt rejection in the same range of crossflow velocity and pressure. Scanning electron microscopy reveals that the dynamic membrane is 8 to 30µ thick, attaining the latter value at a crossflow velocity of 10 ft/s. The SEM also shows that the surface of the membranes is covered by many small circular craters whose diameters vary with the crossflow velocity [244]. The mechanism of formation of Zr(IV)-hydrous oxide membranes deposited under dynamic conditions on a microporous support has been examined. Observations of the decline of transmembrane flux were consistent with a two-step model of membrane formation comprising: a period of pore clogging and bridging, followed by the thickening of a colloidal cake. The selectivity of the dynamically formed Zr(IV)-hydrous oxide membrane was tested with a Rhodamine dye solution. Rejections of 85 to 95% were obtained for over 1 hour with essentially constant filtrate flux [245]. Dynamic Zr(IV)-PAA membrane hyperfiltration was carried out on high temperature dyeing wastewater in a pilot plant. Except for excessive fouling, due to the lack of proper prefiltration, the membrane withstood the high-temperature operation without major deterioration [246].

Conditions to make reproducible dynamically formed membranes were defined. Investigating the properties of these membranes, it became clear that pH affects both water and solute permeabilities, as well as the membrane stability. Membranes that have low water permeability have high rejection at high concentration [247].

Reviews on dynamic membrane formation include papers on dynamic membranes in ultrafiltration and reverse osmosis [248], dynamically formed membranes for ultra- and hyperfiltration [249] and reverse osmosis with dynamic membranes [250].

Patents. Alumina sols, alkali metal aluminates, and a water soluble polymer are deposited on porous substates under pressure and reflux to prepare membranes for reverse-osmosis or ultrafiltration. The membranes have a high-salt rejection [251].

Charge membranes. Thermodynamic effective fixed charge densities of mercuric phosphate and carbonate parchment supported membranes were evaluated by a number of methods. Membrane transport number was calculated and compared with the values determined by the TMS method. The theoretical predictions for membrane potential are borne out quite satisfactorily by experimental results for both membranes [266]. Thermodynamically effective fixed charge densities of parchment-paper-supported mercuric fungstate membranes in contact with various 1:1 electrolyte solutions were evaluated from membrane potential measurements [267]. Further work includes evaluation of thermodynamic parameters and testing of theories of membrane and bi-ionic potential based on nonequilibrium thermodynamics [268, 269, 270, 271].

A theory of thermal membrane potential presented earlier is applied to charge-mosaic membrane systems. The influence of each constituent element of the membrane on thermal membrane potential is made clear. Agreement between the theory and experimental results is shown to be satisfactory [252]. Three membrane series differing in capacity and crosslinkage of both cation- and anion-exchange resins have been studied by means of pressure tests to obtain hydrodynamic permeability coefficient of the membranes, salt enrichment coefficient, volume flux and salt flux for NaCl solution. Permeate concentration was 2.75 to 3.95 times higher than that of feed solution [253].

Porous glass membranes. Porous glass tubes prepared from heat-treated alkaliborosilicate glass show promise as semipermeable membranes in reverse osmosis. The glass tubes must be leached with HNO_3 from one side only if reasonably high salt rejections are to be obtained. This step assures the buildup of a fine structure at the surface opposite that contacting the leachant. Membrane performance is greatly improved by treating the porous glass tubes with aqueous sea salts solutions prior to testing. Initial salt rejections of up to 99% were thus achieved, but the performance deteriorated with time because of the dissolution of SiO_2 [254]. In ultrafiltration at high pressure by porous glass capillary-tube membranes, the degree of concentration of aqueous emulsions depends on the flow rate and the pore size distribution [255].

Glass hollow fiber membranes have a few advantages compared with polymer membranes. These are: high strength under compressive stresses, high stability under low pH and high-temperature conditions and resistance to reactions of bacteria and organic solutions. The two disadvantages of glass membranes are fragility of glass and complication in preparational operations.

Glass membrane is basically an ultrafilter and its performance is dependent on a large variety of parameters, including original glass composition, heat treatment and leaching processes, size and shape of pores, surface chemistry of pores, as well as chemical composition and dielectric properites of the solution. All these parameters are interdependent. Encouraging prospects of glass membrane performance would still be dependent on solving basic problems of mechanical properties of glass [210].

Patents. The formation of hollow fiber membranes from a borosilicate glass melt and the subsequent treatment of the fibers to achieve the correct pore sizes are described [256]. Boron oxide is removed leaving the membranes consisting substantially of silica [257].

LITERATURE TO 3F.

1. D.M. Cavagnaro (NTIS Rept. PS-78/0111/1GA [1978] 188 p)
2. D.M. Cavagnaro (NTIS Rept. PS-79/0090/5GA [1979] 204 p)
3. D.M. Cavagnaro (NTIS Rept. PS-78/0112/9GA [1978] 262 p)
4. D.M. Cavagnaro (NTIS Rept. PS-79/0091/3GA [1979] 261 p)
5. D.M. Cavagnaro (NTIS Rept. PS-79/0092/1GA [1979] 57 p)
6. R. Kretner, G. Dickel (Z. Phys. Chem., Wiesbaden, 105 [1977] 221/224)
7. G. Busch, K.E. Militzer, M. Schubert (Wiss. Z. Tech. Univ. Dresden 27 [1978] 1251/1254)

8. D.D. Spatz, R.H. Friedlander (Water Sewage Works 125, No 2 [1978] 36/40)
9. R. Rautenbach, K. Rauch, R. Abrecht (Verfahrenstechnik 13 [1979] 365/368)
10. P.S. Leung (Sep. Sci. Technol. 14 [1979] 167/174)
11. Anonymous (Hyomen 17, No 2 [1979] 126/137)
12. E. Tatsuata (Hyomen 15 [1977] 483/495)
13. R. Endoh (Kagaku Kogaku 42 [1978] 485/486)
14. S. Kimura (Kagaku Kogaku 42 [1978] 487/488)
15. S. Ishizaka (Seramikkusu 13 [1978] 480/485)
16. Baxter Travenol Laboratories Inc. (Brit. 1.495.670, 21 Dec 1977.- C.A. 88 [1978] 138329)
17. Nippon Oil K.K. (Brit. 1.506.665, 12 Apr. 1978)
18. J.F. Meier, J.D.B. Smith (W. Ger. 2.701.820, 21 Jul 1977.- C.A. 89 [1978] 111657)
19. J.F. Meier, J.D.B. Smith (US 4.076.626, 28 Feb 1978)
20. W. Schultheiss, K. Schmidt (French 2.372.644, 4 Aug 1978)
21. Rhone-Poulenc Industries S.A. (Belg. 867.637, 30 Nov 1978.- C.A. 90 [1979] 73718)
22. S. Prabhakar, R. Nagarajan, K.R. Rao (Proc. 4th Natl. Heat Mass Transfer Conf. [1977] 519/526.- C.A. 90 [1979] 153913)
23. V.I. Cheremisin, I.N. Taganov (Zh. Prikl. Khim. 51 [1978] 233/234.- C.A. 89 [1978] 181703)
24. E.A. Mason, L.A. Viehland (J. Chem. Phys. 68 [1978] 3562/3573)
25. A.R. Cooper, D.S. Van Derveer (Sep. Sci. Technol. 14 [1979] 551/556)
26. S.V. Joshi, D.J. Mehta (Indian Chem. J. 12, No 11 [1978] 28/32.- C.A. 89 [1978] 65916)
27. B. Simon, D. Woermann (Z. Phys. Chem., Wiesbaden, 110 [1978] 99/112)
28. H.G. Burghoff, W Pusch (J. Appl. Polymer Sci. 23 [1979] 473/478)
29. T. Matsuura, S. Sourirajan (I.E.C., Process Des. Devel. 17 [1978] 419/428)
30. C. Kamizawa (J. Appl. Polym. Sci. 22 [1978] 2867/2879)
31. M. Sugahara, Y. Kiso, T. Kitao, Y. Terashima, S. Iwai (J. Chem. Eng. Japan 11 [1978] 366/371)
32. M. Sugahara, T. Kitao, Y. Terashima, S. Iwai (Intern. Chem. Eng. 19 [1979] 322/328)
33. R. Rangarajan, T. Matsuura, E.C. Goodhue, S. Sourirajan (I.E.C., Process Des. Devel. 18 [1979] 278/287)
34. J.L. Halary, C. Noël L. Monnerie (J. Appl. Polym. Sci. 24 [1979] 985/988)
35. J.L. Halary, C. Noël, L. Monnerie (Desalination 32 [1980] 65/76.- C.R. Acad. Sci. C 286, No 25 [1978] 689/691.- 287, No 12 [1978] 491/493)
36. G. Mossa (Quad. Ist. Ric. Acque 22 [1977] 3/22)
37. A.P. Bogdanov, K.M. Saldadze (Dokl. Akad. Nauk SSSR 242 [1978] 860/863)
38. F. Moser, A. Kopp (Proc. Symp. Adv. Sep. Sci. [1978] 132/137)
39. D. Woermann (Chem. Ing. Tech. 50 [1978] 290/293)
40. Y. Nakamura (Sen'i Gakkaishi 35, No 4 [1979] P77/P85)
41. M.C. Porter (Proc. 2nd Pac. Chem Eng. Congr. 2 [1977] 975/982)
42. E. Drioli (Quad. Ist Ric. Acque 22 [1977] 179/214)
43. G. Belfort, B. Marx (Proc. Ind. Waste Conf. 1978, 33 [1979] 377/385)
44. S. Tai, K. Matsushige, S. Hariu, M. Huruichi, K. Tsuchii (Yosui to Haisui 20 [1978] 433/437)
45. S. Tai, K. Matsushige, S. Hariu, M. Huruichi, K. Doi (Yosui to Haisui 20 [1978] 675/681)
46. M. Hollier (Ind. Water Eng. 15, No 3 [1978] 20/21)
47. J.H. MacLeish (Diss. Univ. Calif., Berkeley [1978] 178 p.- Diss. Abst. B 38 [1978] 3788B/3789B)
48. K.S. Spiegler, J.H. MacLeish (AlChE 71st Ann. Meet., Miami Beach, [1978] Paper 109d)
49. V.S. Allen, F. Shippey (NTIS Rept. AD-A055624/1GA [1978] 65p)
50. J. Block (NTIS Rept. PB-287989/8GA [1977] 130p)
51. R.R. Doella, J.R. Smith (Desalination 32 [1980] 113)
52. M.T. Brunelle (Desalination 32 [1980] 127/135)
53. J.C. Schippers, J. Verdoux (Desalination 32 [1980] 137/148)
54. C. Yanagi, K. Mori (Desalination 32 [1980] 391/398)

55. S. Nakamura, Y. Misaka (Jap. 78.28.083, 15 Mar 1978.- C.A. 89 [1978] 220707)
56. M. Koyama, J. Koezuka, Y. Sato (Jap. 78.106.387, 16 Sep 1978.- C.A.90 [1979] 40871)
57. M. Koyama, J. Koezuka, Y. Sato (Jap. 78.106.388, 16 Sep. 1978.- C.A. 90 [1979] 40872)
58. M. Koyama, J. Koezuka, Y. Sato, H. Yoshida (Jap. 78.106.389, 16 Sep 1978.- C.A. 90 [1979] 40873)
59. Y. Otsuka (Jap. 78.119.786, 19 Oct 1978.- C.A. 90 [1979] 61045)
60. T. Yabushita, S. Takadono (Jap. 78.131.980, 17 Nov 1978.- C.A. 90 [1979] 123574)
61. H. Ichinose, Y. Sahara, T. Nakamoto, I. Hazama (Jap. 77.150.388, 14 Dec 1977.- C.A. 89 [1978] 8061)
62. F. Martinola (W. Ger. 2.607.997, 1 Sep 1977.- C.A. 89 [1978] 8047)
63. T. Ikeguchi (Jap. 77.78.677, 2 Jul 1977.- C.A. 88 [1978] 107401)
64. Y. Miyata (Jap. 77.156.176, 27 Dec 1977.- C.A. 89 [1978]148732)
65. Y. Miyata (Jap. 77.156.177, 26 Dec 1977.- C.A. 89 [1978] 148731)
66. S. Takatono, M. Kawasaki (Jap. 77. 156.180, 26 Dec 1979.- C.A.89 [1978] 91490)
67. S. Takatono, M. Kawasaki (Jap. 77.156.181, 26 Dec 1977.- C.A. 89 [1978] 91494)
68. S. Takatono, M. Kawasaki (Jap. 77.156.182, 26 Dec 1977.- C.A. 89 [1976] 91493)
69. S. Takatono, M. Kawasaki (Jap. 77.156.183, 26 Dec 1977.- C.A. 89 [1978] 91492)
70. F. Arai (Jap. 78.17.581, 17 Feb 1978.- C.A. 89 [1978] 91507)
71. K. Ueno, N. Ozawa, H. Oki (Jap. 78.60.380, 30 May 1978.- C.A. 89 [1978] 148507)
72. R.R. Zwack, R.M. Christenson (U.S. 4.136.025, 23 Jan 1979.- C.A. 90 [1979] 170818)
73. R.R. Zwack, R.M. Christenson (U.S. 4.153.545, 8 May 1979.- C.A. 91 [1979] 41212)
74. M. Koyama, J. Koezuka, Y. Sato (Jap. 78.125.274, 1 Nov 1978.- C.A. 90 [1979] 73030)
75. G. Blazejewska, Z. Filipowski, T. Gudra, T. Winnicki (Pol. 97.917, 31 Aug 1978.- C.A. 90 [1979] 192152)
76. K. Kamada, S. Fujii (Jap. 79.23.080, 21 Feb 1979.- C.A. 91 [1979] 62198)
77. M.I. Eman, N.E. Kuz'mitskaya, G.I. Fishman (Proizvod. Pererab. Plastmass Sint. Smol No 8 [1978] 47/51)
78. R.A. Shelden, E.V. Thompson (J. Membrane Sci. 4 [1978] 115/127)
79. R.F. Probstein, W.F. Leung, Y. Alliance (J. Phys. Chem. 83 [1979] 1228/1232)
80. N. Kinjo, M. Sato (Desalination 27 [1978] 71/80)
81. A.K. Ghosh, K.K. Sirkar (J. Appl. Polym. Sci. 23 [1979] 1291/1307)
82. K. Kamide, S. Manabe, T. Matsui, T. Sakamoto, S. Kajita (Kobunshi Kagaku 34 [1977] 205/216)
83. R.M. Livingston (U.K. Inst. Chem. Eng. Symp. Ser. No 51 [1977] 17 p)
84. T.N. Bogdanova, E.L. Akim, T.N. Matveeva, N.P. Novoselov (Izv.Vyssh. Uchebn. Zaved., Khim. Khim. Tekhnol. 21 [1978] 1635/1638.- C.A. 90 [1979] 73520)
85. T. Uragami, K. Fujino, M. Sugihara (Angew. Makromol. Chem. 68 [1978] 39/53.- C.A. 89 [1978] 26283)
86. T.A. Tweddle, S. Sourirajan (J. Appl. Pol. Sci. 22 [1978] 2265/2274)
87. G.R. Gilbert, T. Matsuura, S. Sourirajan (J. Appl. Pol. Sci. 24 [1979] 305/310)
88. R.M. Kava, H.N. Shah, V.P. Pandya, D.J. Mehta (Indian J. Technol. 15 [1977] 342/345)
89. R.M. Kava, V.P. Pandya, D.J. Mehta (Indian J. Chem. 13 No 7 [1979] 17/21)
90. K. Kataoka, T. Sakai, H. Tonami (Sen'i Gakkaishi 34, No 8 [1978] T354/T359)
91. G.J. Johnston (Polymer 19 [1978] 228)
92. L. Broens, F.W. Altena, C.A. Smolders (Desalination 32 [1980] 33/45)
93. K.D. Goebel, G.C. Berry, D.W. Tanner (J. Polym. Sci., Polym. Phys. Ed. 17 [1979] 917/937)
94. V.P. Sribnaya, D.D. Kucheruk (Khim. Tekhnol. No 2 [1979] 47/50)
95. B.M. Brown, S.L. Jones, E.L. Ray (Res. Discl. 176 [1978] 13)
96. R.F. Kesting (NTIS Rept. PB-293144/2GA [1978] 47 p)
97. T. Gulyamov, B.R. Rashidov, S.T. Shaposhnikova, G.M. Kozin, B.I. Aikhodzhaev (Zh. Prikl. Khim. 50 [1977] 2140/2141)
98. G. Mossa (Quad. Ist. Ric. Acque 22 [1977] 155/176)
99. R.E. Kesting (Pure Appl. Chem. 50 [1978] 633/641)
100. H.G. Burghoff, W. Pusch (A.Ch. S. 176th Nat. Meet., Miami Beach [1978] Paper 89)

101. C. Toprak, J.N. Agar (J. Chem. Soc., Faraday Trans. 1, 75 [1979] 803/815)
102. Hoechst A.G. (Brit. 1.486.917, 28 Sep 1977)
103. L.I. Voronkova, Yu. N. Zhilin, B.F. Nikol'skii, A.A. Svittsov, O.F. Shepelev (USSR. 636.007, 5 Dec 1978.- C.A. 90 [1979] 106192)
104. T. Sakamoto, K. Uede, S. Manabe, E. Osafune (Jap. 78.15.104, 22 May 1978.- C.A. 89 [1978] 111646)
105. S. Konomi (Jap. 78.87.982, 2 Aug 1978.- C.A. 89 [1978] 198771)
106. J. Tamura, T. Kawaguchi, T. Ono (Jap. 78.32.885, 28 Mar 1978.- C.A. 89 [1978] 185872)
107. Y. Otsuka, K. Kanki, S. Nomura, K. Okamoto (Japan 78.110.666, 27 Sep 1978.- C.A. 90 [1979] 76357)
108. D. Bartsch, H.G. Hicke, D. Paul, H.H. Schwarz, J. Tscherner, R. Fischer, R. Jaeger, E. Friedrich, D. Tischer, H. Krems (E. Ger. 130.862, 10 May 1978.- C.A. 90 [1979] 73051)
109. D. Bartsch, H.G. Hicke, D. Paul (E. Ger. 131.528, 5 Jul 1978.- C.A. 90 [1979] 55952)
110. J. Borgwardt, H.G. Hicke (E. Ger. 131.529, 5 Jul 1978.- C.A. 90 [1979] 56650)
111. O. Kutowy, W.L. Thayer, S. Sourirajan (U.S. 4.145.295, 20 Mar 1979.- C.A. 90 [1979] 188818)
112. H. Kitano, I. Sumita, Y. Sakamoto (Jap. 78.123.387, 27 Oct 1978.- C.A. 90 [1979] 105216)
113. C. Yanagi (Jap. 78.142.986, 13 Dec 1978.- C.A. 90 [1979] 188815)
114. T. Uemura , M. Kurihara (Jap. 78.144.883, 16 Dec 1978.- C.A. 90 [1979] 206565)
115. L. Nicolais, E. Drioli, C. Migliaresi, H.G. Burghoff, W. Pusch (Int. J. Polym. Mater. 6, No 3-4 [1978] 175/183)
116. H. Ohya, N. Akimoto, Y. Negishi (J. Appl. Polym. Sci. 24 [1979] 663/669)
117. O. Kutowy, T. Matsuura, S. Sourirajan (J. Appl. Polym. Sci. 21 [1977] 2051/2066)
118. A.V. Rao, D.J. Mehta (J. Appl. Polym. Sci. 22 [1978] 3559/3577)
119. T. Uragami, M. Tamura, M. Sugihara (Angew. Makromol. Chem. 66 [1978] 203/220.- C.A. 88 [1978] 91213)
120. T. Uragami, M. Tamura, M. Sugihara (Angew. Makromol. Chem. 68 [1978] 213/217.- C.A. 89 [1978] 26284)
121. S. Takada, H. Hayashi (Jap. 79.16.381, 6 Feb 1979.- C.A. 90 [1979] 205417)
122. B.J. Narola, M.V. Chandorikar (Indian Chem. J. 13, No 6 [1978] 29/31, 13, No 7 [1979] 22/24.- C.A. 90 [1979] 153132, 169791)
123. B.J. Narola, M.V. Chandorikar (Indian Chem. J. 13, No 8 [1979] 26/28.- C.A. 90 [1979] 188107)
124. B.J. Narola, M.V. Chandorikar (Indian Chem. J. 13 No 11 [1979] 23/24.- C.A. 91 [1979] 40544)
125. H.S. Pienaar, R.D. Sanderson (S. Afr. 77.04.960, 9 Oct 1978.- C.A. 90 [1979] 169869)
126. H.S. Pienaar, R.D. Sanderson (S. Afr. 77.04.961, 9 Oct 1978.-C.A. 90 [1979] 138760)
127. M. Koyama, J. Koezuka, Y. Sato (Jap. 78.127.374, 7 Nov 1978.- 90 [1979] 88469)
128. T. Sano, T. Shimomura (U.S. 4.083.904, 11 Apr 1978)
129. P. Meares (Ber. Bnusenges. Phys. Chem. 83 [1979] 342/351)
130. W.D. Munch, L.P. Zestar, J.L. Anderson (J. Memb. Sci. 5 [1979] 77/102)
131. J.K. Beasley (Ind. Water Eng. 15, No 3 [1978] 6/10)
132. C. Rossi (Quad. Ist Ric. Acque 22 [1977] 71/120)
133. A.R. Aksenova, Yu. V. Aleksandrova, E.E. Katalevskii (Plast. Massy No 11 [1978] 58/59)
134. A. Penati, M. Pegoraro (J. Appl. Polym. Sci. 29 [1978] 3213/3223)
135. A.C. Habert, R.Y.M. Huang, C.M. Burns (J. Appl. Polym. Sci. 24 [1979] 489/501)
136. J.M. Dickson, D.R. Lloyd, R.Y.M. Huang (J. Appl. Polym. Sci. 24 [1979] 1341/1351)
137. L.K. Shatayeva, G.V. Samsonov, J. Vasik, J. Kopecek, J. Kalal (J. Appl. Polym. Sci. 23 [1979] 2245/2251)
138. T. Uragami, T. Yono, M. Sugihara (Angew. Makromol. Chem. 75 [1979] 203/213)
139. M. Pegoraro, A. Penati (Quad. Ist. Ric. Acque 22 [1977] 47/70)

142

140. A. Chiolle, L. Credali, G. Gianotti, G. Parrini (Quad. Ist. Ric. Acque 22 [1977] 25/46)
141. H. Ohya, Y. Negishi, T. Yamamoto (Maku 3, No 2 [1978] 143/146)
142. J. Vacik, K. Ulbrich, J. Exner, J. Kopecek (Collect. Czech. Chem. Commun. 43 No 5 [1978] 1221/1226.- C.A. 89 [1978] 110996)
143. H. Sumitomo, K. Hashimoto, T. Ohyama (Polym. Bull., Berlin, 1, No 2 [1978] 133/136.- C.A. 90 [1979] 39486)
144. O.I. Nachinkin, S.I. Kuperman, S.D. Stroganova, N.P. Leksovskaya, N.S. Leonova (Plast. Massy No 3 [1978] 18/19.- C.A. 89 [1978] 110905)
145. S.I. Kuperman, O.I. Nachinkin, I.G. Ruban (Plast. Massy No 6 [1978] 26/27.- C.A. 89 [1978] 110972)
146. J.E. Cadotte, M.J. Steuck, R.J. Petersen (NTIS Rept. 288387/4GA {1978] 50 p)
147. P. Zschocke, H. Strathmann (Angew. Makromol. Chem. 73 [1978] 1/23)
148. V.P. Zhemkov, A.N. Cherkasov, V.S. Soldatov (Vestsi Akad. Navuk BSSR, Ser. Khim. Navuk No 4 [1978] 98/102.- C.A. 89 [1978] 164273)
149. J. M. Rovel, A. Allard (Impianti 10 [1977] 569/571)
150. D. Stevenson, A. Beeber, R. Gandiana, O. Vogl (J. Macromol. Sci., Chem. A 11 [1977] 779/809)
151. M.A. Kraus, M. Nemas, M.A. Frommer (J. Appl. Polym. Sci. 23 [1979] 445/452)
152. I.G. Ruban, O.I. Nachinkin, V.G. Zamoryanskaya, E.T. Peisikhis, I.K. Ivanova, A.N. Shuster (Plast. Massy No 4 [1979] 50/51)
153. S. Calgari, A. Ravazzoli (Quad. Ist. Ric. Acque 22 [1977] 121/131.- C.A. 89 [1978] 152488)
154. S. Calgari, A. Ravazzoli, A. Matera (Quad. Ist. Ric. Acque 22 [1977] 133/154.- C.A. 91 [1979] 21478)
155. N. Yamazaki, A. Hirao, S. Nakahama (J. Macromol. Sci., Chem. A 13 [1979] 21572)
156. Y. Sakai, H. Tanzawa (J. Appl. Polym. Sci., 22 [1978] 1805/1815)
157. T. Uraganni, T. Wakai, M. Sugihara (Angew. Makromol. Chem. 71 [1978] 17/27)
158. T. Uragami, M. Sugihara (Angew. Makromol. Chem. 71 [1978] 43/49)
159. Y. Imai, E. Masuhara, K. Takakura, O. Nakaji, S. Yamashita (Kobunshi Ronbunshu 35 [1978] 765/770.- C.A. 90 [1979] 72953)
160. H. Sato, M. Maeda, A. Nakajima (J. Appl. Polym. Sci. 23 [1979] 1759/1767)
161. A.P. Bogdanov, R.G. Gumen, K.M. Saldadze (Plast. Massy No 3 [1979] 56/58)
162. G. Morel, J. Jozefonvicz, P. Aptel (J. Appl. Polym. Sci. 23 [1979] 2397/2407)
163. O.I. Nachinkin, S.I. Kuperman, I.G. Ruban, L. Ya. Guseva (Plast. Massy No 5 [1977] 27/28)
164. O.I. Nachinkin, T.G. Shubina, I.G. Ruban, S.D. Stroganova (Plast. Massy No 4 [1979] 51/52.- C.A. 90 [1979] 205045)
165. I.A. Tumanova, I.A. Litvinov, S.G. Durgar'yan, O.B. Semenov, Yu.Ya. Podol'skii, N.S. Nametkin (Vysokomol. Soedin., Ser. A. 20 [1978] 1105/1108.- C.A. 89 [1978] 25235)
166. I.C. Cabasso (NTIS Rept. PB-290609/7GA [1979] 72 p).- I. Cabasso, C.N. Tran (J. Appl. Polym. Sci. 23 [1979] 2967/2988)
167. N. Minoura, Y. Fujiwara, T. Nakagawa (J. Appl. Polym. Sci. 24 [1979] 965/973)
168. Teijin K.K (Brit. 1.504.733, 22 Mar 1978)
169. U.O.P. Inc. (Isr. 46.824, 31 Jan 1978)
170. Alfred Rudin Ltd. (Isr.48.013, 10 Mar 1978)
171. W.J. Wrasidlo, S. Spiegelman (W. Ger. 2.743.673, 30 Mar 1978.- C.A. 89 [1978] 25623)
172. J.H. Jensen, L.E. Applegate (W. Ger. 2.747.390, 27 Apr. 1978.- C.A. 89 [1978] 111717)
173. S. Ono, K. Koyama, A. Shimizu (Jap. 78.30.661, 23 Jan 1978.- C.A. 89 [1978] 111632)
174. H. Suzuki, T. Takiguchi, Y. Midorikawa (Jap. 78.95.183, 19 Aug 1978.- C.A. 89 [1978] 198797)
175. H. Hori, H. Sekiguchi, K. Sadamitsu (Jap. 78.96.974, 24 Aug 1978.- C.A. 89 [1978] 198796)
176. Y. Midorikawa (Jap. 78.144.974, 16 Dec 1978.- C.A. 90 [1979] 153176)
177. R.D. Sanderson, H.S. Pienaar, M.J. Mackenzie (S. Afr. 77.04.962, 1 Nov 1978.- C.A. 90 [1979] 169608)

178. J.E. Cadotte (U.S. 4.039.440, 2 Aug 1977.- C.A. 87 [1977] 118989)
179. T. Shinomura (U.S. 4.100.238, 11 Jul 1978.- C.A. 89 [1978] 198739)
180. K. Okita (French 2.363.351, 5 May 1978)
181. T.J. Cochrane (W. Ger. 2.747.408, 27 Apr 1978.- C.A. 89 [1978] 111718)
182. T. Kawaguchi, Y. Hayashi, Y. Taketani, T. Ono, K. Mori (W. Ger. 2.822.784, 30 Nov 1978.- C.A. 90 [1979] 88267)
183. U.O.P. Inc (Jap. 77.127.481, 26 Oct 1977.- C.A. 89 [1978] 111675)
184. K. Yamada, S. Nagata, K. Takakura (Jap. 78.23.873, 4 Mar 1978.- C.A. 89 [1978] 25646)
185. S. Yamashita (W. Ger. 2.838.665, 8 Mar 1979.- C.A. 90 [1979] 169536)
186. F. Kamiyama, A. Nishio, H. Okazaki, O. Harima (Jap. 78.42.185, 17 Apr 1978.- C.A. 89 [1978] 90796)
187. M. Kitano, I. Sumita, Y. Sakamoto (Jap. 78.110.978, 28 Sep 1978.- C.A. 90 [1979] 61039)
188. M. Kitano, I. Sumita, Y. Sakamoto (Jap. 78.110.979, 28 Sep 1978.- C.A. 90 [1979] 61038)
189. M. Kitano (Jap 78.136.099, 28 Nov 1978.- C.A. 90 [1979] 209919)
190. K. Ishii, Z. Honda, K. Sato, H. Tsugaya (Jap. 78.130.285, 14 Nov 1978.- C.A. 90 [1979] 122696)
191. J.D. Grandine (W. Ger. 2.822.265 and 2.822.266, 25 May 1977.- C.A. 90 [1979] 88427 and 105143)
192. S. Munari, F. Vigo, G. Capannelli, C. Uliana, A. Bottino (W. Ger. 2.735.443, 9 Feb 1978.- C.A. 88 [1978] 137565)
193. S. Tokizane, M. Kurihara, T. Tanaka, M. Tanimura, Y. Shimokawa (Jap. 77.152.879, 19 Dec 1977.- C.A. 89 [1978] 185885)
194. S. Hara, K. Mori, K. Iwata (Jap. 78.22.875 17 Aug 1978.- C.A. 89 [1978] 111700)
195. S. Takenishi (Jap. 78.44.486, 21 Apr 1978.- C.A. 89 [1978] 168567)
196. S. Hara, Y. Taketani, Y. Hayashi (Jap. 78.125.980, 2 Nov 1978.- C.A. 90 [1979] 104974)
197. Y. Taketani, Y. Hayashi, T. Kawaguchi, T. Ono, K. Mori (Jap. 78.146.800, 20 Dec 1978.- C.A. 90 [1979] 138797)
198. Y. Hashino, F. Hayano, S. Fuji, K. Ito (W. Ger. 2.829.630, 18 Jan 1979.- C.A. 90 [1979] 105253)
199. J. Latty (U.S. 4.125.462, 14 Nov 1978.- C.A. 90 [1979] 156970)
200. H. Harayama, T. Nakagawa, K. Maejima (Jap. 78.11.884, 2 Feb 1978.- 89 [1978] 113232)
201. I. Sumita, M. Kitano, Y. Sakamoto (Jap. 78.110.997, 28 Sep 1978.- C.A. 90 [1979] 39848)
202. I. Sumita, H. Kitano, Y. Sakamoto (Jap. 78.123.386, 27 Oct 1978.- C.A. 90 [1979] 105215)
203. Daicel K.K. (Brit. 1.516.203, 28 Jun 1978)
204. Sumitomo Chemical Co. (Brit. 1.526.843, 4 Oct 1978)
205. Bayer A.G. (Brit. 1.527.001, 4 Oct 1978)
206. Bayer A.G. (French 2.365.361, 26 May 1978)
207. H. Ito, M. Nagasaki (Jap. 79.01.372, 8 Jan 1979.- C.A. 90 [1979] 170602)
208. S. Fukui, T. Yamamoto, K. Iida (Jap. 78.07.783, 24 Jan 1978.- C.A. 89 [1978] 111678)
209. Y. Murakami, H. Shirane (Jap. 78.26.777, 13 Mar 1978.- C.A. 89 [1978] 111708)
210. D. Bahat, J. Flicstein (J. Appl. Polymer Sci., Appl. Polym. Symp. 31 [1977] 389/395)
211. Y. Sakamoto, M. Kitano, I. Sumita (Jap. 78.132.480, 18 Nov 1978.- C.A. 90 [1979] 123621)
212. T. Sano, T. Shimomura, M. Sasaki, I. Murase (U.S. 4.107.049, 15 Aug 1978)
213. K. Ishii, R. Suzuki, Z. Honda, H. Tsugaya (Jap. 78.149.181, 26 Dec 1978.- C.A. 91 [1979] 6077)
214. N. Kinjo, T. Ishii, Y. Miyadera, H. Yokono (Jap. 79.02.279, 9 Jan 1979.- C.A. 91 [1979] 6079)
215. T. Takemura, K. Kamada, S. Minami, Y. Fujinaga (Jap. 79.04.283, 12 Jan 1979.- C.A. 90 [1979] 188174)

216. S. Onishi, S. Furnya, S. Ito (Jap. 79.08.671, 23 Jan 1979.- C.A. 90 [1979] 188205)
217. Y. Komaki, S. Tsujimura (Jap. 79.11.971, 29 Jan 1979.- C.A. 90 [1979] 188195)
218. M. Horiguchi, S. Isoda, S. Otani (Jap. 79.11.081, 26 Jan 1979.- C.A. 91 [1979] 22046)
219. K. Kamada, S. Minami (Jap. 79.15.478, 5 Feb 1979.- C.A. 90 [1979] 205408)
220. R. Oshiumi, S. Fukuchi, T. Hayashi (Jap. 79.15.479, 5 Feb 1979.- C.A.91 [1979] 22054)
221. T. Ochiumi, Y. Kihara, H. Ichinose, K. Nakamoto (Jap. 79.21.970, 19 Feb 1979.- C.A. 90 [1979] 205410)
222. H. Hori, H. Sekiguchi, T. Komiyama (Jap. 79.25.278, 26 Feb 1979.- C.A. 91 [1979] 6105)
223. K. Yamazaki (Jap. 79.26.283, 27 Feb 1979.- C.A. 91 [1979]6113)
224. V. Groebe, W. Albrecht, D. Bartsch, H.J. Gensrich, P. King, Le Viet Kim Ba, J. Ludwig, W. Makschin, D. Paul (E. Ger. 134.448, 28 Feb 1979.- C.A. 91 [1979] 58316)
225. V. Groebe, W. Albrecht, D. Bartsch, P. Klug, Le Viet Kim Ba, J. Ludwig, W. Makschin, D. Paul, H.J. Purz (E. Ger. 134.447, 28 Feb 1979.- C.A. 91 [1979] 58389)
226. Yu. A. Leikin, A.B. Davankov, V.V. Korshak, T.D. Ignatovich, T.N. Toroptseva (U.S.S.R. 664.973, 30 May 1979.- C.A. 91 [1979] 58211)
227. National Research Council and Montedison SpA, Italy (Isr. 50.135, 12 Mar 1979)
228. J.A. Latty (W. Ger. 2.837.845, 15 Mar 1979.- C.A. 90 [1979] 205398)
229. K. Elfert, H.J. Rosenkranz, G.D. Wolf, F. Bentz (W. Ger. 2.741.669, 22 Mar 1979.- C.A. 90 [1979] 205162)
230. P.V. Hinman, A.T. Bell, M. Shen (J. Appl. Polym. Sci 23 [1979] 3651/3656)
231. J.E. Cadotte, R.J. Petersen, R.E. Larson, E.E. Erickson (Desalination 32 [1980] 25/31)
232. Rhône Poulenc Industries S.A. (Brit. 1.495.887, 21 Dec 1977.- C.A. 89 [1978] 7275)
233. I. Sumita (Jap. 78.39.982, 12 Apr 1978.- C.A. 89 [1979] 185874)
234. R. Oshiumi, S. Fukuchi, Y. Sano (Jap. 78.149.874, 27 Dec 1978.-C.A. 90 [1979] 153204)
235. T. Eguchi, M. Shimokawa (Jap. 78.18.482, 20 Feb 1978.- C.A. 89 [1978] 25622)
236. U.O.P. Inc. (Isr. 49.261, 17 Dec 1978)
237. U.O.P. Inc. (Isr. 50.384, 12 Mar 1979)
238. F. Hayano, Y. Hashino, K. Ichikawa (Jap. 78.132.478, 18 Nov 1978.- C.A. 91 [1979] 23229)
239. Yu. I. Dytnerskii, G.V. Terpugov, N.M. Trapeznikov, V.A. Ovchinin (Tr. Mosk. Khim.- Tekhnol. Inst. im D.I. Mendeleeva 93 [1977] 107/110.- C.A. 90 [1979] 209581)
240. M. Igawa, M. Seno, H. Takahashi, T. Yamabe (J. Appl. Polym. Sci. 22 [1978] 1607/1618)
241. T.V. Knyaz'kova, A.A. Kavitskaya, L.A. Kul'skii (Dopov. Akad. Nauk Ukr. RSR. Ser. B. Geol., Khim. Biol. Nauki No 1 [1979] 33/35.- C.A. 90 [1979] 139241)
242. S. Antoniou, J. Springer, A. Grohmann (Desalination 32 [1980] 47/55)
243. K. Haraya, T. Nakane, H. Yoshitome (Nippon Kaisui Gakkaishi 32, No 4 [1978] 197/205)
244. D. Freilich, G.B. Tanny (Desalination 27 [1978] 233/251)
245. J.A. Mikhlin, G.B. Tanny (J. Coll. Interface Sci. 68 [1979] 157/162)
246. A.M. El-Nashar (NWSIA 5th Ann. Meet., San Diego [1977] 21p)
247. T. Nomura, S. Kimura (Desalination 32 [1980] 57/63)
248. G.B. Tanny (Sep. Purif. Methods 7 [1978] 183/220)
249. E. Drioli (Recent Dev. Sep. Sci. 3, Part B [1977] 343/354)
250. M. Igawa (Nippon Kaisui Gakkaishi 31, No 2 [1977] 52/60)
251. H. Takahashi, M. Igawa (Jap. 79.02.276, 9 Jan 1979.- C.A. 90 [1979] 209920)
252. M. Tasaka (J. Membr. Sci. 4 [1978] 51/59)
253. T. Winnicki, G. Blazejewskaya, A. Mika-Gibala (Desalination 32 [1980] 77/89)
254. T.H. Elmer (Am. Ceram. Soc. Bull. 57 [1978] 1051/1053, 1060)
255. R. Schnabel (Fette, Seifen, Anstrichm. 81, No 2 [1979] 83/88)
256. Jenaer Glaswerk Schott & Gen. (Brit. 1.487.109, 28 Sep 1977)

257. Janaer Glaswerk Schott & Gen. (Isr. 48.373. 31 Aug 1978)

258. H. Ohshima, T. Mitsui (Seibutsu Butsuri 18, No 3 [1978] 120/129.- C.A. 90 [1979] 157501)

259. M. Nakagaki, S. Kitagawa (Yakugaku Zasshi 98 [1978] 840/849.- C.A. 89 [1978] 95486)

260. H. Strathmann (Chem. Ztg. 103, No 6 [1979] 211/219)   =

261. M.G.A. Khedr (Chem. Ing. Tech. 51 [1979] 756/757)

262. M. Tamura, T. Uragami, M. Sugihara (J. Membrane Sci. 4 [1979] 305/314.- C.A. 91 [1979] 6621)

263. M. Tamura, T. Uragami, M. Sugihara (Angew. Makromol. Chem. 79 [1979] 67/77.- C.A. 91 [1979] 58995)

264. T. Uragami, K. Maekawa, M. Sugihava (Desalination 27 [1978] 9/20)

265. T. Uragami, K. Maekawa, M. Sugihava (Polymer 19 [1978] 1437/1440.- C.A. 90 [1979] 187705)

266. F. Siddiqi, P. Prakash, S.P. Singh (Colloid Polym. Sci. 256 [1978] 552/562)

267. F. Siddiqi, M. Nasim-Beg, P. Prakash, S.P. Singh (Indian J. Chem. A 16, No 1 [1978] 7/11.- C.A. 89 [1978] 31302)

268. F.A. Siddiqi, M. Nasim-Beg, M.I.R. Khan, A. Haq, S.K. Saksena, B. Islam (Can. J. Chem. 56 [1978] 2205/2215.- C.A. 89 [1978] 186474)

269. F.A. Siddiqi, M. Nasim-Beg, P. Prakash (J. Polym. Sci., Polym. Chem. 17 [1979] 539/550)

270. F.A. Siddiqi, M. Nasim-Beg, A. Haq, M. Aqueel-Ahsan, M.I.R. Khan, S.A. Khan (J. Chim. Phys. 76 [1979] 57/60)

271. M. Nasim-Beg, F.A. Siddiqi, M.I.R. Khan, M. Aqueel-Agcan, B. Islam (J. Membrane Sci. 4 [1978] 275/281)

3G. Reverse Osmosis Process

An empirical modeling method has been suggested for the reverse osmosis process. Least-square fitting of data to a third-order polynomial has resulted in the accurate modeling of Du Pont's hollow fiber B-10 module for water desalination. The general modeling methodology is not specific to this system but can readily be used to model other available membrane configurations and solute-water systems. The simple mathematical form of the empirical model allows its use in scale-up, process optimization and economic studies of the reverse osmosis process [1]. A correlational dependecne, which determines the relative value of the coordination number of solution-component hydration, was derived by using experimental data of reverse-osmosis separation of multicomponent aqueous concentrated solutions [2].

In the technological and economical evaluation of various reverse osmosis units, the major structural systems and materials tested were one hollow fiber unit, two spiral wound units and one tubular unit. The process was reliable. The performances of both hollow fiber and spiral wound elements were close to predictions and that of the tubular element was promising [3]. Physico-chemical criteria of reverse osmosis separation of inorganic compounds from waste water were established, based upon extensive experimental studies on five different types of membrane materials with a large number of inorganics. Because of problems encountered with membrane fouling and deterioration, a pretreatment scheme was proposed prior to the reverse osmosis removal of inorganics [4].. In a morphological approach to desalination by reverse osmosis the relation of the physical properties of membranes to their structure and of the latter to their permeation and selection properties were discussed [5].

Plant site, water intake, pretreatment, choice of materials of construction, design alternatives and energy recovery are important variables to be considered in the design of seawater desalination plants employing reverse osmosis modules [6]. Four commercial modules (Roga-4000, Westinghouse, Raypak and Du Pont B9) were studied for specification and prediction of their performance. Three sets of calculated data were reported for the operating pressure of 400 psig. The first shows good agreement between calculated and experimental results on the performance of the modules with aqueous sucrose feed solutions. The second set of data shows the

variations in solute separation and membrane productivity for each of the four modules, as functions of volumetric fraction product water recovery and membrane compaction for a 3000 ppm NaCl-water feed solution. The third set of data shows the
variations in the solute separation as a function of solute transport parameter at
different levels of mass transfer coefficient on the high pressure side of the membrane for very dilute aqueous feed solutions [7]. Tables and graphical presentations
of data on the performance of various membranes, lines of research and development
which are or ought to be followed and the economic benefits of their application in
various industries were presented [8].

Low pressure desalination for various brackish and seawater feeds were carried
out with improved cellulose acetate membranes. The replacement of acetone by dioxane improves the performance of the membranes [9]. A reverse osmosis cluster module producing as much as 70000 gpd from a single 12 inch diameter shell has been
developed, tested and placed into service. One pressure vessel with a single feed
inlet and single product and brine outlets contains seven parallel streams. The
relatively small individual membrane elements can be tested for productivity and
salt rejection without removal from the vessel. Thus, membrane can be replaced in
75 ft^2 elements rather than more costly large diameter elements. The module can be
used for both brackish and seawater treatment [10]. The selection, design, construction and operating procedures for a portable demonstration water treatment system
using reverse osmosis and electrodialysis to provide information, including operating conditions and costs, to small communities for improving their drinking water
supplies to meet existing regulations were described [11].

For the desalination of seawater by means of reverse osmosis, factors influencing the design, shipment, operation and maintenance of systems for offshore installations were presented. In addition, field data concerning the technical and economical aspects of the operation of reverse osmosis systems on offshore installations
are given. The significance and practicality of seawater desalination via reverse
osmosis technology are illustrated [12]. An experimental reverse osmosis seawater
system has been in operation for two years at Eilat on the Red Sea shore. Simplified seawater pretreatment, comprising only sand filtration and acid dosing, was
applied. Hollow fiber and spiral wound membranes were tested. Investment and
water cost analysis of a 4000 m^3/day seawater desalting system under various operating conditions is given. The resulting water production cost is approximately
Dollar 1.1 per m^3 [13].

A comparison of piston pumps, centrifugal pumps and rotary positive displacement pumps for use in low-capacity reverse osmosis plants was given with detailed
information on Worthington, Mustang, Gardner-Denver, Goulds, Sunflo and Moyno designs, based on actual performance tests. The recommended materials of construction for such pumps were given [14].

The cost for seawater desalting by reverse osmosis is approximately Dollars 4.0
per 1000 gallons of product water. Present technological product, process and engineering capability can be used in high recovery full two-stage systems to reduce
these costs by 40% to approximately Dollars 2.60 per 1000 gallons of product water,
meeting the 500 mg/ l total dissolved solids upper limit. The above approach produces substantial savings in power requirements, membrane replacement and pretreatment capital and operating costs, which far outweigh the added capital amounts to
be amortized [15].

In the foreseeable future reverse osmosis is expected to grow at a rate of between 20 and 30% and will constitute the major portion of new brackish water desalination capacity. Seawater reverse osmosis is expected to grow at an even faster
rate. The reverse osmosis growth will be sustained in at least two ways: expansion
of installed capacity based on the current state of the art and innovations leading
to better membranes and plant design, which will make reverse osmosis even more
economically attractive [16].

Other work, mostly reviews, includes present status of application of the reverse
osmosis water treatment process [17], state of the art of seawater desalination by
hyperfiltration [18], present status of membrane technology, especially the reverse
osmosis method [19], state of the art, experience and perspectives of water desalting by reverse osmosis [20], general design of experimental reverse osmosis desalination units [21], current work of the Reverse Osmosis Division of KIWA, Netherlands [22], economics of boiler feed pretreatment by reverse osmosis versus ion

exchange [23], thorough desalination of water by electroionization after reverse osmosis [24], combined process of reverse osmosis and electrodialysis methods [25], use of membrane processes for separating molecular mixtures [26], reverse osmosis for preconcentration [27], processing of river water by reverse osmosis [28], high temperature reverse osmosis technology [29], operating experience in desalting brackish water by means of reverse osmosis [30] and four years experience in preparing water for pharmaceutical preparations with reverse osmosis [31].

Patents. The arrangements of solution flow channels across the supports and of inlet and outlet channels to and from an apparatus for membrane filtration are described [32]. A reverse osmosis system with automatic pressure relief valve was described [33]. A desalination system with by-product recovery consists of a first reverse osmosis unit, the reject stream of which is passed through a second reverse osmosis unit, operating at a much higher pressure, and then to a distillation unit [34]. Pretreated brackish water is processed in an electrodialysis sybsystem and then to a reverse osmosis subsystem. Water of high salt content is first passed through the reverse osmosis subsystem and then processed in the electrodialysis subsystem [35]. Water is passed through a magnetic field prior to reverse osmosis to increase membrane life and to improve desalination [36].

A fabric structure is presented for supporting two semipermeable membranes in a sandwich structure that is wrapped around a tube [37]. An apparatus, giving a high surface area and easy replacement, consists of osmosis membranes arranged around the axis of rotation of a rotatable housing [38].

In submarine seawater desalination by means of reverse osmosis, a relative velocity between the membrane and the surrounding seawater created or existing, natural velocities are increased [39]. Seawater is desalinated by using spirally wound membranes in a small, narrow, lightweight apparatus, which is connected to a pump and a freshwater collector by long flexible lines. The apparatus can be removed to the surface for maintenance without disconnection from the pump [40]. Seawater is desalinated in an underground reverse osmosis installation, using the hydrostatic head of the influent seawater to provide the differential pressure required for reverse osmosis [41].

Membranes are mounted on the inner and outer surfaces of a cylindrical porous support, inside a pressure vessel [42]. Porous disks support semipermeable membranes on both sides. Details of construction are given [43]. Porous disks are stacked within a pressure vessel, each disk supporting a semipermeable membrane on each face [44]. Strips of cellulose acetate are affixed to both sides of long bands of porous material [45]. Modular system of membranes, porous carriers and turbulization inserts are described [46]. An ultrafiltration system uses a simple or multistage jet pump to feed the suspension or solution into the closed recirculation circuit [47]. An apparatus for use on ships is designed to produce permeate of constant salt content, independently of variations in the salinity of the feed [48]. An installation for production of potable water, comprising a fixed component and a replaceable component, consisting of an ion exchange unit, is described [49].

A reverse osmosis or ultrafiltration apparatus is used for water desalination or for purifying water containing colloids [50]. A reverse osmosis apparatus prepares water containing 3000 to 15000 ppm NaCl for treatment of fish stored aboard a fishing vessel [51]. A solution is subjected to reverse osmosis to recover fresh water and the effluent is evaporated to produce more water and simultaneous recovery of byproducts [52]. A reverse osmosis apparatus consists of at least two stages at 50 kg/cm^2 each. A portion of the treated water is used in successive treatments and the water from the last unit is recovered entirely. The concentrated solutions from each stage are returned to the previous stage and the concentrated solution from the first stage is discarded [53]. The design of an apparatus avoids damage to the membranes during assembly and use [54].

A reverse osmosis unit has a portable water container, which receives the treated water from the modules via a diameter-reduced channel providing fluid back pressure on the reverse osmosis modules [55]. An apparatus consists of a hollow cylinder, semipermeable folded membrane, coaxial channel for collecting liquid, membrane spacers and seals [56]. Water, desalted by reverse osmosis and then irradiated with

UV in the presence of H_2O_2 to oxidize organic compounds, is further demineralized by a mixed bed ion exchange resin to prepare ultrapure water for electronic or pharmaceutical uses [186].

PRETREATMENT OF FEEDWATER

Pretreatments of seawater for reverse osmosis desalination by manganese zeolite filtration, coagulation/flocculation in an upflow sludge blanket precipitator and by in-line coagulation were tested. In-line coagulation was the most effective treatment but required air scouring to prevent mudball formation in the sand filter [57]. Ion-exchange softening as a practical pretreatment method for membrane desalting of water was conducted. The concept demonstrated was one of reducing Ca in raw SO_4^{2-} fouling concentration by ion exchange, desalting by reverse osmosis and using the recovered reject brine for regeneration of the exhausted ion-exchange resin. The test results demonstrated that ion exchange pretreatment of water can be operated as a closed-loop exhaustion-regeneration cycle. Only the reject brine from reverse osmosis need be used to regenerate the resin [58].

A 10000 gal/day skid-mounted alumina-lime-soda pilot plant has been constructed and operated to establish operating parameters for removal of dissolved silica from reverse osmosis brines. Removal of silica foulant permits further water recovery in additional stages. Results showed that dissolved silica could be routinely reduced to less than 12 mg/l with the alumina-lime-soda process. Common heavy metals were nearly quantitatively removed by the process under optimum conditions for silica removal. Costs of the process range from $0.30 to $0.35 per 1000 gal for waters initially containing 30 to 50 mg/l of silica [59].

The UVOX process, which is essentially the retention of chlorinated or ozonated seawater for a period of about 3 days in open ponds, has been proved to be effective in improving the treatability of nutrient-rich recalcitrant seawater to the extent that simple sand and/or diatomaceous earth filtration will yield a water appropriate for reverse osmosis desalination. The process eliminated the need for chemical coagulation and relatively expensive clarification systems [60].

The possibilities of removing salts from Rhine water by means of reverse osmosis depend considerably on the efficiency of the pretreatment techniques. Experiments showed that this water can be so purified by coagulation with ferric salts that tubular membrane systems can be used. An additional treatment is necessary for spiral-wound and hollow fiber membranes. Experiments with in line coagulation have shown that interesting resulcts cen be achieved. This process is seriously disturbed by the presence of humic substances, because they react with the cation-active polymer added. Humic substances can be removed by adjusting the addition of ferric salts in the preceding coagulation process [61].

The filtration mechanism in pretreatment of seawater for desalination by reverse osmosis was discussed and analytical methods for understanding the suspended matter, which affects the quality of seawater, were evaluated [62]. Off the shelf reverse osmosis systems have proved to be expensive to operate on brackish waters in Saudi Arabia, unless proper pretreatment techniques were employed. Improper pretreatment can raise the operational and maintenance costs of reverse osmosis systems from Dollars 0.65 to 11.57 per 1000 gallons for the worst operational instance [63].

Patents. The seawater is pretreated with chelating ion-exchange resins containing functional groups to remove Ca and Mg components. Desalination efficiency and service life of membranes are improved [64]. Waste water containing calcium and fluorine ions is pretreated with an appropriate aluminum or iron compound to form insoluble complexes with the scale forming constituents. After their separation, the water is treated before it undergoes reverse osmosis [65]. To prevent scale formation on membranes, the feed water is treated by multistage reverse osmosis at 15 to 60 kg/cm^2 and the brine is softened and then recycled to the reverse osmosis treatment process [187].

ENERGY CONSIDERATIONS

The performance of each of six large operating reverse osmosis systems in the U.S. for desalting water was analyzed to determine the desalting capabilities, operating costs, energy efficiency and methods for improving the energy efficiency. The units investigated used from 9 to 18 kWh per 1000 gallons of product water. The energy efficiency of reverse osmosis as currently operated is superior to other forms of commercially available unit processes to achieve 90 to 95% salt reduction. The energy efficiency of reverse osmosis could be improved by 30% or more by improving the pumping systems, using higher performance membranes, using waste heat and energy recovery devices and by optimizing the ratio of permeate produced to concentrate discarded [66].

In reverse osmosis desalination plants energy is wasted with the discharged flow of concentrated brines. Schemes of hydraulic turbine coupled with electric generator, tied to the electric power supply, were considered for various plant sizes from 1 to 10 Mgd. The analysis presents the interrelationships between the cost of money, cost of electrical energy, and recovery factors for the differnet plant sizes and for seawater and brackish water systems. The results serve as a guide to determine when such a power recovery system should be seriously considered [67]. Similar considerations aim to recuperate the largest part of energy in the discharge by means of an hydraulic turbine, use this energy at the axis of the pump and consider pump and turbine as two compartments of the same machine with one axis [68].

An attempt has been made to develop a desalting unit employing reverse osmosis and a photoelectric generator to actuate the pump. Experimental results have been obtained and the operation of individual elements of the unit are analyzed [69].

Commercially available wind turbines appear promising for producing electrical energy at costs ranging from 5 down to 1.8 cents/kWh. It is possible to reach costs for desalting brackish water at less than $1.00 per 1000 gal with large wind energy/desalination systems. The coupling of wind energy turbines to electrodialysis and reverse osmosis also offers technological advantages, such as variable energy outputs from wind turbines, to optimize variable fluid flows in the desalination processes for maximum.economic production of potable water.[70]. A conceptual design of solar assisted reverse osmosis plant was developed, wherein the necessary power is provided by the combination action of a diesel engine and a solar thermal energy driven Sterling engine. The design parameters for this desalting plant were identified and a preliminary economic analysis was made [71].

Patents. In an energy conserving system, means are provided that the hydrostatic head of the salt water creates a differential pressure across the membrane sufficient for osmotic separation of the salt water [72]. The use of a pressure pump in the feeding conduct and a separation pump in the circulation system enables separation control of the pressure and the flow velocity [73]. An additional liquid-jet pump connected to the ultrafiltration apparatus is made in such a way that the pump is driven by the circulating stream, thereby reducing the power requirements and the maintenance costs [74]. A device for performing reverse osmosis is described, characterized by low energy consumption. The apparatus can be powered manually or by wind energy [75].

PLATE AND FRAME SYSTEM

The first reverse osmosis systems developed in the U.S. were plate and frame systems, which were not very successful. Similar systems were developed in Europe with satisfactory performance. De Danske Sukkerfabrikker has marketed the DDS-RO system for industrial use, mainly in food, dairy and pharmaceutical industries, as well as in the pulp and paper industry. A review of the development of plate and frame systems is given. The system was applied for the desalination of brackish water and seawater for drinking purposes, desalination of high-salt content ground water for market gardens with a special demand on water quality and for purification of bleaching effluents from the sulfate process [76].

Another module was developed at the Gesellschaft für Kernenergieverwertung in Schiffbau und Schiffahrt. A program, ESPRO-I, computes structure and balances of water desalting plants using the Series G of plate modules, but might also be used for other systems. The plant has a single stage and the modules are arranged to one, two and three series. A new version, computer program ESPRO-II, includes additional parameters, such as investment, capital and operational costs [77].

Separation of multi-component solutions by reverse osmosis in flat chamber type apparatus was also reported in the U.S.S.R. [78].

The effect of slip-velocity at a porous surface was studied in detail for a parallel flat membrane system assuming fully-developed flow. The effect of slip coefficient on velocity profiles, pressure gradient and concentration polarization in ultrafiltration was examined. The equations of motion are solved by the regular perturbation method. The coupled diffusion equation in the boundary layer is solved using a finite difference scheme [188].

Patents. The apparatus consists of module plates, membrane plates and connecting channels in a stack [79]. A light-weight unit, that can be used on board aircraft, consists of a stack of cells, each consisting of two flat membranes, separated by a porous flexible support [189].

TUBULAR MODULE

Two membranes, 2-cellulose acetate and 2,3-cellulose diacetate, were tested in a tubular module for relation of flow rate-pressure-concentration-membrane characteristics in NaCl solution and seawater. The increase of NaCl concentration of feed water resulted in decrease of product flux and salt rejection. There was a good correlation between desalination characteristics of seawater and NaCl solution. By using 2-stage reverse osmosis of seawater for 400 h continuous operation, the Cl^- remaining was 1820 ppm in the 1st stage and 128 ppm in the 2nd stage from an initial 16300 ppm. The membrane was coated with Na, Si, K, Fe, and Cr after continuous operation. To remove this coating Na dithionate, oxalic acid and citric acid were effective [80].

The free convective motion in a horizontal circular pipe, which is superimposed upon the main axial flow, is caused by buoyancy forces arising from the buildup of a dense solute boundary layer near the membrane surface. The three-dimensional convective diffusion problem is solved by dividing it into a perturbation part accounting for the buoyancy effects present and a nonperturbation part for the intrinsic convective flow pattern present. An approximate solution to the nonperturbation equations is obtained from the literature, and the perturbation equations are solved using a stream function-vorticity scheme valid for high Schmidt numbers. Correlations are developed for the asymptotic Sherwood number and the effective axial length at which free convection becomes significant. The numerical results are in reasonable agreement with limiting analytical solutions and with the experimental asymptotic Sherwood numbers reported in the literature [81].

Calculation of actual fluxes in reverse osmosis desalination through membrane tubes involves considerable iteration due to the existence of concentration polarization. Explicit flux expressions have been developed for such cases. Low levels of solvent flux in existing membranes allow approximations leading to such explicit expressions. The computer flow diagram for reverse osmosis plant design with membrane tubes in series has been simplified by means of the explicit flux expressions [82]. The effect of slip velocity at a membrane surface was studied in detail for a tubular membrane system. A second-order perturbation solution of the equations of motion is found to be very satisfactory. As in the case of channel flow system, the effect of slip coefficient on concentration polarization is identical to that of Peclet number. It augments diffusive transport of solute molecules from the membrane surface to the bulk solution [83].

In the fabrication process of asymmetric tubular cellulose acetate membranes the chief determining factor of membrane performance is the amount of evaporation of solvent in the casting solution. To measure the evaporation amount a test sample of a membrane was collected by using a device at each section of the tube. The evaporation amount of acetone and membrane performance were measured with the membranes fabricated by various methods of casting and the relationship between them studied from their distribution curve. The vapor of acetone evaporated from a membrane caused a downward stream in the tube. Therefore a casting bob was modified to prevent the stream and maintain the atmosphere in the tube constant. In the fabrication process using a modified bob, the rate of evaporation and evaporation time were kept constant over the whole region of a tube to obtain a uniform tubular membrane [84].

A 4.2 gpm (about 6000 gpd) tubular reverse osmosis pilot plant was operated at Pomona Advanced Wastewater Treatment Research Facility for over one year to study the membrane life. Repeated membrane failure resulting from poor mechanical design of the system obviated drawing significant conclusion [85]. Reverse-osmosis units using tubular cellulose acetate, spiral-wound cellulose acetate and spiral-wound polyamide membranes were tested for wastewater treatment for recycle. The control of temperature and pH was important. The membranes were restored by soaking in demineralized water. Water recovery of over 70% will be difficult with wastewater containing total dissolved solids of over 4 g/l [86].

Other work includes reverse osmosis tubular plant performance under continuous 30 days operation [87] and a description of a tubular reverse osmosis plant [88].

Patents. A tubular membrane on the inside of a porous support is wound into a helical form. The conditions for heat treating the membrane during the winding step are specified [89] . A membrane is coiled on the outer surface of a porous tubular support [90]. The semipermeable tubular element comprises a core, covered by a layer or layers of threads or fabric, onto which the membrane is applied [91]. The tubular membrane elements move within the pressure vessel to minimize the effects of concentration polarization [92]. The membrane filter apparatus is composed of many parallel tubes that are connected in series. The membranes, placed inside the tubes, are flat strips rolled to form a long cylinder with overlapping edges [93]. The tubular membranes are mounted together as a number of small bundles, which are fixed in end plates [94]. A tubular membrane filtration device comprises a semiporous core with channels for the feed water and channels for the purified water. The core is enclosed in a pressure box. The channels for the incoming water are covered with semipermeable membranes. Rotational movement of the core causes turbulence and elimates concentration polarization [95].

Several tube-type reverse osmosis elements are placed in a pressure vessel and the adjacent tubes are connected to form fluid paths [96]. A high pressure feed solution is supplied to the surface of these elements to yield a permeated solution [97]. A tubular membrane is prepared from cellulose acetate [98, 99]. A tubular membrane apparatus consists of several modules connected from low to high concentrations in multiple stages [100]. A helically wound tubular membrane has a corrugated spacer, which provides space for the flow of the feed solution [102]. Several layers of membranes are formed inside tubular supports and treated to yield cylindrical filtration modules [103]. A tubular membrane prepared from a PVC and polyethylene glycol solution, showed a water permeation of 8 cm^3/min at 0.5 kg/cm^2 [104]. The membrane is supported on the outside of a narrow (1.5 to 8 mm) extruded tube of PVC [105]. A porous support for a tubular membrane is formed from a strip of fibrous material [106]. The edge of a tubular fibrous layer, on which a membrane is mounted, is heat sealed to prevent fluid flow across the edge of the layer [107]. Membrane tubes comprise an inner fibrous non-woven tube, an outer support tube and the membrane itself [108]. Tubular membranes are cleaned by forcing sponge balls through the tubes using a carrier liquid[109].

Each layer of porous filler material is sandwiched between two semipermeable membranes. The sealing and connecting arrangements to achieve serpentine flow path are described [110]. Tubular ultrafiltration membranes with improved durability were prepared by coating the inside of a tubular polyester nonwoven fabric with mixtures containing polysulfone [111]. Spherical foams of polyvinyl formal having 10 to 30% OH-groups are used for washing the internal surface of tubular membranes [112]. Seawater is filtered with a 50μ pore filter to have less than 0.1 ppm turbidity and then processed in a tubular or other reverse osmosis apparatus [113].

SPIRAL WOUND MODULE

A detailed engineering and economic study was undertaken on spiral wound and hollow fiber reverse osmosis module designs and a tortuous path electrodialysis

stack design used in membrane desalting plants with product water capacities of
1,5 and 10 million gallons per day. The minimum pretreatment required, based on
each of the six brackish waters, was determined to achieve water recoveries of 70,
80 and 90%. The preferred pretreatment was acidification. Plant layout arrangements, process flow and piping and instrumentation diagrams were prepared in detail
for use as a basis for preparing capital and annual cost estimates. The impact of
reduced pressure operation, interest rate, electric power cost and membrane life on
product water costs was determined. The cost data indicated that membranes and
modules averaged 26.5% of total capital costs and 19.4% of annual costs. Operating
and maintenance labor required 23.4% of annual costs while electricity and chemicals
were 13.8% and 11.9% respectively. A research and development program was outlined
to reduce costs and improve reliability [114].

New thin film composite spiral-wound membrane system, designated PEC-1000, formed
by the acid catalyzed polymerization on the surface of a reinforced-porous supporting membrane, make it possible to produce potable water from seawater by reverse
osmosis in a single-stage with a high recovery operation. TBS rejection over 99.9%
and stable water fluxes of 0.20 to 0.30 m^3/m^2-day have been attained with 3.5% synthetic seawater at an applied pressure of 56 kg/cm^2. For brackish water, sodium
chloride rejections of 99.6 to 99.9% and fluxes of 0.61 to 0.81 m^3/m^2.day have been
attained with 5000 ppm sodium chloride feed at an applied pressure of 40 kg/cm^2.
TDS rejection of 99.8% and water flux of 0.30 m^3/m^2.day have been attained with two-
or four-inch diameter PEC-1000 composite membrane elements at an applied pressure
of 56 kg/cm^2 in a single-stage synthetic seawater desalination test. This performance is kept for more than 1500 hours in PEC-1000 thin film composite membrane and
two-inch diameter element. 280 ppm in TDS and water flux of 0.11 m^3/m^2.day are
observed at an applied pressure of 56 kg/cm and 40% water recovery. This membrane
shows high selectivity for low molecular weight valuable organic materials such as
ε-caprolactam, dimethylsulfoxide, dimethylformamide. The thickness of ultrathin
salt barrier of the composite membrane was found to be 300 Å [115].

A 56.8 m^3/day pilot plant was operated at the Pomona Advanced Wastewater Treatment Research Facility on secondary effluent to establish the effective membrane
life, to determine the reliability of the process performance and to derive a realistic process cost estimate. A cost estimate for a 10 Mgd plant indicated that for
membranes with only one-year life the process cost was about 57.4 cents per 1000 gal.
However, the cost could be substantially reduced to 41.3 cents per 1000 gal for membranes with two-year life. Both cost estimates did not include the costs for carbon adsorption pretreatment and brine disposal.[116].

Patents. A module for reverse osmosis and ultrafiltration consists of spirally
wrapped membranes with passageway for the solution and permeate on a hollow mandrel
that has a transverse portion [117].

A cylindrical housing contains convoluted spirally wound membrane elements,
arranged in parallel. The model is designed to give short channel lengths and hence
low pressure drops [101].

HOLLOW FIBER MODULE

The manufacture of hollow fibers is described, stressing the importance of the
furan resin components which acts as the rejecting barrier of the membrane and which
is chemically very stable. Reverse osmosis using hollow fiber membranes has several
industrial uses [118]. This membrane is comprised of a furan-based rejecting barrier
deposited on and near the surface of a microporous polysulfone hollow fiber of dimensions typically 250μ O.D. and 75μ I.D. The rejection barrier is a negatively charged, cross-linked furan resin formed in situ. The composite membrane system generally resembles the so-called NS-200 type, but significant changes in both form and
chemistry have resulted from work carried out [119].

In the development of hollow fiber membranes, new applications to problems in
agricultural chemistry, in ion exchange and in biology and medicine are reviewed.
Both cation- and anion-exchange fibers are evaluated for continuous ion exchange.
Medical applications which are examined include the use of fibers as sorbent carriers and as vascular supplies for tissue culture. Finally, some new developments in

reverse osmosis and ultrafiltration fibers are summarized [120]. Pretreatment of feed solution to hollow fiber modules is important, especially elimination of organic and inorganic colloidal substances [121].

In previous work composite plasma-polymer hollow fibers were prepared in a batch operation using an inductively coupled plasma polymerization reactor. In the present work a coupled plasma polymerization reactor has been used to coat polysulfone hollow fibers in a continuous fashion. Experimental data are presented which show that composite reverse osmosis membranes for salt rejection can be made by deposition of plasma polymer on hollow fibers in a continuous manner and at high fiber take-up rates [122]. A study of the effects of changing plasma polymerization variables on the properties of the resultant plasma polymer has been carried out. Salt rejections of 87 to 93% have been achieved at fluxes of 1.5 to 4.0 gal/ft^2.day. The reverse osmosis properties obtained were sensitive to the nature of the hollow fiber, presumably because of surface pore size [123].

Composite polysulfone hollow fibers consisting of a polysulfone porous substrate coated with crosslinked polyethyleneimine or furan resin are analogous to the flat-sheet composite membranes known as NS-100 and NS-200. Scanning electron microscope observations and reverse osmosis transport studies showed that the support fiber must have surface pore diameters of less than 0.2μ to obtain a durable composite hollow fiber membrane. Since both the dense layer and surface of the porous substrate contract when exposed to the curing temperature, it was found to be profitable to cure the hollow fiber before applying the coating. When tested in a reverse osmosis rig, the hollow fiber bundles displayed 98% salt rejection and a flux of 5 to 7 gal/ft^2.day for a feed solution of 10000 ppm NaCl at a hydraulic pressure of 400 psi. A new method of depositing furan resin on the polysulfone hollow fiber is described. The furfuryl alcohol is instantaneously polymerized by exposing the alcohol-soaked fiber to a 60% solution of concentrated sulfuric acid. It has been demonstrated that in such a polymerization procedure a dense, semipermeable layer is formed on top of the porous substrate; the resulting composite hollow fiber membrane yields salt rejections higher than 98% when tested under the above reverse osmosis conditions [124].

Cellulose triacetate hollow fiber membranes with high performance, especially high salt rejection, have been developed for one pass seawater desalination. High salt rejection makes it possible to operate under severe operating conditions of high product water recovery and high salinity seawater [125].

A 50000 gpd pilot plant based on cellulose triacetate hollow fibers was constructed to demonstrate the economics and feasibility of reverse osmosis for seawater desalination. Development and testing of 8 inch modules (2500 gpd) was successful and the product cost of Dollars 3.40 to 4.00 per 1000 gallons, based on three year membrane life, was low enough to assure that reverse osmosis will gain acceptence in seawater desalting [126].

Patents. A process for making hollow fibers is described [127]. An assembly of permeable hollow fibers is supported at one end by a tube sheet, cast from resin [128]. The hollow fiber permeator consists of an elongate, ordered bundle of hollow fibers, which pass through a central tubesheet and are formed around a perforated feed tube containing a coaxial permeate conduit, which is connected to permeate collecting bores in the tubesheet. The assembly is enclosed in a casing. A typical permeator with cellulose triacetate fibers produces 172 to 245 gal/min, operating at 250 to 400 psig [129]. Hollow fiber membranes comprising ethylene copolymers with 5 to 20% vinyl monomer show superior permeabilities to organic compounds, when compared to polyethylene hollow fibers [130].

Manufacture of the polymer to spun hollow fibers and the control of the wall porosity are described [131]. Porous reinforcement is used in a separation apparatus with hollow fibers [132]. A desalination hollow fiber membrane from acronitrile-vinyl chloride copolymer, without salt rejection capacity, was treated to give a membrane having salt rejection of 93% CaCl$_2$ [133]. The body of a filter module is formed by winding resin-impregnated fabric directly onto a bundle of hollow semipermeable fibers [134]. Hollow fibers were prepared by spinning a cellulose acetate solution into a precipitating bath [135].

Preparation of hollow fibers from cellulose diacetate and treatment gave water transmission 2.64×10^{-4} ml/h.mm.cm^2, compared with 1.95×10^{-5} for the untreated fibers [136]. A fluid separator contains a bundle of hollow fibers [137]. In the multistage separation, hollow fiber membranes must be used at least in the first stage, while tubular apparatus should be used as the last stage [138]. Hollow acetate fibers were prepared to give water permeation 130 l/m^2.day and solute removal ratio 96% [139]. Hollow fibers from ethylene-vinyl alcohol copolymer gave water permeation 40×10^{-2} ml/cm^2.h.atm [140]. Hollow fibers from cellulose acetate and Me carbitol gave water permeability 350 l/m^2.day and 90% salt rejection [141].

The hollow fibers, which may be wound around a perforated support tube, are covered with a finely divided filter material [142]. At least one constriction in its transverse cross-section is applied in a hollow fiber permeator [143]. The components of the casting solution [144] and an improved process for making membranes with selective permeability are specified [145]. The tube plate for hollow fine fiber apparatus is formed of a solid cylindrical block of hardened resin [146].

A selfsupporting unitary structure comprises rigid, porous, inorganic hollow filaments arranged in a three-dimensional crisscrossing network. The structure can be made by coating a yarn with a paste of sinterable composition, e.g. Al_2O_3, mullite, cordierite etc. [147]. An apparatus is described for removing gas from interior of hollow fiber permeability apparatus [148]. A hollow fiber bundle separatory device exhibits improved fluid flow characteristics and more uniform wetting of the available membrane surface [149]. The selectivity of the separation of fluid mixture by hollow polysulfone fibers is increased by storing them in unsterilized tap water over one day [150]. The membranes are cast in the presence of organic or inorganic microfilaments and then dissolving out the filaments [151]. Preparation of hollow fiber membranes from cellulose ester is described [152]. Ethylene copolymer hollow fiber membranes are useful for the separation of phenols from wastewater [153]. Hollow rinal fibers were prepared by spinning together a liquor containing poly(vinyl alcohol) and a coagulating gaseous medium [190].

HYGIENIC CONSIDERATIONS

From the results of the control experiment in identifying organic micropollutants, it can be concluded that only a small part of the organics isolated from the samples originated in the membrane modules. Most of the compounds detected are plasticizers and these compounds are ubiquitous in environmental samples [154].

OPERATING EXPERIENCE

France. Operating experience is reported on various plants erected in the Balearic Islands, on board seagoing vessels and yachts, as well as on platforms [155].

Germany W. A seawater desalination test station has been installed on board the nuclear research vessel Otto Hahn [156]. A reverse osmosis pilot plant was erected at the island of Helgoland. The plant has 600 units in the first stage and 100 units in the second totalling 157 m^2 [157]. Two pilot plants were erected to investigate the optimum connection, the load factor and means to overcome operation troubles [158].

India. One year operating experience is reported for a 10 m^3/day capacity tubular reverse osmosis plant [159]. Research and development effort at Central Salt and Marine Chemicals Research Institute on reverse osmosis and electrodialysis during the past decade has been reviewed. Some of the current activities in tackling the pressing needs of potable water to the rural masses in India are outlined [160].

Italy. Pilot plants for desalination by reverse osmosis are reviewed [161].

Japan. Highly satisfactory and stable performances have been recorded for the two-stage reverse osmosis desalination process using modules developed in Japan. Construction of a demonstration plant of 500 m^3/day using domestic modules is being planned [162]. For the purpose of constructing a reverse osmosis seawater desalination plant of 800 m³/day, a series of tests have been carried out in the Chigasaki laboratory. Domestic modules showed good and stable performance during long term operation. The water recovery ratio of these modules has been raised to 40% [163]. Reverse osmosis plants erected in Japan were reviewed [164]. The 800000 gpd ROGA spiral plant was put on stream in late 1971 and it has been expanded to the present 3.5 Mgd. Development of the most appropriate pretreatment process and application of specially determined indexes to evaluate the pretreated feed water quality have greatly contributed to maintaining the membrane performance at satisfactory levels. A 98% plant availability, module replacement of some 5 to 6% per year and flux decline slope of -0.02 can be achieved with present engineering techniques [165]. A comprehensive review of the operation, maintenance, steps taken for trouble-shooting and performance data of this plant were presented [166].

Netherlands. Hyperfiltration of ground and surface water, prevention of membrane fouling and membranes used were reported [167].

Saudi Arabia. Lacking supplies of potable water the country depends upon the desalination of brackish well water or seawater to satisfy the increasing demands, which are expected to reach 430 Mgd in 1980 and 700 Mgd in 1985 [168]. Pretreatment and reverse osmosis units in Riyadh with a total capacity of 2730 m³/h inflow were described. In order to make maximum use of water resources, the water losses were limited to 12% [169]. The design, testing, installation and commissioning of the Ras Al Mishab seawater desalination plant with a capacity of 75000 gpd are summarized [170]. The technical design and construction of the new water treatment plants in Riyadh, with a total capacity of about 254000 m³/day, were presented. The treatment of Minjur deep-well water includes not only chemical softening but also demineralization with reverse osmosis as a final purification stage [171]. A description is given of the world's largest seawater desalination plant at Jeddah with a capacity of 12000 m³/day [172]. A hollow fiber desalination plant with a capacity of 568 m³/day was erected in a concrete plant [173].

Sweden. Membrane technology research at Lund Institute of Technology is reported. Experimental investigations performed are mainly in the following fields: dairy projects, dewatering of liquid foods, effluent treatment and water purification. Basic research is mainly devoted to concentration polarization studies [174].

United Kingdom. Field performance of spaghetti reverse osmosis modules in the application to brackish water, primary and secondary settled sewage and seawater was discussed. Potable water was produced from brackish waters with TDS up to 3000 ppm. With seawater, modules were operated for over 3000 h with little feed pretreatment. but flux declined 40% because of compaction and fouling. With sewage, flux declines were up to 70% within 24 h but were restored partially with detergent flushing [175].

U.S.A. Design details are described for both the treatment and supply components of the 3 Mgd hollow fiber reverse osmosis plant at Cape Coral, Florida [176]. The history of the 1 Mgd reverse osmosis brackish water desalination plant at Venice, Florida, is presented. This plant is significant because it is operating well after almost three years with the original membranes [177]. A reverse osmosis plant with a capacity of 700000 gpd is operating at Lake Killarney, Bahamas, to desalt the lake water with a salinity of 17000 ppm [178]. Case histories of three reverse osmosis seawater plants with Permasep permeators are presented. Critical operating data are included for each case [179]. Experiences with reverse osmosis demineralizing for boiler feed water were reported at Willow Glen Power Station [180] and for water control and recycle using ERDA's Rock Flats Plant in Golden, Colorado [181]. The purpose and results of the test activities both for the evaluative testing of the various membrane desalting units and for the development of the most suitable and economic pretreatment were outlined. In the desalting area the methods used to evaluate long term reverse osmosis flux data were specifically discussed. In the pretreatment area process selection and the importance of reactor pH in the partial lime softening process were emphasized [182]. The design and construction approach and the primary mechanical and control features of the Yuma plant were described.

156

Items covered are plant size, control block size, recovery, plant factor, plant split and amount and size of equipment to be furnished. Also covered are plant design and layout, plant piping, pumping, acid handling and energy recovery equipment. The plant equipment arrangement is shown and the plant computer and manual control system is described [183]. Increased product water recovery was obtained using interstage ion exchange and a Spiractor with simulated reverse osmosis concentrate of high calcium-high sulfate water, similar to projected Yuma Desalination Facility reject. The self-sustaining system consisted of ion exchange softening of interstage concentrate using as regenerant only the final reject from secondary reverse osmosis and no imported salt. No antiscalant was used for protection of either the ion exchange or reverse osmosis systems. The level of regeneration was enhanced by reuse of previously used regenerant, desupersaturated with respect to gypsum by a fuidized seed bed crystallizer, a Spiractor [184].

Venezuela. A review of the design and operation of the 800000 gpd reverse osmosis seawater system installed at the Punta Moron Station is given [185].

LITERATURE TO 3G.

1. M.N. Aschauer, E.S.K. Chian (AlChE Symp. Ser. 74, No .172 [1978] 209/217)
2. L.S. Aksel'rod, V.I. Fedorenko, B.K. Sokolov, O.Yu. Timofeeva (Zh. Fiz. Khim. 53 [1979] 1030/1032)
3. P. Glueckstern, M. Greenberger (Israel Nat. Coun. R & D Rept. NCRD 8-76 [1977] 85/109)
4. H.H.P. Fang, E.S.K. Chian (AlChE Symp. Ser. 73, No 166 [1977] 137/143)
5. R.D. Sanderson, H.S. Pienaar (NCRD Pub. 8-78 [1978] 405/427)
6. H.W. Pohland (Desalination 32 [1980] 157/167)
7. T.A. Tweddle, W.L. Thayer, T. Marsuura, F.H. Hsieh, S. Sourirajan (Desalination 32 [1980] 181/198)
8. K.C. Channabasappa (NCRD Pub. 8-78 [1978] 377/402)
9. A.K. Ghosh, K.K. Sirkar (J. Appl. Polym. Sci. 23 [1979] 1291/1307)
10. N.S. Call, W.M. King, R.A. Tidball (NWSIA 5th Ann. Conf. San Diego, Calif. [1977])
11. D.B. Wilson, H.G. Folster, G. Kramer, S. Hanson, W. Boyle (NTIS Rept. PB-285963/5GA [1978] 74 p)
12. D.L. Lyftogt, R.R. Callaway (Offshore Technol. Conf. 4 [1978] 2507/2512)
13. P. Glueckstern, M. Wilf, Y. Kantor (Desalination 30 [1979] 235/245)
14. L. Spinosa, A. Rozzi (Quad. Ist. Ric. Acque 22 [1977] 333/347)
15. S.S. Kremen (Desalination 30 [1979] 59/68)
16. J.K. Beasley (Desalination 30 [1979] 69/74)
17. E. Chiriea (Energetica, Bucharest, 25 [1977] 367/375)
18. H.U. Demisch, W. Pusch (Fortschr. Ber. VD1-Z, 47 [1977] 148/168)
19. A. Watanabe, K. Umeda (Japan Food Sci. 17, No 10 [1978] 26/35)
20. K.W. Böddeker (Chem. Ztg. 103, No 6 [1979] 221/228)
21. R.M. Abu-Eid, A.M. Hassan, A. Fakhoury, A.L.A. Malik, F. Butt, B. Al- Sederawi (Ann. Res. Rep. Kuwait Inst. Sci. Res. [1977] 70/74)
22. J.C. Schippers (KIWA res. progr. rep. No 5 [1977] 6 p)
23. J.W. Mason (Triton Tech. Rep. No 1 [1977] 12 p)
24. A.K. Reshetnikova, A. Sh. Shayakhmetor, A.V. Severin, N.S. Kobeleva (Nauch. Tr. Kuban Un-t 232, No 2 [1977] 108/110)
25. T. Yamabe (Kagaku Kogaku 42, No 9 [1978] 491)
26. H. Strathmann (Chem. Tech., Heidelberg 7, No 8 [1978] 333/347)
27. D. Pepper (Chem. Eng., London, No 339 [1978] 916/918)
28. N.J. Ray, M.A. Jenkins, A. Coates (VGB Kraftwerkstechnik 58 No 3 [1978] 213/220)
29. Y. Hayashi (Kagaku Kogaku 42, No 9 [1978] 489)
30. A. Weise (Tech. Mitt. Krupp 36, No 2 [1978] 33/40)
31. B. Certain, C. Lavaux, Y. Le Cudonnec, G. Petit (Sci. Tech. Pharm. 7 [1978] 201/206)
32. W. Saupe (U.S. 4.062.721, 13 Dec 1977)
33. D.T. Bray (U.S. 4.077.883, 7 Mar 1978.- C.A. 88 [1978] 176978)
34. F.E. Conger (U.S. 4.083.781, 11 Apr 1978)

35 F.E. Conger (U.S. 4.141.825, 27 Feb 1979.- C.A. 91 [1979] 27146)
36. G. Mach (W. Ger. 2.750.783, 24 May 1978.- C.A. 89 [1978] 117491)
37. R. Bairinji, T. Tanaka (W. Ger. 2.829.893, 1 Feb 1979.- C.A. 90 [1979] 139430)
38. T.L. Siwecki, G.B. Andeen (W. Ger. 2.837.489, 15 Mar 1979.- C.A. 90 [1979] 156967)
39. B.C. Drude, E. Klapp, T. Peters (W. Ger. 2.719.907, 9 Nov 1978.- C.A. 90 [1979] 92217)
40. B.C. Drude, E. Klapp, T. Peters (W. Ger. 2.722.975, 23 Nov 1978.- C.A. 90 [1979] 92219)
41. D. Hebden (S. Afr. 77.00.499, 28 Jul 1978.- C.A. 90 [1979] 156954)
42. Hoechst A.G. (Isr. 43.135, 31st Jan 1978)
43. Hoechst A.G. (Brit. 1.504.464, 22 Mar 1978)
44. U.O.P. Inc (Isr. 48.121, 15 Jan 1978)
45. U.O.P. Inc (French 2.325.405, 22 Apr 1977.- C.A. 88 [1978] 138322)
46. M. Skrabak, E. Vavrik, S. Kolarik, M. Mazak (Czech. 175.008, 15 Nov 1978.- C.A. 90 [1979] 106252)
47. D. Hauffe (Belg. 866.503, 27 Oct 1978.- C.A. 90 [1979] 139410.- W. Ger. 2.718.882)
48. I.M. Tseitlin, A.K. Orlov, R.G. Milovidov (USSR. 548.293, 21 Mar 1977)
49. C. Roiz Noriega, J.A. Garcia Arroyo (Span. 455.286, 1 Jan 1978.- C.A. 89 [1978] 152537)
50. T. Maeda, H. Hotta, H. Tejima, Y. Yoshii (Jap. 77.119.481, 6 Oct 1977.- C.A. 89 [1978] 30540)
51. S. Itoi, H. Kimura (Jap. 77.123.545, 17 Oct 1977.- C.A. 89 [1978] 30532)
52. Stone and Webster Engineering Corp. (Jap. 78.09.278, 27 Jan 1978.- C.A. 89 [1978] 203960)
53. I. Niitsu (Jap. 78.22.876, 2 Mar 1978.- C.A. 89 [1978] 135626)
54. A.H. Clemens, P.H. Chang (U.S. 4.092.223, 30 May 1978)
55. P.D. Maples (U.S. 4.110.219, 29 Aug 1978.- C.A. 90 [1979] 192338)
56. Y. Sakaguchi (W. Ger. 2.825.698, 21 Dec 1978.- C.A. 90 [1979] 206323)
57. A.C. Epstein (Proc. Int. Water Conf. Eng. Soc. West. Pa. 39 [1978] 149/161)
58. R.J. Eisenhauer, R.W. Schiller (NTIS Rept. PB-286972 [1977] 23 p)
59. A.D. Tippit, E.P. Shea, J.W. Nebgen (NTIS Rept. 290755 [1978] 80 p)
60. D. Hebden, G.R. Botha (Desalination 32 [1980] 115/126)
61. J.C. Schippers, J. Verdouw, J.M. Hofman (Desalination 32 [1980] 103/112)
62. T. Matsumura, I. Furuta, M. Takeda, H. Tsuge, Y. Sugino (Desalination 32 [1980] 93/101)
63. R.R. Doelle (Desalination 30 [1979] 317)
64. H. Oya, H. Uejima, T. Kudo (Jap. 78.30.482, 22 Mar 1978.- C.A. 89 [1978] 203962)
65. Hitachi K.K. (Brit. 1.521.362, 16 Aug 1978)
66. Larson and Associates (Rept. ORNL/TM-6735.- NTIS Rept. PC A12/MF A01 [1979] 257 p)
67. R. Singh, S.V. Cabibbo (Desalination 32 [1980] 281/296)
68. L. Goulvestre, C. Lepert (Desalination 32 [1980] 297/302)
69. I.G. Savchenko, B.V. Tarnizhevskii, T.M. Kotina, A.M. Rozhdestvenskii (Gelio-tekhnika 13,No 4 [1977] 54/59.- Appl. Solar En. 13, No 4 [1977] 42/46)
70. E.A. Cadwallader, J.E. Westberg, W.R. Williamson (NTIS Rept. 276174 [1977] 68 p)
71. A.M. El-Nashar, A.A. Husseiny (Desalination 32 [1980] 239/256)
72. J.W. Chenoweth (U.S. 4.125.463, 14 Nov 1978.- C.A. 90 [1979] 156953)
73. B.J. Eiselein (W. Ger. 2.622.461, 24 Nov 1977.- C.A. 88 [1978] 138321)
74. D. Hauffe (W. Ger. 2.718.882, 2 Nov 1978.- C.A. 90 [1979] 139413)
75. B.G. Keefer (W. Ger. 2.812.761, 28 Mar 1978.- C.A. 90 [1979] 43640)
76. W.K. Nielsen, R.F. Madsen, O.J. Olsen (Desalination 32 [1980] 309/326)
77. G. Gassenmaier, J. Kaschemekat (GKSS Rept. 79/E/10 [1979] 52 p)
78. L.S. Aksel'rod, V.I. Fedorenko (Khim. Mashinostroenia No 8 [1977] 131/136)
79. Gesellschaft fuer Kernenergieverwertung in Schiffbau und Schiffahrt m.b.H. (Jap. 77.73.189, 18 Jun 1977.- C.A. 89 [1978] 65105.- W. Ger. 2.556.210)
80. K. Mori, H. Tsuge, C. Yanagi, M. Takeda, S. Takahashi (R & D Kobe Seiko Giho 28, No 3 [1978] 62/67)
81. C.Y. Chang, J.A. Guin (AlChE J. 24 [1978] 1046/1054)
82. G.H. Rao, K.K. Sirkar (Desalination 27 [1978] 99/116)

158

83. R. Singh, R.L. Laurence (Intern. J. Heat Mass Transfer 12 [1979] 731/737)
84. M. Matsuda, C. Kamizawa, H. Yoshitome (Desalination 27 [1978] 41/50)
85. C. Chen, R.P. Miele (NTIS Rept. PB-287117/6GA [1978] 37 p)
86. M. Scherm, H.E. Mynhier (Proc. Int. Water Conf., Eng. Soc. West. Pa. 38 [1978] 211/228)
87. M.V. Chandorikar, A.V. Rao (Chem. Concepts 5, No 12 [1978] 19/25)
88. P.C. Freschi (Chim. Ind., Milan, 60 [1978] 138/140)
89. Oxy Metal Industries Corp. (Isr. 45.370, 30 Dec 1977)
90. Rhône-Poulenc Industries (Isr. 47.646, 30 Dec 1977)
91. R. Brien. M. Pages (USSR. 528.020, 31 Jan 1977)
92. V.D. Volgin (USSR. 528.101, 1 Oct 1976)
93. Fr. Krupp G.m.b.H (Neth. 76.10.852, 13 Apr 1977.- C.A. 88 [1978] 107201)
94. C. Jonas, M. Karl (E. Ger. 129.043, 28 Dec 1977.- C.A. 90 [1979] 40607)
95. R.F. Connelly (W. Ger. 2.754.627, 5 Oct 1978.- C.A. 90 [1979] 28859)
96. N. Fukushima, K. Nakamura, T. Okada, H. Totani (Jap. 77.124.478, 19 Oct 1977.- C.A. 89 [1978] 45438)
97. N. Fukushima, K. Nakamura, T. Okada, H. Totani (Jap. 77.124.479, 19 Oct 1977.- C.A. 89 [1978] 48768)
98. S. Takedono, H. Iwahori, M. Ohta (Jap. 78.00.948, 13 Jan 1978.- C.A. 89 [1978] 26338)
99. H. Iwahori, S. Ishii, S. Takedono (Jap. 78.103.982, 9 Sep. 1978.- C.A. 89 [1978] 216510)
100. T. Ogawa, K. Ebara, S. Takahashi, S. Komori (Jap. 78.06.106, 4 Mar 1978.- C.A. 89 [1978] 113350)
101. N.S. Call (U.S. 4,083.780, 11 Apr 1978)
102. Daicel K.K. (Brit. 1.508.156, 19 Apr 1978)
103. Y. Inukai, S. Tada, M. Suefuji (Jap. 77.156.778. 27 Dec 1977.- C.A. 89 [1978] 91491)
104. A. Shimizu, Y. Hirose, N. Tayama (Jap. 77.137.466, 16 Nov 1977.- C.A. 89 [1978] 30543)
105. G.M. Gale (U.S. 4.100.064, 11 Jul 1978)
106. Wafilin B.V. (Brit. 1.519.991, 1.519.992, 2 Aug 1978)
107. Wafilin B.V. (Brit. 1.521.045, 9 Aug 1978)
108. Wafilin B.V. (Brit. 1.521.287, 16 Aug 1978)
109. Hitachi K.K. (Brit. 1.516.792, 5 Jul 1978)
110. U.O.P. Inc. (Brit. 1.517.220, 12 Jul 1978)
111. H. Yoshino, S. Takada, M. Goto, M. Iznmi (Jap. 79.14.376, 2 Feb 1979.- C.A. 90 [1979] 205406)
112. S. Takahashi, K. Ebara, H. Murakami, S. Kitagawa (Jap. 79.32.181, 9 Mar 1979.- C.A. 91 [1979] 59490)
113. K. Mori, H. Tsuge, T. Masaki (Jap. 79.51.982, 24 Apr 1979.- C.A. 91 [1979] 62538)
114. S.V. Cabibbo, D.B. Guy, A. Ko, A. Ammerlaan, R. Singh (NTIS Rept. PB-294080/7GA [1979])
115. M. Kurihara, N. Kanamaru, N. Harumiya, K. Yoshimura, S. Hagiwara (Desalination 32 [1980] 13/23)
116. C. Chen, R.P. Miele (NTIS Rept. PB-288197/7GA [1978] 71 p)
117. W.J. Schell (Can. 1.047.413, 30 Jan 1979.- C.A. 90 [1979] 170609)
118. R.B. Davis, M.J. Coplan, A.E. Allegrezza, R.D. Bruchesky (Inform. Chimie No 163 [1977] 171/174)
119. M.J. Coplan, R.B. Davis, J.H. Beale, A.E. Allegrezza, G. Lopatin (NTIS Rept. PB-287990/6GA [1978] 124 p)
120. E. Klein (J. Appl. Polym. Sci., Appl. Polym. Symp. 31 [1977] 361/381)
121. H. Ludwig (Fortschr. Ber. VDI.-Z - R3, No 47 [1977] 169/184)
122. N. Morosoff, B. Hill, H. Yasuda (A. Ch. S. 176th Nat. Meet., Div. Polym. Chem [1978] 549/552)
123. N. Morosoff (NTIS Rept. PB-285383/6GA [1978] 119 p)
124. I. Cabasso, A.P. Tamvakis (J. Appl. Polym. Sci. 23 [1979] 1509/1525)
125. T. Ukai, Y. Nimura, K. Hamada, H. Matsui (Desalination 32 [1980] 169/178)
126. R.D. Ammons (NTIS Rept. PB-291158/4GA [1978] 284 p)
127. F.M. Aspin (U.S. 4.056.418, 1 Nov 1977)

128. G.B. Clark (U.S. 4.061.574, 6 Dec 1977)
129. G.B. Clark (U.S. 4.080.296, 21 Mar 1978.- C.A. 89 [1978] 94857.- W.Ger. 2.806.
     222, 17 Aug 1978.- C.A. 90 [1979] 56838)
130. A.K. Fritzsche, R.L. Leonard (U.S. 4.082.658, 4 Apr 1978.- C.A. 89 [1978] 44849)
131. Asahi Kasei Kogyo K.K. (Brit. 1.506.785, 12 Apr 1978)
132. E.A. McLain, D.N. Dean (W. Ger. 2.814.326, 12 Oct 1978.- C.A. 90 [1979] 40622)
133. E.J. Kiser, J.A. Latty (W. Ger. 2.816.088, 2 Nov 1978.- C.A. 90 [1979] 122675)
134. F.N. Karelin, A.A. Askerniya, A.B. Kosminskii (USSR. 523.699, 17 Nov 1976)
135. L.P. Perepechkin, B.L. Biber, M. Ya. Ivan, G.A. Budnitskii, R.A. Kutnova,
     T.A. Drozdova, V.G. Volkova (USSR. 614.135, 5 Jul 1978.- C.A. 89 [1978] 112117)
136. K. Isawa, E. Murakami (Jap. 77.123.983, 18 Oct 1977.- C.A. 88 [1978] 106424)
137. K. Watanabe, T. Kikuchi (Jap. 77.126.681, 24 Oct 1977.- C.A. 88 [1978] 107207)
138. T. Tsukamoto (Jap. 77.128.888, 28 Oct 1977.- C.A. 89 [1978] 113347)
139. K. Hamada, J. Takada, K. Numata (Jap. 78.19.423, 22 Feb 1978.- C.A. 89 [1978]
     25625)
140. K. Yamanouchi, T. Tanaka, S. Kawai, H. Tanii (Jap. 78.82.669, 21 Jul 1978.-
     C.A. 89 [1978] 130714)
141. E. Kuzumoto, H. Matsumoto, T. Ukai (Jap. 78.82.670, 21 Jul 1978.- C.A. 89 [1978]
     185876)
142. Bayer A.G. (Brit. 1.526.183, 27 Sep 1978)
143. Nippon Zeon Co. Ltd (French 2.367.520, 16 Jun 1978)
144. T.J. Cochrane (French 2.368.289, 23 Jun 1978)
145. J.H. Jensen, L.E. Applegate (French 2.368.290, 23 Jun 1978)
146. Dow Chemical Co. (French 2.380.051, 13 Oct 1978)
147. R.A. Baker, G.D. Forsythe, K.K. Likhyani, R.E. Roberts, D.C. Robertson (U.S.
r   4.105.548, 8 Aug 1978.- C.A. 90 [1979] 56857)
148. N. Kaneko, Y. Joh (U.S. 4.108.764, 22 Aug 1978)
149. C.W. Walter (U.S. 4.140.637, 20 Feb 1979.- C.A. 90 [1979] 192580
150. R.C. Chang, R.R. Ward (U.S. 4.157.960, 12 Jun 1979.- C.A. 91 [1979] 59319)
151. R. Stemme, B. Schilling (W. Ger. 2.750.897, 17 May 1979.- C.A. 91 [1979] 59306)
152. M. Mishiro, S. Kasai (W. Ger. 2.827.012, 21 Dec 1978.- C.A. 91 [1979] 22914)
153. Monsanto Co. (French 2.380.053, 8 Sep 1978.- C.A. 90 [1979] 205443)
154. F.C. Koplfer, W.E. Coleman, R.G. Melton, R.G. Tardiff, S.C. Lynch, J.K. Smith
     (Ann. N.Y. Acad. Sci. 298 [1977] 20/30)
155. H. Lerat (Desalination 32 [1980] 201/210)
156. K.W. Böddeker, W. Hilgendorff, J. Kaschemekat (GKSS Rept. 77/E/17 [1977] 4 p)
157. Anonymous (Blick Dch. D. Wirtsch., Beil. Frankf. Allg. 20 [1977] 5)
158. A. Weise (Tech. Mitt, Krupp Werksber. 36, No 2 [1978] 33/40)
159. B.M. Misra, K.C. Thomas (Proc. 4th Natl. Heat Mass Transfer Conf. [1977] 537/545)
160. D.H. Mehta, A.V. Rao, K.P. Govindan (Desalination 30 [1979] 325/335)
161. C. Carrieri, P. Mappelli, G. Boari, M. Santori (Quad. Ist. Ric. Acque 22 [1977]
     217/230)
162. Y. Kunisada, Y. Murayama (Desalination 27 [1978] 333/344)
163. Y. Kunisada, H. Kaneda, M. Hirai, Y. Murayama (Desalination 30 [1979] 337/345)
164. Y. Taniguchi (Chem. Econ. Eng. Rev. 10, No 1 [1978] 18/25)
165. Y. Taniguchi, K. Horio (NWSIA J. 5, No 1 [1978] 29/37)
166. K. Horio (Desalination 32 [1980] 211/220)
167. J.C. Schippers (PT-Procestech. 33 [1978] 503/512)
168. R.R. Doelle, C.K. Wojcik (Desalination 30 [1979] 315)
169. P. Treille (Proc. Int. Water Supply Assoc. Congr. 12 [1978] S9/S12)
170. T. Dupree, B. Andrews (NWSIA 5th Ann. Conf., San Diego, Calif.[1977])
171. H.E.A. Ghulaigah, B. Ericsson (Desalination 30 [1979] 301/314)
172. C.E. Hickman, I. Jamjoom, A.B. Riedinger, R.E. Seaton (Desalination 30 [1979]
     259/281)
173. Anonymous (Chem. Rundschau 32, No 3 [1979] 5)
174. G. Eriksson, P. Eriksson, B. Hallström, R. Wimmerstedt (Desalination 27 [1978]
     81/89)
175. R.H. Knibbs, R.G. Gutman (Inst. Chem. Eng. Symp. Ser. 54 [1978] 125/138)
176. E.E. Shannon, T.R. Smallwood, F.A. Eidsness (AWWA Proc 97th Ann. Conf. 1,
     10-4 [1977] 25 p.- R.M. Quinn (NWSIA 5th Ann. Conf., San Diego, Calif. [1977]
177. M. Lamendola, C.S. Miller (NWSIA 6th Ann. Conf., Sarasota, Florida [1978] 7 p)

160

178. R.A. Tidball (NWSIA 6th Ann. Conf., Sarasota, Florida [1978] 12 p.- Water
     Sewage Works 125, No 2 [1978] 69)
179. D.C. Brandt (NWSIA 6th Ann. Conf., Sarasota, Florida [1978] 11 p)
180. M. Hollier (Proc. Int. Water Conf., Eng. Soc. West. Pa. 1977, 38 [1978] 277/282.-
     Ind. Water Eng. 15, No 3 [1978] 20/21)
181. R.T. Heizer, C.E. Plock (NWSIA J. 5, No 2 [1978] 21/26)
182. C. Van Hoek, J.D. Mavis (NWSIA 5th Ann. Conf. San Diego, Calif. [1977])
183. F. Engstrom, I. Taylor, L. Haugseth, D. Bell, W. Sattler, R. Hogg (NWSIA 5th
     Ann. Conf. San Diego, Calif. [1977])
184. A.B. Mindler, S.T. Bateman (NTIS Rept. PB-287920/3GA [1978] 131 p)
185. R. Quinn (Desalination 32 [1980] 179)
186. T. Ogoshi (Jap. 78. 149.873, 27 Dec 1978.- C.A. 90 [1979] 192358)
187. K. Tsukamoto (Jap. 79.11.882, 29 Jan 1979.- C.A. 91 [1979] 27147)
188. R. Singh, R.L. Laurence (Int. J. Heat Mass Transfer, 22 [1979] 721/729)
189. J. Fourcas, G. Rodet (W. Ger. 2.811.826, 21 Sep 1978.- C.A. 90 [1979] 56847)
190. A. Sueoka, T. Okamoto, A. Omori, S. Kawai (Jap. 79.27.025, 1 Mar 1979.-
     C.A. 91 [1979] 22101)

3H. Other Applications of Reverse Osmosis and Ultrafiltration

Reverse osmosis is not used only for the preparation of potable water. A great deal of other applications have made reverse osmosis and ultrafiltration a generally accepted unit operation process, especially in the treatment of wastes. With the existing know-how and development of membranes of improved quality, membrane processes might now solve many problems in production, recycling and environmental protection. As these activities are out of the main scope of present literature survey, only a very condensed summary of recent publications will be given in the following.

DOMESTIC AND MUNICIPAL WASTES

A bibliography with 213 abstracts has been compiled with citations from the NTIS data base covering the treatment of sewage and industrial wastes for the period 1964 to July 1978 [1]. A similar bibliography from Engineering Index data base contains 97 abstracts in Volume 1 on the treatment of sewage for the period 1970 to July 1978 [2] and 95 abstracts in Volume 2 on the use of membranes in industrial waste water treatment for the period 1976 to July 1978 [3].

The renovation of waste water for injection to provide a fresh water barrier against seawater infiltration into the ground water basin at the Orange County, known as Water Factory 21, was described [4, 5].

Other work includes treatment by tubular reverse osmosis membranes of waste treatment plant effluents [6], demineralization of carbon-treated secondary effluent by spiral-wound reverse osmosis process [7], development and evaluation of a two-step membrane filter method for fecal coliform recovery in chlorinated sewage effluents [8], recycling for nonpotable use of complex wastewaters by means of ultrafiltration [9], anaerobic digestion and membrane separation of domestic waste water [10], ultrafiltration to meet discharge regulations and to reduce waste pump-out [11], potential of permeate from the reverse osmosis treatment of sewage effluents as a source of potable water [12], reverse osmosis renovation of secondary effluent [14], application of reverse osmosis to reclamation of municipal wastewaters [15] and development of water reuse treatment system for municipal wastewater [16].

Patents. A suspension of a feed solution, e.g. sewage, and surface-active solid microparticles is ultrafiltered while it is being circulated. Compressed air is bubbled into the filtration tank for stirring [153].

SPACECRAFT WASTES

The effect of operating parameters on the performance of the hyperfiltration membrane when operating on spacecraft washwater was examined. Data taken included rejections of organic materials, ammonia, urea and an assortment of ions. The membrane used was a dual layer, polyacrylic acid over zirconium oxide, deposited in situ on a porcelain ceramic substrate [17].

RECYCLING OF INDUSTRIAL WASTES

A review was given of wastewater treatment plants in various chemical industries [18]. Reviews and papers on the subject include ultrafiltration in industrial waste water technology [19], treatment of industrial wastewater by reverse osmosis for reuse [20], ultrafiltration in the treatment of chemical wastes [21], ultrafiltration for coal gasification processes [22], reverse osmosis pilot plant for treating washing waste solutions from nuclear power plants [23], processing of liquid radioactive wastes of nuclear power plants by reverse osmosis [24], evaluation of ultrafiltration membranes for treating low-level radioactive contaminated liquid waste [25], reusable water from electronics waste [26, 27], waste water treatment at the Mooka plant of Kobe Steel [28], development of a high performance reverse osmosis membrane for reclaiming agricultural waste water [29], reuse and treatment of waste water in agriculture [30], reclamation of brackish waters for irrigation with emphasis on optimization of process parameters for fertilizer-driven osmosis [31], cellulose acetate membranes for mine water purification [32], effective treatment of acid mine wastes by a low pressure charged membrane ultrafiltration process [33], charged membrane ultrafiltration of multisalt systems as applied to acid mine waters [34] and desalination and utilization of saline mine waters in Poland [35].

Applications of reverse osmosis and ultrafiltration in various industrial branches refer to membrane filtration of spent sulfite liquor for recovery of by-products and pollution control [36], combined reverse osmosis and freeze concentration process in the treatment of bleach plant effluents [37], application of reverse osmosis and ultrafiltration for industrial water from pulp washing waste water [38] and for treatment of bleaching plant effluents [39], membrane processes for pulp mill pollution control [40], membrane filtration of neutral sulfite semichemical process wastes [41], hyperfiltration for renovation of composite waste water at eight textile finishing plants [42], textile size and water recovery by means of ultrafiltration [43], textile waste water renovation for reuse with reverse osmosis [44], ultrafiltration in the treatment of dehairing wastes in the tanning industry [45] and ultrafiltration of effluents from the degreasing of skins in the tanning industry [46].

RECYCLING IN THE FOOD INDUSTRY

A fundamental approach to reverse osmosis offers quantitative methods of predicting membrane performance for several separation problems involved in the concentration of fruit juices and food sugars [47]. Fouling of membranes is particularly prevalent in the food industry [48]. Membrane apparatus for food plants are reviewed [49].

Specific applications of reverse osmosis and ultrafiltration include membrane systems for whey concentration [50], water and solute transport across cellulose acetate membranes in the treatment of soybean whey [51], performance of a membrane type enzyme reactor utilizing ultrafiltration [52] and use of soluble enzymes in continuous reactor operation by means of hollow fiber ultrafiltration membranes [53].

Patents. A recycling apparatus with concentration control is used to concentrate a dairy waste water from 3.5 to 5% up to 18 to 22% [54].

HEMODIALYSIS AND ARTIFICIAL KIDNEY

One of the most important applications of reverse osmosis and ultrafiltration might be hemodialysis and the construction of artificial kidneys. Recent work re-

162

fers to structure and mechanism of membranes for artificial kidney [55], preparation and evaluation of optimized hemodialysis membranes [56, 57], asymmetric cellulose acetate fibers for the dialytic separation of urea and salt [58], evaluation
of membranes for use in hemofiltration [59], transport characteristics and blood
compatibility of acrylonitrile copolymer membranes [60], poly(vinyl alcohol)-based
membranes for hemodialysis [61], transport and thromboresistant properties of hemodialysis membranes based on hydrated cellulose [62], sieving properties of hemodialysis membranes [63], polyelectrolyte membrane permeability using low-molecular-
weight metabolites [64], mass transfer in regular arrays of hollow fibers in countercurrent dialysis [65], water removing apparatus for blood filtration [66], membrane system to remove urea from the dialyzing fluid of the artificial kidney [67],
fluid dynamics of blood cells and applications to hemodialysis [68] and analytical
study of ultrafiltration in a hollow fiber artificial kidney [69].

Patents. The multi-stage dialyzer is composed of concentrical units consisting
of two chambers partitioned by a semipermeable membrane [70]. Preparation of a
hydrophilic membrane for dialysis is described [71]. A plate for a stack of plates
and semipermeable membranes can be made by injection molding and is suitable for
blood dialyzers [72]. Preparation of diurethane, useful as an adhesive in the manufacture of haemodialysis apparatus, is described [73]. Ascites containing blood was
filtered through two hollow fiber type membranes of pore size 0.2 to 0.01 μ, the
first filtering out bacteria and tumor cells and the second retaining proteins [74].
Aqueous urea was passed through a cuprammonium rayon hollow fiber dialyzer, while
pure water was being circulated outside the fibers. A higher degree of dialysis was
obtained [75]. A dilution apparatus [76] and a hollow fiber mass transfer apparatus
[77], especially useful for blood dialysis for kidney patients are described. Bovine
skin was treated to obtain a dispersion containing collagen fibers, which can be
used for preparing dialysis membranes for artificial kidneys [78]. Membranes for
dialysis of blood were prepared by treatment of cellulose acetate [79], ethylene-
vinyl alcohol copolymer [80], methacrylate or diethylene glycol monomethacrylate [81],
by spinning a compound containing cellulose acetate into a coagulating bath [82].
High molecular weight substance is dissolved in solvents and nozzle-sprayed to form
hollow fibers for hemodialysis [83]. Hollow fibers with improved filtration efficiency, useful for blood dialysis, are prepared by wet-spinning solutions acrylonitrile-
sulfonate salt polymers [84]. Hollow acetate fibers were prepared without yarn
breakage by spinning together a mixture of cellulose acetate in organic solvent and
formamide [85].

TREATMENT AND SEPARATION OF INORGANICS

The use of reverse osmosis and ultrafiltration in reclaiming useful materials
from waste streams is another application of these processes. The characteristics
of waste streams suitable for such treatment were summarized [86]. Relevant work
includes treatment of ammonium sulfate effluents by reverse osmosis [87], aluminum
sulfate recovery and recycling process [88], removal of dissolved substances from
waste water [89], use of membrane processes for the recovery of valuable materials
[90], material recovery from waste streams by membrane processes [91], synthesis
of membranes for the ultrafiltration of acids and alkalis [92], removal of sodium
chloride from waste water from an electric desalination unit of a petroleum processing plant [93], application of reverse osmosis in a photographic film laboratory
[94], coupled transport membranes for metal separations [95], recovery of metal
salts from metal-finishing wastes [96], permeative separation of metal ions through
polymer membranes containing the gluconate group [97], charged membrane ultrafiltration of toxic metal oxyanions and cations from single and multiple aqueous solutions
[98], charged membrane ultrafiltration of heavy metals from nonferrous metal [99],
purification of chromium-containing plating effluents [100], PBI membranes from
chromium plating rinse water [101], use of membrane processes in nickel plating
baths[102], reverse osmosis treatment of electroplating rinse waters [103], treatment of gold plating rinse by reverse osmosis [104] and potable water preparation
from contaminated brackish waters [154].

Patents. Ultrafiltration and reverse osmosis are used for treating waste electrolytes, coating solutions etc [105]. Waste solution or wastewater containing inorganic or organic compounds are separated to a concentrated portion and a diluted portion [106]. A waste solution containing 8% NaCl, from regeneration of ion exchange resin, and having absorbancy 1.4 at 420 mμ, when passed through a cellulose acetate membrane at 20 kg/cm^2, gave a solution containing 7.6% NaCl with 0.04 absorbancy at 420 mμ [107]. In a multistep solution-concentrating process a 26.5 g/l Na$_2$SO$_4$ solution was separated to give water containing 1.95 g/l Na$_2$SO$_4$ and a concentrated solution with 114.5 g/l Na$_2$SO$_4$ [108]. Removal of ammonia was 59.4% when CO$_2$ was added in the feed solution, versus 6% when CO$_2$ was not dissolved [109].

TREATMENT AND SEPARATION OF ORGANICS

Theoretical and experimental results are reported for the ultrafiltration of a solution which contains micro-ions and a single macro-ion which is rejected by the membrane. In the experimental portion of this study, bovine serum albumin served as the macro-ion [110]. A method is presented for determining the gel concentration and diffusivity at gelling of macromolecular solutions by comparing measured ultrafiltration limiting fluxes in plane, laminar and turbulent channel flow with theoretical fluxes obtained from analytical mass transfer solutions. The diffusivity of bovine serum albumin solution as a function of concentration is found at pH 4.7 and discrepancies among existing literature values are analyzed [111].

A selection is made from the literature on treatment and separation of organic liquids by means of reverse osmosis and ultrafiltration, which includes use of ultrafiltration for the separation and fractionation of organic ligands in fresh waters [112], removal of organic substances and clarification of drinking water by ultrafiltration [113], removal of organics in sewage and secondary effluent [114], transport properties of thermally conditioned sludge liquors [115], selectivity of a membrane based on n-vinylpyrrolidone with methyl methacrylate for dissolved organic substances [116], treatment of waste waters from adhesives and sealants manufacture [117], permeability of membranes made from high-pressure polyethylene in relation to liquid hydrocarbons [118], removal of hexamethylenediamine from waste waters by reverse osmosis and ion exchange methods [119], transport of methanol-water and formamide-water mixtures through cellulose acetate membranes [120], waste water treatment of waste water from a synthetic alcohol plant [121], separation of organic and inorganic solutes in alcoholic solutions [122], separation of the three-component water-diethylene glycol-sodium chloride mixture in a tubular reverse osmosis apparatus [123], analysis of reverse osmosis data for the system polyethylene glycol-water-cellulose acetate membrane at low operating pressures [124], reverse osmosis separations of polyethylene glycols in dilute aqueous solutions using porous acetate membranes [125], ultrafiltration, a new technique for separation of oil-water emulsions [126], a case history of treatment of oily wastes by ultrafiltration and reverse osmosis [127], ultrafiltration of vegetation waters from olive oil extraction plants [128], membrane separation of emulsified oil in waste water [129], removal of nonionized detergents by ultrafiltration [130], ultrafiltration characteristics of oil-detergent-water system, especially membrane fouling mechanisms [131], a comparison of stirred cell and hollow fiber techniques in ultrafiltration of fulvic and humic acids [132], ultrafiltration and hyperfiltration of phenolic compounds in coal gasification wastewater streams [133], purification of waste waters from the manufacture of olefins by a membrane method [134], separation of solutions of low-molecular-weight organic substances by reverse osmosis [135], permselectivities of some aromatic compounds in organic medium through cellulose acetate membranes [136], permeability of phenol through cellulose acetate membranes by reverse osmosis in various alcoholic systems [137], treatment of aniline plant waste water by means of semipermeable membranes [138], ultrafiltration separation of dyes from aqueous solutions [139], reverse osmosis treatment of dyeing waste water [140], purification of polymeric dyes by ultrafiltration [141], ultrafiltration of electrodeposition paint [142], reverse osmosis as an alternative for concentration of a thermolabile antibiotic and a waste stream from a paper mill [143] and purification of effluents of biological purification installations and IJs-salt water by reverse osmosis with dynamically formed membranes [144].

Patents. Organic components can be separated from aqueous solutions, e.g. MeOH with a polybutadiene membrane at 1 atm [145]. Poly(vinyl alcohol) membranes are useful for separating phenol from wastewaters [146]. Washing of ultrafiltration membranes, used for the filtration of microorganism culture solutions, is reported [147]. Preparation of membranes suitable for separation of unsaturated compounds from organic mixtures [148, 149] and of kerosine [150]. Water-oil emulsions are separated by a gravity flotation separator and the remaining mixture is passed through an ultrafilter [151]. A waste water from dye manufacture is treated in a reverse osmosis module to recycle dye containing solution [152].

LITERATURE TO 3H.

1. D.M. Cavagnaro (NTIS Rept. PS-78/0811/6GA [1978] 221 p)
2. D.M. Cavagnaro (NTIS Rept. PS-78/0812/4GA [1978] 103 p)
3. D.M. Cavagnaro (NTIS Rept. PS-78/0813/2GA [1978] 101 p)
4. D.G. Argo (NWSIA 5th Ann. Conf. San Diego, Calif. [1977])
5. Anonymous (Public Works 108, No 9 [1977] 166)
6. H.K. Johnston, H.S. Lim (Res. Rep., Res. Program Abatement Munic. Pollut. Provis. Can.-Ont. Agreement Great Lakes Water Qual. No 84 [1978] 122 p.- C.A. 90 [1979] 156521)
7. C.L. Chen, R.P. Miele (NTIS Rept. PB-288197 [1978] 71 p.- C.A. 91 [1979] 62016)
8. S.D. Lin (NTIS Rept. PB-293146/7GA [1978] 21 p)
9. D. Bhattacharyya, A.B. Jumawan, R.B. Grieves (J. Water Pollut. Control Fed. 50 [1978] 846/861)
10. H.E. Grethlein (J. Water Pollut. Control Fed. 50 [1978] 754/763.- C.A. 89 [1978] 135090)
11. L.R. Harris, P. Schatzberg, D. Bhattacharyya, D.F. Jackson (Water Sewage Works 125, No 8 [1978] 66/71.- C.A. 90 [1979] 126887)
12. T. Jobling (Water Pollut. Control 77 [1978] 460/461.- C.A. 90 [1979] 76143)
13. Y. Taniguchi (Chem. Econ. Eng. Rev. 10, No 1 [1978] 18/25)
14. J.E. Beckman (NTIS Rept. PB-293761/3GA [1979] 61 p)
15. C.K. Wojcik, J.G. Lopez, J.W. McCutchan (Desalination 32 [1980] 353/364)
16. T. Sawa, M. Kubota, S. Takahashi, Y. Masaki (Desalination 32 [1980] 373/382)
17. J.L. Gaddis, C.A. Brandon (NTIS Rept. NASA- 151689 [1978] 134 p)
18. W.J. Brenner (Chem.-Anl. Verfahren No 10 [1978] 147/148, 152)
19. W. Guetling (Fachber. Huettenprax. Metallweiter-Verarb. No 11 [1978] 969/972.- Masch. Markt 84 [1978] 1510/1511)
20. Y. Kojima, A. Komura, M. Kambara (PPM 9, No 2 [1978] 20/26)
21. Nguyen Quang Trong (Inf. Chim. 184 [1978] 139/142)
22. G.L. Anderson, W.G. Bair, J.A. Hudziak (NTIS Rept. CONF-780611-12 [1978] 26 p.- C.A. 91 [1979] 62041)
23. Anonymous (Mitsubishi Genshiryoku Giho 11 [1978] 9/11)
24. V.A. Mamet, A.A. Svittsov, G.A. Shchapov, B.F. Nikol'skii, N.M. Khromchenko, V.V. Chernov, Yu. N. Zhilin (Teploenergetica No 4 [1978] 52/54.- C.A. 89 [1978] 94529)
25. J.W. Koenst, R.C. Roberts (NTIS Rept. MLM-2448 1978 13 p.- C.A. 89 [1978] 203670)
26. C. Caprio, M.D. Beasley, L. Luttinger (Ind. Water Eng. 14, No 6 [1977] 24/30)
27. M.D. Beasley, L. Luttinger, C. Caprio (Proc. Ind. Waste Conf. 1977, 32 [1978] 630/638.- C.A. 89 [1978] 48410)
28. Anonymous (Mech. Eng. 99, No 9 [1977] 58.- Water Waste Treat. 20, No 9 [1977] 54)
29. M.C.S. Chan (D. Thesis Univ. Calif. Los Angeles [1978] 188 p.- Diss. Abst. B 39 [1978] 306/307)
30. D.B. Brice, R.R. Lindholm (NWSIA 5th Ann. Conf. San Diego, Calif. [1979])
31. J.O. Kessler, C.D. Moody (NTIS Rept. PB-277709/2GA [1977] 56 p.- G.R.A. 78, No 11 [1978] 126/127)
32. A.I. Kovalenko, D.D. Kucheruk, V.M. Bagnyuk, A.A. Kul'skii (Khim. Tekhnol. No 2 [1978] 46/48.- C.A. 89 [1978] 135549)
33. D. Bhattacharyya, S. Shelton, R.B. Grieves (Proc. Ind. Waste Conf. 1978, 33

[1979] 869/876.- C.A. 90 [1979] 192076)

34. D. Bhattacharyya, S. Shelton, R.B. Grieves (Sep. Sci. Technol. 14, No 3 [1979] 193/208.- C.A. 90 [1979] 91911)

35. J. Kepinski (Desalination 32 [1980] 399/408)

36. P.H. Claussen (Prepr. Pap. 63rd Ann. Meet. Techn. Sect. C.P.P.A., B [1977] 125/130.- C.A. 89 [1978] 151987)

37. A.J. Wiley, L.E. Dambruch, P.E. Parker, H.S. Dugal (TAPPI 61, No 12 [1978] 77/80.- Environ. Prot. Technol. Ser. E.P.A. No 600/2-78-132 [1978] 156 p.- NTIS Rept. PB-285912/2GA [1978] 158 p.- C.A. 90 [1979] 141870)

38. M. Bodzek, O. Kominek (Przegl. Papier 34, No 5 [1978] 164/168.- C.A. 91 [1979] 26679)

39. M. Bodzek, O. Kominek, E. Kowalska, I. Tanistra (Przegl. Papier 34, No 7 [1978] 248/251.- C.A. 90 [1979] 60644)

40. G. Maples, E.W. Lang (TAPPI Proc. Environ. Conf. [1978] 71/82)

41. G. Jonsson, S. Kristensen (Desalination 32 [1980] 327/339)

42. C.A. Brandon, J.J. Porter, D.K. Todd (NTIS Rept. PB-279451/9GA [1978] 247 p.- C.A. 90 [1979] 28577)

43. G.R. Groves, C.A. Buckley, G.L. Dalton (Progr. Water Technol. 10 [1978] 469/477.- C.A. 89 [1978] 152486)

44. A. El-Nashar (NWSIA Ann. Conf., Sarasota, Forida [1978] 28 p)

45. B. Cortese, E. Drioli (Inquinamento 20, No 6 1978 51/54.- C.A. 90 [1979] 191952)

46. B. Cortese, E. Drioli (Cuoio, Pelli, Mater. Concianti 54, No 2 [1978] 167/171.- C.A. 90 [1979] 174069)

47. T. Matsuura, S. Sourirajan (AlChE Symp. Ser. 74 No 172 [1978] 196/208)

48. W. Eykamp (AlChE Symp. Ser. 74, No 172 [1978] 233/235)

49. K. Hashimoto (Japan Food Sci. 17, No 10 [1978] 36/41)

50. Anonymous (Chem. Eng., New York, 85, No 9 [1978] 89/90)

51. E.C. Baker, G.C. Mustakas, M.D. Moosemiller, E.B. Bagley (J. Appl. Polym. Sci. 24 [1979] 135/145)

52. S. Katoh, T. Yanagida, E. Sada (J. Chem. Eng. Japan 11, No 2 [1978] 143/146)

53. C. Wandrey, E. Flaschel, K. Schuegerl (Germ. Chem. Engg. 1, No 1 [1978] 39/43)

54. Y. Sakaguchi, H. Koi (Jap. 77.131.978, 5 Nov 1977.- C.A. 88 [1978] 107211)

55. H. Strathmann (Chem. Ing. Tech. 49 [1977] 412/417)

56. A.J. Rosenthal, H.J. Davis, L.C. Sawyer (NTIS Rept. PB-288228/0GA [1978] 49 p)

57. A.J. Rosenthal, D.B. Clark, H.J. Davis (NTIS Rept. PB-288123/3GA [1978] 158p)

58. G. Tanny (J. Appl. Polym. Sci. 31 [1977] 407/413)

59. E. Klein, F.F. Holland, K. Eberle (NTIS Rept. PB-288555/6GA [1978] 33 p)

60. J.M. Courtney, J.D.S. Gaylor, R.M. Lindsay, I. Martin (Proc. Conf. Tech. Aspects Renal Dial. [1978] 126/132.- C.A. 90 [1979] 174645)

61. S. Yamashita, K. Takakura, Y. Imai, E. Masuhara (Kobunshi Ronbunshu 35 [1978] 283/289.- C.A. 89 [1978] 111592)

62. E.R. Levitskii, L.B. Baeva, V.G. Panov, A.K. Chepurov, N.S. Makhortov, V.A. Landysheva, I.D. Shamolina (Probl. Gematol. Pereliv. Krovi 23, No 6 [1978] 55/57.- C.A. 89 [1978] 152681)

63. R.P. Wendt, E. Klein, E.H. Bresler, F.F. Holland, R.M. Serino, H. Villa (J. Membrane Sci. 5 [1979] 23/49)

64. A.R. Rudman, N.A. Vengerova, R.I- Kalynzhnaya, B.S. El'tsefon, A.B. Zezin (Khim.- Farm. Zh. 13, No 3 [1979] 82/85.- C.A. 90 [1979] 192514)

65. I. Noda, C.C. Gryte (AIChE J. 25, No 1 [1979] 113/122.- C.A. 90 [1979] 154011)

66. K. Sasaki, T. Suehiro, N. Fujino, N. Nakabayashi (Sen'i Kobunshi Zairyo Kenkyu-sho Kenkyu Happyokai Sanko Shiryo 52 [1977] 141/152.- C.A. 89 [1978] 220858)

67. W.J. Kolff, D.E. Gregonis, E. Klein, R. Wendt, J. Walker (NTIS Rept. PB-288125/8GA [1978] 72 p)

68. P.L. Blackshear, S.V. Patankar, R.W. Heil, T. Nippoldt, A. Rosenstein (NTIS Rept. PB-288587/9GA [1978] 38 p)

69. H.D. Papenfuss, J.F. Gross, S.T. Thorson (AlChE J. 25, No 1 [1979] 170/179)

70. O. Hitomi, T. Yamaguchi, T. Iijima (Jap. 77.38.985, 1 Oct 1977.- C.A. 88 [1978] 172375)

71. T. Nomura (Jap. 78.120.688, 21 Oct 1978.- C.A. 90 [1979] 104898)

72. Rhone-Poulenc S.A. (Belg. 846.604, 24 Mar 1977.- C.A. 88 [1978] 172391)

73. G. Graber (French 2.361.452, 10 Mar 1978.- C.A. 89 [1978] 180777)

74. N. Tsuda, N. Kominami, K. Inagaki, T. Imamiya (Brit. 1.525.177, 20 Sep 1978.-
    C.A. 90 [1979] 157099)
75. Y. Ono (Jap. 77.144.380, 1 Dec 1977.- C.A. 89 [1978] 131562)
76. H. Tashiro (Jap. 78.10.383, 30 Jan 1978.- C.A. 89 [1978] 131594)
77. Y. Jo, M. Yamazaki, T. Sakai (Jap. 78.22.164, 1 Mar 1978.- C.A. 89 [1978] 91510)
78. T. Sode, A. Goto, K. Iwamoto, Y. Okamoto (Jap. 78.65.358, 10 Jun 1978.- C.A. 89
    [1978] 152702)
79. R. Nakatsuka, S. Suzuki, Y. Matsui, J. Takeda (Jap. 78.92.867, 15 Aug 1978.-
    C.A. 89 [1978] 198795)
80. T. Tanaka, K. Yamauchi, S. Kawai (Jap. 78.126.319, 4 Nov 1978.- C.A. 90 [1979]
    105259)
81. J. Aoyagi, T. Ichikawa (W. Ger. 2.834.716, 15 Feb 1979.- C.A. 90 [1979] 169570)
82. M. Mishiro, M. Kaneko (Jap. 79.15.476, 5 Feb 1979.- C.A. 90 [1979] 169918)
83. Y. Jo, M. Yamazaki, N. Kaneko, S. Oikawa, Y. Makuta, C. Hayashi (Jap. 79.15.030,
    3 Feb 1979.- C.A. 90 [1979] 192582)
84. Y. Jo, N. Kaneko, Y. Makuta, M. Yamazaki (Jap. 79.34.416, 13 Mar 1979.- C.A. 91
    [1979] 40650)
85. Y. Jo, M. Yamazaki, Y. Makuta, N. Kaneko, C. Hayashi, S. Oikawa (Jap. 79.42.419,
    4 Apr 1979.- C.A. 91 [1979] 40831)
86. D.D. Spatz (Chem Tech. No 11 [1977] 696/699)
87. H.I. Shaban, A.M. Akbar, M.A. Fahim (J. Environ. Sci. Health, A 13 [1978] 315/
    324.- C.A. 89 [1978] 168431)
88. E.E. Lindsay, W.L. Short (NTIS Rept. PB-278353/8GA [1976] 61 p)
89. H.H. Hahn, K.P. Kiefhaber (Umwelt No 5 [1978] 359/363)
90. H. Strathmann, U. Von Mylius (Chem. Ing. Tech. 50 [1978] 113/115)
91. H. Strathmann (Chem. Product. 8, No 3 [1979] 16/21)
92. Z.V. Borisenko, L.A. Dubova, L.I. Borovskaya, T.D. Elizarova, V.G. Ivanenko
    (Khim. i Khim. Tekhnol. v Tekstil'n. i Legk. Prom-sti [1977] 21/23.- Ref. Zh.
    Khim. [1979] 1 T 291)
93. G.A. Enaki, E.M. Kaliniichuk, V.M. Tkachenko (Neftepererab. Neftekhim. 16
    [1978] 77/80.- C.A. 89 [1978] 113381)
94. E. Dobolyi, F. Dobos (Hidrol. Kozl. 58, No 3 [1978] 122/130.- C.A. 90 [1979]
    60643)
95. W.C. Babcock, R.W. Baker, D.J. Kelly, H.K. Lonsdale (NTIS Rept. PB-293029/5GA
    [1978] 93 p)
96. W. Pusch, A. Walch (Recent Dev. Sep. Sci. 4 [1978] 1/12.- C.A. 90 [1979] 190261)
97. K. Kobayashi, H. Sumitomo (Polym. Bull., Berlin, 1, No 2 [1978] 121/125.- C.A. 90
    [1979] 29532)
98. D. Bhattacharyya, M. Moffitt, R.B. Grieves (Sep. Sci. Technol. 13 [1978] 449/463)
99. D. Bhattacharyya, A.B. Jumawan, R.B. Grieves (J. Water Pollut. Control Fed. 51,
    No 1 [1979] 176/186.- C.A. 90 [1979] 156569)
100. M.M. Mardanyan, V.A. Minaev, G.V. Makarov (Tr. Mosk. Khim. Tekhnol. Inst. im
    D.I. Mendeleeva 93 [1977] 98/102.- C.A. 91 [1979] 62068)
101. H.J. Davis, F.S. Model, J.R. Leal (NTIS Rept. PB-280944/OGA [1978] 38 p)
102. M. Rovel, M. Dangeon (Trait. Surf. 164 [1978] 7/10)
103. P.S. Cartwright (AlChE 71st Ann. Meet., Miami Beach [1978] Paper 51f)
104. C. Kamizawa, H. Masuda, M. Matsuda, T. Nakane, H. Akami (Desalination 27 [1978]
    261/272)
105. M. Kanematsu, H. Sato, M. Fushijima (Jap. 78.25.265, 8 Mar 1978.- C.A. 89 [1978]
    152052)
106. K. Tsukamoto (Jap. 78.28.082, 15 Mar 1978.- C.A. 89 [1978] 117288)
107. T. Kawabata, E. Kawaguchi, H. Okada, K. Ikeda (Jap. 78.39.272, 11 Apr 1978.-
    C.A. 89 [1978] 91818)
108. K. Tsukamoto (Jap. 78.58.974, 27 May 1978.- C.A. 89 [1978] 203965)
109. H. Ono, T. Hayashi, S. Inoue (Jap. 78.66.897, 14 Jun 1978.- C.A. 89 [1978]
    181915)
110. A.J. Di Leo, K.A. Smith, C.K. Cotton (AlChE 71st Ann. Meet., Miami Beach [1978]
    Paper 126a)
111. R.F. Probstein, W.F. Leung, Y. Alliance (J. Phys. Chem. 83 [1979] 1228/1232)
112. J. Buffle, P. Deladoey, W. Haerdi (Anal. Chim. Acta 101 [1978] 339/357.- C.A. 90
    [1979] 141968)

113. A.A. Zaborskii, N.G. Skvortsov, I.A. Donetskii, G.M. Kolosova, M.M. Senyavin (Vodosnabzh. Sanit. Tekh. No 10 [1978] 20/22.- C.A. 90 [1979] 92167)
114. E.S.K. Chian, S.S. Cheng, F.B. Dewalle, P.K.P. Kuo (Progr. Water Technol 9 [1978] 761/776.- C.A. 89 [1978] 168480)
115. M. Sugahara, T. Kitao, Y. Terashima, S. Iwai (Kagaku Kogaku Rombunshu 4, No 1 [1978] 43/48.- C.A. 90 [1979] 60683)
116. A.A. Zaborskii, G.M. Kolosova, I.A. Donetskii, S.S. Surzhenko (Zh. Prikl. Khim. 51 [1978] 2115/2117.- J. Appl. Chem. USSR 51 [1979] 2007/2010)
117. M.H. Kleper, R.L. Goldsmith, T.V. Tran, D.H. Steiner, J. Pecevich (NTIS Rept. PB-287823 [1978] 135 p.- C.A. 91 [1979] 62108)
118. E.P. Ageev, G.F. Vasygova (Deposited Doc. VINITI 2367-77 [1977] 11p.- C.A. 90 [1979] 72744)
119. G.L. Fishman, N.E. Ku'mitskaya, I.D. Pevzner, M.I. Eman, S.L. Khazitonova, N.M. Lyutikova (Plast. Massy No 2 [1979] 48/49.- C.A. 90 [1979] 209543)
120. S. Duckwitz, H. Moraal (Z. Naturforsch. 32 A [1977] 1077/1083.- C.A. 87 [1977] 207070)
121. N.G. Muratova, V.B. Shevchenko, G.F. Stychinskii, A.V. Dolganova, I.N. Smirnov (Intem. Chem. Eng. 19 [1979] 350/352)
122. H. Cohen, M. Ventura (Ben Gurion Univ. Rept. BGUN-RDA-164-77 [1979] 14 p)
123. V.I. Fedorenko, B.K. Sokolov, L.S. Aksel'rod (Khim. Mashinostr. 7 [1977] 112/ 117.- C.A. 90 [1979] 40697)
124. F.H. Hsieh, T. Matsuura, S. Sourirajan (I.E. C., Process Des. Dev. 18 [1979] 414/423.- C.A. 91 [1979] 21542)
125. F.H. Hsieh, T. Matsuura, S. Sourirajan (J. Appl. Polym. Sci. 23 [1979] 561/ 573.- C.A. 90 [1979] 105939)
126. J. Olszewski, A. Annusewicz (Powloki Ochr. 64, No 4-5 [1978] 89/95.- C.A. 90 [1979] 174081)
127. M.K. Sonksen, F.M. Sittig, E.F. Maziarz (Proc. 33rd Ind. Waste Conf. 33 [1979] 696/705.- C.A. 90 [1979] 156599)
128. C. Carrieri (Oli, Grassi, Deriv. 14, No 4 [1978] 29/32.- C.A. 90 [1979] 126957)
129. M. Kanematsu, K. Imasu (PPM 9, No 4 [1978] 56/62.- C.A. 90 [1979] 191901)
130. Y.Y. Kusov, N.A. Kalinchuk, L.M. Gogilashvili (Bioorganiceskaja Chimija 4 [1978] 832/835)
131. D. Bhattacharyya, A.B. Jumawan, R.B. Grieves, L.R. Harris (Sep. Sci. Technol. 14 [1979] 529/549.- C.A. 91 [1979] 44096)
132. J.C.T. Kwak, R.W.P. Nelson, D.S. Gamble (Geochim. Cosmochim. Acta 41 [1977] 993/996.- C.A. 88 [1978] 63607)
133. S.L. Klemetson, M.D. Scharbow (Prog. Water Technol. 10, No 1-2 [1978] 479/491.- C.A. 90 [1979] 109335)
134. V.B. Shevchenko, G.F. Stychinskii, N.G. Muratova, L.I. Gamaga (Neftepererab. Neftekhim. No 7 [1978] 31/33.- C.A. 90 [1979] 60636)
135. A.A. Yasminov, V.T. Kalgina, A.V. Kozhevnikov (Khim. Prom-st. No 10 [1978] 745/750.- C.A. 89 [1978] 221475)
136. H. Nomura, S. Yoshida, M. Seno, H. Takahashi (J. Appl. Polym. Sci. 22 [1978] 2609/2620.- C.A. 89 [1978] 164576)
137. H. Nomura, M. Seno, H. Takahashi, T. Yamabe (J. Appl. Polym. Sci. 24 [1979] 1191/1203)
138. S.V. Rjabchuk, E.I. Cyrlin, A.A. Evert (Khim. Prom. 2 [1977] 110/111)
139. L. Ya. Kukushkina, E.V. Migalatii, A.F. Nikiforov, V.V. Pushkarev (Zh. Prikl. Khim. 50 [1977] 1847/1852.- C.A. 89 [1978] 11533)
140. Kwangtung College of Chemical Engineering (Huan Ching K'o Hsueh No 3 [1978] 60/62.- C.A. 90 [1979] 191976)
141. A.R. Cooper, R.G. Booth (J. Appl. Polym. Sci. 23 [1979] 1373/1384)
142. Y. Hasegawa (Sumitomo Keikinzoku Giho 19, No 1-2 [1978] 35/48.- C.A. 89 [1978] 91152)
143. J. Wagner (Inst. Chem. Eng. Symp. Ser. 54 [1978] 147/152.- C.A. 89 [1978] 168453)
144. W.H. Rulkens, H.J. Van Veen, J.W. Van Heuven (PT-Procestech. 33 1978 746/ 753.- C.A. 90 [1979] 209513)
145. E. Perry (W. Ger. 2.654.296, 8 Jun 1977.- C.A. 89 [1978] 45677)
146. S. Peter, R. Stefan (W. Ger. 2.730.528, 25 Jan 1979.- C.A. 90 [1979] 422486)

147. K. Tako, K. Matsumoto, K. Iizuka (Jap. 77.120.978, 11 Oct 1977.- C.A. 88 [1978] 123256)
148. T. Sumie. R. Kobavashi, Y. Kivomatsu. M. Takahashi (Jap. 77.122.278,77.122.279 14 Oct 1977.- C.A. 88 [1978] 138615, 89 [1978] 45680)
149. K. Motegi, K. Kamimura, J. Morimoto, S. Yamazaki (Jap. 78.113.276, 3 Oct 1978.- C.A. 90 [1979] 39849)
150. M. Sato, Y. Hashino (Jap. 78.137.876, 1 Dec 1978.- C.A. 90 [1979] 139615)
151. A.A.J. Lefeuvre (W. Ger. 2.818. 127, 16 Nov 1978.- C.A. 90 [1979] 206315)
152. S. Takahashi, T. Nagaretani, H. Matono (Jap. 78.136.034, 28 Nov 1978.- C.A. 90 [1979] 174286)
153. J. Sasaki (Jap. 77.134.882, 11 Nov 1977.- C.A. 89 [1978] 8057)
154. W. Oehler (Techn. Mitt. Krupp 36, No 2 [1978] 41/44)

4. Other Desalting Processes

ICE FORMATION AND MELTING

In studying the performance characteristics of a scale model annular flow ice-water heat sink, comparisons were made between experimental results and computer predictions of outlet water temperature and melting time. The computer program gave a reasonable simulation of actual behavior and could be used as the basis for design calculations [1]. Data on the ice formation rate were presented for a salt water drop suspended by drag forces in a flowing cold organic liquid. The effects of refrigerant under cooling, salt concentration, drop size and time were studied. Ice formation rates in drops of 3 wt. % NaCl solution were two to three times lower than in pure water drops. A parallel-plate model was used to correlate the data and predict ice formation rates for other drops and refrigerants [2].

An experimental determination was made of the velocity profiles which result from the free convective melting of a vertical ice sheet into fresh water at temperatures in the range from 2.0°C to 7.0°C. The results suggest that upward flows exist for water temperatures below 4.7°C. Entirely downward flowing boundary layers are suggested for temperatures above 7.0°C. For intermediate temperatures, an oscillatory dual flow regime is indicated [3].

FREEZING PROCESSES

Thermodynamic properties of normal butane over the temperature range -30 to + 50°C were presented in tables giving the values of the volume, enthalpy and entropy for the subcooled liquid, the saturated liquid and vapor and the superheated vapor [4]. The effect of butane gas agitation on the crystallization of ice was investigated in an experimental crystallizer. The specific production rate of ice was 0.55 kg/h.kg solution and superheating for evaporation of butane was between 0.8 and 1.6° [5].

The mixing was studied experimentally in a model horizontal crystallizer prior to the construction of a seawater desalination plant by freezing [6]. While operating a direct contact crystallizer at the mixing condition by the buoyancy force of vaporized refrigerant, the mean diameter of volume-surface obtained from the measured permeability and porosity of the ice bed coincided with that calculated from the distribution of ice particles counted by microphotos. The characteristics of the crystallizer agitated by the buoyancy force of refrigerant vapor showed almost the same characteristics as the mechanically agitated crystallizer, but the rate of crystallization/unit volume of crystallizer was lower, the rate of nucleation was higher and the size of the crystals was smaller [7].

Operational testing, maintenance and modification were conducted for nine months on a 75000 gpd secondary refrigerant freezing desalination plant at the Wrightsville Beach Test Facility. The operational parameters of a 13 foot in diameter counterwash column for this process were evaluated. Changes were made to various parts of the pilot plant to improve operations. Design defficiencies in the process and the pilot plant equipment were identified and material analysis of various aluminum components in the plant was made [8].

Further trials were conducted in Canada for desalination of brackish water by spray freezing of tap water and a synthetic sodium chloride solution at the outdoor laboratory and on brackish groundwater at several locations [9]. A desalination unit with a heat-utilizing lithium bromide absorption refrigerating machine combines a heating unit with a compressor condenser. The working fluid in the heat engine is a concentrated LiBr solution and that in the condenser is water vapor obtained from the feed solution. The product is obtained both from melting ice and from condensing vapor. A thermodynamic analysis and detailed typical heat and mass balances indicate that, if 5% of the product is uded for washing ice crystals, 1 kg of fresh water can be obtained from 2.1 kg of 3.5% TDS saline water with a heat expenditure of 90 kcal [10].

A design study was conducted of the absorption freezing vapor compression process. This process overcomes one of the major problem areas in previous freezing processes by using a closed cycle refrigerant loop to power the adsorption cycle, thus eliminating compressor problems caused by direct compression of refrigerants contaminated with seawater. The major advantages of the process are: low power consumption of freezing compared to heat processes: less corrosion, no pretreatment is needed to prevent scaling even with brackish water saturated with calcium sulfate. Power consumptions of 42 and 38 kWh per 1000 gal were calculated for 100000 and 1 million gpd plants, respectively [11].

A suggestion was made to develop a freezing process based upon the freezing of ice from a saline solution through a heat transfer surface as a slurry without plug or ice sheet formation on the surface. A conceptual design for 1 Mgd desalting plant was described [12].

Other work includes a monograph on freshening of water by freezing [13], experimental finishing for crystallization of ice and its separation from brine [14], crystallization method for the desalting of sea and saline waters and for purification of waste waters [15], dynamics of seawater desalting through zone freezing [16] and a detailed review on freeze crystallization [17].

Patents. In a saline water desalination system, the water is vaporized in a evaporator and the vapor passed to a condenser. Heat is supplied to the evaporator and extracted from the condenser by a refrigeration system, having its condensing coils in the water evaporator and its evaporative coils in the water condenser [18]. A two-stage crystallizer uses a water-immiscible refrigerant which is boiled in the feedwater in the first stage and is disentrained in the second stage [19]. An apparatus comprises a vertical, cylindrical vessel, a vertical annular heat exchanger on whose outside and inside surfaces ice is formed, ice removing scrapers and annular ice melter placed above the level of the top of the freezing heat exchanger, an annular melt receiver and a centrally located screw conveyor for raising the scraped ice into the melter. The apparatus is intended for use on ships. The still is mounted in the same casing as the freezer [20]. Water is purified by freezing and melting, then electrolyzed in the presence of electrolytes to prepare a mineralized alkaline water [21].

HYDRATES

The desalination of sea water, based on the formation of a solid crystalline H_2S hydrate, requires a considerable expenditure of energy, which can theoretically be obtained from a reverse cycle operating on the difference in temperature of the surface and lower water layers. The water temperature in the Black Sea at 30 to 50 m depth is about 281°K throughout the year. This ensures a temperature difference of about 15° and, although this gives a very low Carnot cycle efficiency, the large volumes of water handled by desalination plants makes the process economically attractive. The lower layers of water in the Black Sea contain H_2S, which may be recovered in addition to the fresh water [22]. Water contents for methane, ethane, propane hydrates were calculated and are temperature dependent. The maximum value is attained at the quadruple point where liquid water, ice, gas and hydrate coexist [23].

BIOCONVERSION

Experiments were conducted on the use of algae for removal of salt from water in an attempt to develop an economically feasible, low-energy method for the conversion of saline water. Techniques for monitoring cell sodium levels were evaluated for the most efficient means of removing external sodium and for digestion of the cells prior to sodium analysis. Environmental changes indicated in the literature search and which could be easily manipulated were made in attempts to induce uptake/release of sodium from algal cells. Factors selected for experimental manipulation were: pH salinity, temperature, light versus dark conditions, addition and depletion of energy sources, addition of toxic compounds, and addition of sewage sludge. No significant uptake or excretion of Na or Cl⁻ was obtained in the algal species studied, when the selected environmental manipulations were made, with the exception of a possible uptake of sodium when the algal culture was studied in the presence of activated sewage sludge [24].

CHELATE TREATMENT

The low-molecular weight fraction of black liquor forms an insoluble chelate when treated with $KMnO_4$. This chelate is used for the purification of waste effluents and for desalting seawater. The chelate reduces the Na^+ content of seawater by 10% in single-stage treatment or by 26% in 2-stage treatment [25].

LITERATURE TO 4.

1. J. Stubstad, W.F. Quinn, Y.C. Yen (AlChE Prepr. 17th Nat. Heat Transfer Conf. [1977] 188/198.- C.A. 88 [1978] 9008)
2. S.T. Bustany, P. Harriott, H.F. Wiegandt (AlChE J. 25 [1979] 439/446)
3. N.W. Wilson, B.D. Vyas (J. Heat Transfer 101 [1979] 313/317)
4. R.T. Kurnik, A.J. Barduhn (Desalination 26 [1978] 211/283)
5. M. Oowa, S. Kawasaki (Nippon Kaisui Gakkaishi 32, No 4 [1978] 176/182)
6. Y. Nagashima, S. Maeda (Kagaku Kogaku Rombunshu 3 [1977] 371/375.- C.A. 90 [1979] 40682)
7. Y. Nagashima, S. Maeda (Kagaku Kogaku Rombunshu 4 [1978] 509/514.- C.A. 90 [1979] 109706)
8. R.S. Robinson (NTIS Rept. PB-296811/3GA [1978] 110 p)
9. J. W. Spyker (Saskatchewan Res. Coun. Publ. E 78/3 [1978] 114 p)
10. L.M. Rozenfel'd, Yu. V. Kuz'mitskii (Kholod. Tekh. No 2 [1978] 17/24)
11. Concentration Specialists Inc. (NTIS Rept. PB-288276/9GA [1977] 130 p)
12. W.E. Johnson (Desalination 31 [1979] 417/425)
13. M.V. Kolodin (Ylym. Ashkhabad, Turkm. SSR [1977] 244 p)
14. V.P. Chursin, V.A. Odintsov (Kristallizatsiya No 2 [1976] 150/151)
15. V.P. Alekseev, L.F. Smirnof (Visn. Akad. Nauk Ukr. SSR. No 2 [1978] 41/54)
16. D.M. Lupu (Desalination 31 [1979] 415)
17. J.A. Heist (Chem. Eng., New York, 86, No 10 [1979] 72/82)
18. N.L. Foley (Aust. 495.414, 10 Aug 1978.- C.A. 90 [1979] 100763)
19. A.I. Lloyd (U.S. 4.092.834, 6 Jun 1978)
20. L.F. Smirnov, V.M. Parkhitko, V.I. Zverkhovskii, O.A. Burtov, N.I. Razuvaev, F.E. Dovzhko, M.G. Kleiman, V.I. Dzyan (U.S. 4.112.702, 12 Sep 1978.- C.A. 90 [1979] 106183.- French 2.341.344, 16 Sep 1977.- C.A. 90 [1979] 8071.- Jap. 77.119.470, 6 Oct 1977.- C.A. 89 [1978] 91486)
21. S. Sakamoto (Jap. 78.146.270, 20 Dec 1978.- C.A. 90 [1979] 192357)
22. E.I. Kleshchunov, A.I. Kleshchunov, V.B. Ermakov (Tr. Krasnodar. Politekh. Inst. 72 [1976] 17/20.- C.A. 89 [1978] 48706)
23. S. Sh. Byk, V.I. Fomina (Zh. Fiz. Khim. 52 [1978] 1306/1308)
24. Biospherics Inc. (NTIS Rept. PB-283768/0GA [1978] 58 p)
25. M.G. Morekhin (Khim. Drev. No 4 [1978] 68/71.- C.A. 89 [1978] 204917)

5. Economic Considerations

The reliability and availability of water systems including desalination units were examined, using fault tree logic and field data whenever possible. An assessment is made of factors impacting availability and safety of MSF desalination plants. The emphasis has been on integrating the plant into a water supply system such, that the district fresh water requirements are met or exceeded. Required equipment reliability is a function of the system design characteristics and the expected demand. Failure models of MSF desalination systems and reliability data on desalination equipment can be conveniently incorporated into estimates of unit effectiveness. MSF desalination equipment reliability and maintainability problems can be analyzed using existing availability methodology [1].

An investigation was made to obtain values for specific costs and in particular their mean values and spreads, taking as a basis the data contained in tenders from MSF plant suppliers, in order to be able to make statements on the reliability of investment cost estimates. The possibility of technical reasons being responsible for this spread in tender prices and the nature of these were examined. Some interesting aspects result with regard to the state of the act in the fields of construction and design [2].

In a comparative investigation of the economics of desalting, based on current and projected technology, current operating cost of various plant types operating in Israel are reported. These costs range from less than Doll. 4 per m^3 for membrane plants desalting brackish water to more than three times as much for thermal plants desalting seawater. For new systems, two plant sizes were evaluated: 4000 m^3/day plants applying current technology and 100000 m^3/day plants applying projected technology. The water costs obtained for the various plant types and applied economic parameters, especially energy prices, range between Doll. 0.2 and 0.6 per m^3 for brackish water desalting and from Doll. 0.5 to 2.4 per m^3 for seawater desalting [3].

Costs of desalting seawater and brackish water were originally presented in document ORNL/TM-5070 based upon first quarter 1975 prices [4]. A first update was based upon 1977 costs [5]. Since 1975, desalted water costs using MSF distillation, have risen from about $3.60 to $4.90 per 1000 gal for Mgd plants and from about $1.70 to approximately $2.15 per 1000 gal for 100 Mgd plants. The major change in water costs is attributed to fixed charge rate increasing from 11.5% to 15.5%. The cost of brackish water desalting by reverse osmosis has increased from about $1.05 per 1000 gal to $1.25 per 1000 gal for 1 Mgd plants and from roughly 80 to 90 cents per 1000 gal for 25 Mgd plants. The increases for electrodialysis are of the same order as for reverse osmosis. Since 1975, the estimated costs of reverse osmosis desalting of seawater have changed from $6 to $8 per 1000 gal from a 20000 gpd plant and from $1.50 to $4 per 1000 gal from a 1 Mgd plant. This is attributed to the development of more realistic data [6].

Cost estimates were given for desalting seawater by distillation and reverse osmosis and for desalting brackish water by reverse osmosis and electrodialysis. The energy costs for both reverse osmosis and electrodialysis are based upon the availability of electricity at a fixed rate. Cost data were computed as a function of plant size and include both capital costs and construction costs, which are considered as typical. The presentation is such, that the costs can be easily adjusted to reflect local conditions [7].

Rising energy prices will influence the water conversion market drastically in the years to come. To enable an estimation of the consequences, the cost structures of the most potential water conversion alternatives were compared and the effect of energy demand, energy price, feed water salinity, operating temperature range and plant capacity on the production cost of fresh water were presented [8].

The preparation of make-up water for thermal plants may be integrated with a saline water distillation unit introduced in the turbine system. Heat extracted from a bleedstream of the turbine is given back to the feedwater in the heater [9].

In predicting costs for membrane desalination systems, careful estimation should be made for the amounts of electricity, labor, chemicals etc. required. The assumptions made in establishing unit costs can greatly affect the predicted operating cost and hence the ultimate investment decision. The fact that 0.2 pounds of acid per 1000 gal of permeate is required in a Florida reverse osmosis plant is more useful for predicting operating costs than is the fact of the cost of acid. Data based on units rather than costs are presented for seven reverse osmosis facilities and one electrodialysis facility [10].

Economic aspects of multi-stage flash distillation and seawater reverse osmosis plants were compared. The design parameters and cost data are based on the operation and construction of these facilities on the U.S. Gulf Coast. The reference designs used in the study may be used as either single or dual purpose facilities, depending upon specific site requirements [11]. Cost data are reported for unit sizes up to 10 Mgd of the two main processes, multi-stage flash distillation and reverse osmosis. Both single-pass and double-pass systems were considered for the reverse osmosis unit, while the cost of water from MSF plants was calculated for different performance ratios. Controlling factors such as fuel prices, pretreatment costs, membrane life, energy recovery and terms of financing were considered. It was concluded that reverse osmosis is the more economical of the two processes for samll unit sizes and regions of high fuel costs. There is a break-even point, depending on these two parameters, beyond which the MSF process yields lower water costs. Attention is focused on a third process, vertical tube evaporation combined with vapor compression as a low energy consuming system. This process is potentially competitive with reverse osmosis, even for small unit sizes [12].

The comparison between desalination and other water supply methods is usually based on a conventional economic evaluation, which is adequate when one of the alternative water supply methods is clearly superior. A methodology was presented for comparing water desalination to fresh water transportation and treatment when the alternatives are competitive. The methodology considers the complete water supply and disposal chain, and takes into account technical, environmental, economic and political, as well as legal aspects of the problem. Actual examples, mainly from the experience of the Southwest Florida Water Management District, are provided to illustrate the methodology. The economic analysis is based on a present-value life-cycle cost model, which accounts for interest, tax, insurance and escalations in energy, labor and material costs [13]. Estimation of costs of product water conveyance from desalination plants was also reported [14].

Further work includes a procedure for technical-economical analysis of methods for desalting natural waters [15] and a discussion of problems arising from misguided economies at the planning and desgin stages of water plants [16].

RECOVERY OF BYPRODUCTS FROM THE SEA

A phase diagram was given showing the main five components found in seawater: sodium, potassium, magnesium, sulfate, chloride and water. The sequence of salts formed is: $CaCO_3$, halite + anhydrite(trace), halite + picromerite + epsomite, halite + epsomite + sylvite + picromerite(trace) + kainite(trace), halite + $MgSO_4$ hydrates + carnallite, and halite + $MgSO_4$ hydrates + carnallite + bischofite [17]. Successive evaporation of sea bittern, after removing NaCl crystals, gives sels mixtures between 34 and 36° Baumé, mixed salt between 36 and 38° Baumé and leaves behind a mother liquor rich in $MgCl_2$. The fraction sels mixtures is a mixture of NaCl and $MgSO_4$ in widely varying proportions in the range NaCl 55 to 65% and $MgSO_4 \cdot 7H_2O$ 35 to 45% [18].

The evaporation of concentrated seawater, made by electrodialysis, in vacuum pans was discussed on the basis of the amounts of water evaporated and NaCl deposited and total salts concentrations of mother liquid [19].

Other work includes extraction of certain elements from seawater [20], a review of the recovery of chemical products from marine nodules and seawater [21] and of minerals from seawater [22].

Patents. The described facility sequentially precipitates $CaCO_3$, $Mg(OH)_2$ and $Ca(OH)_2$ with NaOH, then evaporates the seawater and condenses the vapors, and elec-

trolyzes the brine to NaOH and chlorine [23]. Seawater and other salt solutions are
treated for recovery of the salts by adding Me_2CO [24]. Carbonates, sulfates, borates
and chlorides of sodium and potassium are recovered from natural underground brines.
The brine of predetermined concentration is evaporated to form crystal salts and a
hot brine. The salts from the hot brine are carbonated to recover Na_2CO_3, K_2SO_4 and
NaCl, and the hot brine is flash cooled to recover KCl and $Na_2B_4O_7 \cdot 5H_2O$ [25].

Common salt. The physical and chemical qualities of NaCl obtained by solar evaporation from various parts of the world were reviewed [26]. A simplified model was
presented for studying the evaporation of sea brine in solar ponds [27]. Caking
tendency of common grade salt made by ion-exchange membrane method was compared with
that made by solar evaporation. Generally common grade salt made by membrane method
showed a greater caking tendency than that made by salt-field method [28]. Design
calculations were presented for producing NaCl from the brine of the Sivash Bay of
the Azov Sea by evaporation in a 4-effect system [29].

Patents. Brine is prepared by dissolving NaCl and mirabilite obtained by crystallization from seawater [30]. The degree of recovery of common salt is increased
by mixing the solution with soda-industry wastes containing $CaCl_2$ and NaCl [31].
High purity salt is prepared from brine by controlling the Ca^{2+}/SO_4^{2-}- equivalent concentration ratio at over 1 and crystallizing while using fine salt crystals as
crystallization nuclei [32]. A salt solution is concentrated in a series of evaporators, by direct contact of hot gas to crystallize the salt [33]. Sea salt was
prepared by removing the water from a brine containing sodium and magnesium chlorides with heating to a residual moisture content of 0.1 to 0.2% followed by drying
and granulating the product [34].

Halogens. In an industrial process for bromine separation from brine waters, the
oxidation of Br^- by chlorine proceeds in two stages, the first of which is rate determining. A procedure was developed for calculating the desorption rate and the
excess air needed to remove bromine from the reaction mixture [35]. Schemes for the
recovery of useful components, including bromine, from geothermal brines were developed [36].

Patents. Iodide salts contained in brines are separated by repeated cycles of
passing the brine through an anion exchange resin to absorb the iodide and oxidizing
it to iodine [37]. Iodine is adsorbed from brine by fluidized countercurrent
strongly basic anion exchange. The exhausted resin is removed from the bottom of
the apparatus [38]. Molecular iodine and bromine are obtained by adding an oxidizing liquid to underground brine or seawater [39]. Iodine in natural brines is
oxidized by NaClO to molecular form [40].

Lithium. A review was presented of the recovery of lithium from seawater [41].
The recovery of lithium from Dead Sea brines as aluminate precipitate was examined
Yields of about 90% were obtained from diluted brines processed with $AlCl_3 \cdot 6H_2O$ [42].
Structural properties and adsorbability of Li^+ from seawater were studied comparatively for three types of adsorbents: hydrated Al_2O_3, a composite of hydrated Al_2O_3
and activated C and a mixture of hydrated Al_2O_3 and MnO_2 . Amorphous adsorbents
were highly effective in absorbing Li from seawater. The adsorption was due to
amorphous Al_2O_3 and activated C and MnO_2 played no role in the adsorbtion process [43].

Patents. $Al(OH)_3$ suspended in a particulate anion exchange resin is reacted with
LiOH for a time and at a temperature sufficient to form microcrystalline $LiOH \cdot$
$2Al(OH)_3$ [44]. The anion exchange resin is impregnated with $AlCl_3$, which is con-
verted to $Al(OH)_3$ with NH_4OH. The $Al(OH)_3$ containing resin is treated with an
aqueous lithium halide solution to form a lithium aluminate complex. The resulting
mixture is heated at a temperature sufficient to form microcrystalline $LiX \cdot 2Al(OH)_3$,
where X is a halide [45]. Seawater, containing 0.17 ppm Li, is concentrated to
specific gravity 1.13 to 1.14, the NaCl, KCl and other salts are removed, and the
solution finally concentrated to specific gravity 1.4, diluted with a 5-fold amount
of fresh water. The Li and Mg salts are absorbed by a cation exchanger and the
LiCl separated by dissolving it in hydrochloric acid. The yield is about 70% [46].

Magnesium. Magnesium hydroxide was precipitated from seawater by means of roasted dolomite under varying conditions. A decisive parameter of precipitation was pH, which should be 11.5 to 12.0. Addition of NaOH improved the purity of the precipitate. Of several flocculants tested the most effective were the polyacrylamide resins [47]. The recovery of magnesium from Inowroclaw natural brines was effected after acidification to pH 4 in order to remove the calcium present in the brine. Magnesium was precipitated by addition of NaOH [48] . A similar procedure was used for the recovery of magnesium from coal mine brines [49].

Magnesium hydroxide suspensions obtained by reaction with CaO were investigated using cationic, anionic and neutral high molecular weight polyacrylamide flocculants. The main impurities were $CaCO_3$ and $CaSO_4$ in the precipitates from seawater and the bitterns from salt marshes of Araya, respectively [50]. In a seawater magnesite plant, heat conservation was increased by improving the efficiency of the mechanical dehydrators, using the waste gases from the dryer for $Mg(OH)_2$, and using the waste heat from product cooling in the plant. Precompressing the $Mg(OH)_2$ by briquetting and improving the calcination process and the heat exchange during calcination improved the quality of the product [51]. Other work includes the extraction of magnesia from seawater [52], the preparation of magnesium oxide from magnesium chloride brines [53] and investigation of sedimentation rate of magnesium hydroxide precipitated from seawater [54].

Patents. Microcrystalline $MgX_2.2Al(OH)_3$, where X is a halogen, suspended in a particulate anion exchange resin was used to recover Mg^{2+} from brines [55]. $Mg(OH)_2$ containing minimal amounts of B and Ca can be obtained from seawater by adding $Mg(OH)_2$ seed crystals and alkali in an amount greater than that required to precipitate all of the magnesium ions [56]. Brines or bitterns were treated with milk of calcined dolomite to precipitate contaminated $Mg(OH)_2$ and obtain a saturated $Ca(OH)_2$ solution, which is used in a second stage of the treatment of fresh brines or bitterns. $Mg(OH)_2$ precipitated in the second stage is more pure and can be used in the manufacture of refractory materials [57].

Bittern from electrodialysis is treated with $MgCO_3.3H_2O$ slurry, separated from $CaCO_3$, added with $(NH_4)_2CO_3$ or Na_2CO_3 and filtered to obtain $MgCO_3.3H_2O$, part of it being recycled as a slurry [58]. Underground brine is decarbonated with HCl and then electrodialyzed, using the brine as the catholyte and a concentrated NaCl solution as the anolyte, to give $Mg(OH)_2$ floc in the catholyte [59]. In the manufacture of $MgSO_4$ by reacting seawater with milk of lime to precipitate $Mg(OH)_2$ and treating the resulting $Mg(OH)_2$-based slurry with waste gases containing SO_2 and NO_x, the NO_x in the waste gases is effectively removed by adding an iron chelate and powdered iron to the reaction system. Thus, $MgSO_4$ production and waste gases purification are performed simultaneously [60]. A solution containing Mg^{2+} is reacted with alkali containing CaO or $Ca(OH)_2$ to give $Mg(OH)_2$, which is washed and calcined to obtain MgO clinker [61].

Potassium. Potassium is selectively recovered from concentrated brine from a seawater desalination plant by precipitation with dipicrylamine [62].

Strontium. Patents. To recover Sr from a brine containing Ca and Sr, $SrSO_4$ seed crystals are added and then Na_2SO_4 or other water-soluble sulfate is added [63].

Uranium. A bibliography contains 471 references pertaining to the evaluation of U.S. territorial ocean waters as a potential uranium resource and to the selection of a site for a plant designed for the large scale extraction of uranium from seawater. The literature cited is listed by author with indices to the author's countries, geographic areas of study and to a set of keywords to the subject matter [64]. A selected annotated bibliography of 521 references was prepared as a part of a feasibility study of the extraction of uranium from seawater. For the most part, these references are related to the chemical processes whereby the uranium is removed from the seawater [65].

The U.S. coastal waters were evaluated as a uranium resource, as well as the selection of a suitable site for construction of a large-scale plant for uranium extraction.

Evaluation of the resource revealed that although the concentration of uranium is quite low, about 3.3 ppb in seawater of average oceanic salinity, the amount present in the total volume of the oceans is very great, some 4.5 billion metric tons. Of this, perhaps only that uranium contained in the upper 100 meters or so of the sur-face wellmixed layer should be considered accessible for recovery, some 160 million tonnes. The study indicated that open ocean seawater acquired for the purpose of uranium extraction would be a more favorable resource than rivers entering the sea, cooling water of power plants or the feed or effluent streams of existing plants pro-ducing other products such as magnesium, bromine or potable and/or agricultural water from seawater. Various considerations led to the selection of a site for a pumped seawater coastal plant at a coastal location. Puerto Yabucoa, Puerto Rico was sel-ected [66].

A major assessement was made of the uranium resources in seawater. Several con-cepts for moving seawater to recover the uranium were investigated, including pumping the seawater and using natural ocean currents or tides directly. From the various processes for extracting uranium from seawater, the adsorption process is the most promising at the present time. Of the possible adsorbents, hydrous titanium oxide was found to have the best properties. A uranium extraction plant was conceptually designed. Of the possible methods for contacting the seawater with the adsorbent, a continuous fluidized bed concept was chosen as most practical for a pumped system. A plant recovering 500 tonnes of U_3O_8 per year requires 5900 cubic meters per second of seawater to be pumped through the adsorbent beds for a 70% overall recovery efficiency. Total cost of the plant was estimated to be about $6.2 billion. A com-puter model for the process was used for parametric sensitivity studies and economic projections. Several design case variations were developed. Other topics addressed were the impact of co-product recovery, environmental considerations, etc [67].

Nineteen adsorbents were tested for their ability to concentrate uranium from seawater. Most of them proved to be not sufficiently effective. Hydrated titanium oxide showed a fast pick-up rate in the batch studies and a good yield of 92 ± 3% was obtained by the passage of 75 l of seawater spiked with uranium to 4 times the con-centration of open ocean water. Maximum amount of uranium that could be obtained on this bed was 1.4 mg/g U. This adsorbent could be utilized to concentrate uranium from seawater if the practical problems of compaction and algal growth are elimin-ated [68]. Uranium absorption from seawater on browncoal is performed quantitatively in short contact times. However saturation is attained already at 3 ppm uranium, because of large amounts of Ca^{2+} present [69].

Uranium sorption kinetics and sorbent volume stability have been studied in a con-tinuous operation process. From several variants of uranium extraction, the most satisfactory results were obtained with the highly basic anion exchanger AM-Khp, produced in the USSR. A sorption capacity of 5.3 mg/g uranium was reached and the uranium content in the eluate was 600 to 1600 mg/l [70]. The kinetics of sorption of micro-amounts of uranium on phosphoric acid cation exchanger KRF from model solution of seawater depend on exchanger particle size, concentration of binder in the exchanger and on ionic strength of solutions [71].

In studying the adsorption of uranium, dissolved in seawater, hydrous TiO_2 was prepared by precipitation from $TiCl_4$ and aqueous NH_3 at various temperatures. The amount of OH groups on the surface of hydrous TiO_2 decreased with increasing pre-paration temperature. Potassium adsorption increased linearly with the amount of surface OH groups of hydrous TiO_2. The reverse relation existed between the amount of uranium adsorbed and the surface OH groups [72]. Another process consists of adsorption of uranium from seawater on hydrous titanium oxide bound with polyacryla-mide in a floating bed column, desorption of uranium with a Na_2CO_3 solution and con-centration of uranium in the eluate by anion exchange resin or by extraction of trio-ctylphosphine oxide in naphthalene [73].

A sudy was made on the adsorption of uranium from seawater by means of a compo-site adsorbent prepared by impregnating basic zinc carbonate on activated carbon [74]. Stable foams collected at the seashore returned a small volume of seawater when kept standing overnight. This water contained 36 mg/l uranium, hence ten times greater than normally found in seawater. Sands collected near the same shore also contained 10 to 100 times more uranium than found in sands collected in other places [75].

Further work on uranium extraction includes a technical problem of uranium collection from seawater [76], possibility of an industry for extracting uranium from seawater [77], developments of absorbents for the extraction of uranium from seawater [78] and a review of extraction of uranium from seawater [79].

Patents. Heavy metals, such as uranium and vanadium, are recovered from seawater by adsorption on humic acid containing peat. The metals are eluted from the peat with an acidic solution and recovered by neutralization [80]. A matrix for adsorbing substances consists of a permanently magnetic substrate and adsorbent or catalytic particles that are attached magnetically to the substrate. Seawater was passed through Co-ferrite fabric, which absorbed 10 mg/kg uranium. Particles of magnetite coated with TiO_2 gave a fabric which absorbed 300 mg V and 200 mg U per kg TiO_2 [81].

A cation-exchange resin containing COOH or SO_3H, coated with hydroxides or carbonates of titanium or other metals, exhibited an absorption of 3.1 to 3.8 mg/g versus 0.2 without titanium [82]. A titanium compound forming titanic acid by hydrolysis is bonded on an inorganic porous material having hydroxide groups on its surface to obtain a uranium adsorbing medium [83]. A mixture of polymer-uranium adsorbent-solvent and a coagulating solution are contacted at a high shear to obtain a uranium adsorbing medium [84]. Uranium in seawater is adsorbed on a magnetic adsorbate, consisting of spheres of a magnetic iron oxide core and titanic acid adsorbate exterior [85]. A porous substrate, impregnated with a titanic acid solution, is used as a heavy metal adsorbent, especially uranium [86]. Aminostyrene copolymers adsorbed 4 mg/g uranium versus 1,55 with titanic acid [87]. Aminated styrene-divinylbenzene copolymer was diazotized and coupled with 2,4-dihydroxyhippuric acid [88] or coupled with 3-nitrocatechol [89] to prepare a UO_2 adsorbent. The reaction product of chlorosulfonated polystyrene and 3-nitrocatechol had uranium adsorption 3.65 mg/g adsorbent or 73% of uranium was adsorbed [90].

A metal hydroxide is adsorbed on porous activated carbon beads, prepared by carbonization of porous spherical organic polymers. It adsorbed 2.2 mg/g uranium from 2 liters of seawater containing 1.2 mg/l uranium [91]. Metatitanic acid, high density polyethylene and CH_2Cl_2 were treated to give a product used as uranium adsorbent [92]. Resins having nitrile groups were treated with hydroxylamine to give resins useful for absorbing uranium [93].

LITERATURE TO 5.

1. A. Unione, E. Burns, A. Husseiny (Desalination 32 [1980] 225/237)
2. H.E. Hoemig, B.R. Soeltner (Desalination 30 [1979] 515/523)
3. P. Glueckstern (Desalination 30 [1979] 223/234)
4. S.A. Reed, W.F. Savage (ORNL Rept. TM-5070 [1976].- NWSIA J. 3, No 7 [1976] 11/15)
5. S.A. Reed, J.V. Wilson (ORNL Rept. TM-5926 [1977].- NTIS Rept. PC-A03/MF-A01 [1977])
6. W.F. Savage, S.A. Reed (NWSIA 5th Ann. Conf., San Diego [1977] 15 p)
7. T.J. Larson, G. Leitner (Desalination 30 [1979] 525/539)
8. F. Kranebitter (Desalination 30 [1979] 501/514)
9. W. Bongard (Desalination 30 [1979] 583/593)
10. O.K. Buros (Desalination 30 [1979] 595/603)
11. O.J. Morin, S.R. Latour (NWSIA 6th Ann. Conf., Sarasota, Florida [1978] 26 p)
12. I. Kamal, D.K. Emmermann, G.F. Tusel (Desalination 30 [1979] 499)
13. S.P. Kasper, N. Lior (Desalination 30 [1979] 541/552)
14. S.A. Reed, M.L. Marsh (ORNL Rept TM-6260 [1978] 19 p)
15. G.M. Davydor (Uch. Zap. Azerb. In-t Nefti: Khimii No 5[1977] 79/85.- Ref. Zh., Teploenerg. 5 R 143)
16. G.T. Jones (NWSIA 5th Ann. Conf., San Diego, Calif. [1977] 6 p)
17. J.A. Fernandez Lozano (Acta Cient. Venez. 27, No 1 [1976] 18/22.- C.A. 90 [1979] 139676)
18. K.P. Patel, O.K. Srivastava, N.N. Sharma (J. Indian Chem. Soc. 54 [1977] 658/ 660.- C.A. 89 [1978] 45770)
19. M. Murakami (Nippon Kaisui Gakkaishi 31 [1977] 191/201)
20. Yu. A. Afanas'ev, A.I. Ryabinin (Izv. Sev.- Kavk. Nauchn. Tsentra Vyssh. Shk., Ser. Estestv. Nauk No 4 [1977] 89/95)
21. J. Lopez Ruiz (Ciencias, Madrid, 43 [1978] 186/191)

22. B. Seetharara, D. Srinivasan (Chem. Eng. World 13, No 11 [1978] 63/65)
23. K. Kuenstle, C. Koch, I. Erdody, K. Schuh (W. Ger. 2.653.649, 1 Jun 1978.- C.A. 89 [1978] 152536)
24. Agnew Clough Ltd. (Brit. 1.538.239 (17 Jan 1979.- C.A. 91 [1979] 26815)
25. A. Sadan (U.S. 4.088.451, 9 May 1978, continuation of 3.966.541.- C.A. 89 [1978] 91829)
26. T. Masuzawa (Nippon Kaisui Gakkaishi 31 [1978] 241/255)
27. S. Pancharatnam (Salt Res. India 14, No 2 [1978] 13/20.- C.A. 90 [1979] 189212)
28. T. Masuzawa, K. Takenaka, Y. Fujimoto, K. Kagiwada (Nippon Kaisui Gakkaishi 31 [1978] 228/240.- C.A. 90 [1979] 123928)
29. A.F. Shakhova, S.G. Fridman, V.P. Plekhova (Khim. Tekhnol. No 2 [1979] 12/14.- C.A. 91 [1979] 7029)
30. E.E. Marenich, L.V. Meshkova, N.S. Starchikov, S.A. Petrenko, Yu. G. Okatyi, V.M. Shishkine (USSR 659.526, 7 May 1979.- C.A. 91 [1979] 23432)
31. S.G. Fridman, A.F. Shakhova (USSR 653215, 24 Mar 1979.- C.A. 91 [1979] 23387)
32. I. Ekawa, R. Ehara, T. Oda, S. Ogawa (Jap. 78.124.634, 31 Oct 1978.- C.A. 90 [1979] 57390)
33. K. Izumi, I. Okouchi, S. Takahashi, H. Yamazaki (Jap. 78.21.081, 27 Feb 1978.- C.A. 89 [1978] 94678)
34. A.B. Mazurkevich, S.E. Lyandres, V.A. Budkov, Yu. G. Dubrovskii, B.G. Baldin, L.G. Prishlyak (USSR 655.647, 5 Apr 1979.- C.A. 91 [1979] 59598)
35. D.S. Stasinevich (Zh. Prikl. Khim. 51 1978 998/1001.- C.A. 89 [1978] 91785)
36. M. Zlatanova (Rudodobiv 33, No 2 1978 31/24)
37. K.A. Keblys, J.M. McEven (U.S. 4.131.645, 26 Dec 1978.- C.A. 90 [1979] 89498)
38. T. Ayabe, K. Yoshino (Jap. 78.28.038, 11 Aug 1978.- C.A. 90 [1979] 28866)
39. M. Suzuki, M. Kanai (Jap. 78.73.491, 29 Jun 1978.- C.A. 90 [1979] 57287)
40. N. Mineshima, N. Ueda, H. Uzawa (Jap. 79.21.987, 19 Feb 1979.- C.A. 91 [1979] 59573)
41. T. Ishimori (Genshiryoku Kogyo 24, No 4 [1978] 57)
42. I. Pelly (J. Appl. Chem. Biotechnol. 28 [1978] 469/474.- C.A. 90 [1979] 90447)
43. T. Kitamura, H. Wada (Nippon Kaisui Gakkaishi 32, No 2 [1978] 78/81.- C.A. 90 [1979] 209875)
44. J.M. Lee, W.C. Bauman (U.S. 4.116.856, 26 Sep 1978.- C.A. 90 [1979] 57413)
45. J.M. Lee, W.C. Bauman (U.S. 4.116.858, 26 Sep 1978.- C.A. 90 [1979] 57415)
46. T. Ishimori, K. Ueno (W. Ger. 2.847.864, 10 May 1979.- C.A. 91 [1979] 41526)
47. G. Bienkiewicz, M. Starczewski (Zesz. Nauk. Politech. Slask., Chem. 80 [1977] 55/66.- C.A. 89 [1978] 135560)
48. G. Bienkiewicz, M. Starczewski (Zesz. Nauk. Politech. Slask., Chem. 80 [1977] 67/77.- C.A. 89 [1978] 45752)
49. G. Bienkiewicz, M. Starczewski (Zesz. Nauk. Politech. Slask., Chem. 80 [1977] 79/88.- C.A. 89 [1978] 45753)
50. A. Acosta, S.A. Cho, E. Romero D (Acta Cient. Venez. 29 [1978] 274/279.- C.A. 90 [1979] 209899)
51. H. Dahmen (Tonind. Ztg. 102 [1978] 701/703.- C.A. 90 [1979] 154165)
52. S.R. Khanna (Indian Refract. Makers Assoc. J. 11, No 1 [1978] 10/13.- C.A. 90 [1979] 91251)
53. A.V. Sofronova, L.A. Yukseva (Khim. y Khim. Tekhnol., Perm [1978] 81/84.- Ref. Zh. Khim [1979] 9 L 52)
54. B. Petric, N. Petric (Chem. Ing. Tech. 51 [1979] 745)
55. J. M. Lee, W.C. Bauman (U.S. 4.116.857, 26 Sep. 1978.- C.A. 90 [1979]57414)
56. A.J. Brown, N. Heasman, J. Williamson (Brit. 1.502.422, 1 Mar 1978.- C.A. 89 [1978] 131789)
57. I. Motyka, W. Gunia, H. Szczypa, J. Chrzaszcz (Pol. 73.061, 30 Apr 1977.- C.A. 90 [1979] 106546)
58. T. Imamoto, H. Kataoka, T. Kawamura (Jap. 77.143.999, 30 Nov 1977.- C.A. 88 [1978] 123352)
59. S. Inoue (Jap. 78.28.155, 12 Aug 1978.- C.A. 90 [1979] 28855)
60. H. Arita (Jap. 78.55.497, 19 May 1978.- C.A. 89 [1980] 181905)
61. K. Sasaki, Y. Oda, T. Shiba, Y. Suzuki (Jap. 79.03.884, 28 Feb 1979.- C.A. 91 [1979] 8717)
62. M. Matsuda, H. Matsuda, S. Ishizaka (Nippon Kaisui Gakkaishi 32, No 2 [1978] 82/88.- C.A. 90 [1979] 206627)

63. W.C. Bauman, J.M. Lee, J.D. Watson (U.S. 4.110.402, 29 Aug 1978.- C.A. 90 [1979] 57422)
64. A.C.T. Chen, L.I. Gordon, M.R. Rodman, S.E. Binney (NTIS Rept. PC A13/MF A01 [1979] 279 p)
65. S.E. Binney, S.T. Polkinghorne, R.R. Jante, M.R. Rodman, A.C.T. Chen, L.I. Gordon (NTIS Rept. PC A14/MF A01 [1979] 320 p)
66. M.R. Rodman, L.I. Gordon, A.C.T. Chen (NTIS Rept. PC A08/MF A01 [1979] 163 p)
67. M.H. Campbell, J.M. Frame, N.D. Dudey, G.R. Kiel, V. Mesec, F.W. Woodfield, S.E. Binney, M.R. Jante, R.C. Anderson, G.T. Clark (NTIS Rept. PC A08/MF A01 [1979] 169 p)
68. V.N. Sastry, G.R. Doshi, T.M. Krishnamoorthy, T.P. Sarma (Indian J. Mar. Sci. 6, No 1 [1977] 55/58)
69. L. Astheimer, H.J. Schenk, K. Schwochan (Chem. Ztg. 101 [1977] 544/546)
70. B.M. Laskorin, S.S. Metal'nikov, G.I. Smolina (Atomn. Energ. 43 [1977] 472/476.- Sov. At. En. 43 [1978] 1122/1126)
71. O.T. Krylov, P.D. Novikov (Zh. Fiz. Khim. 53 [1979] 1267/1270)
72. H. Yamashita, Y. Ozawa, F. Nakajima, T. Murata (Nippon Kagaku Kaishi No 8 [1978] 1057/1061.- C.A. 89 [1978] 150104)
73. Y. Shigetomi, T. Kojima, H. Kamba (Nippon Genshiry-oku Gakkaishi 20 [1978] 789/795)
74. Y. Miyai, N. Takagi, T. Kitamura, S. Katoh, H. Miyazaki (Nippon Kaisui Gakkaishi 32 [1978] 89/94, 141/149.- C.A. 90 [1979] 190251, 209876)
75. S. Seno (Nippon Genshiryoku Gakkaishi 20 [1978] 578/580)
76. M. Kanno (Nippon Kaisui Gakkaishi 31 [1977] 155/163)
77. N. Ogata (Genshiryoku Kogyo 24, No 1 [1978] 27/31)
78. H. Miyazaki (Shikoku Kokenkai 28 [1977]16/20)
79. H. Miyazaki (Nippon Kikai Gakkaishi 81 [1978] 475/480)
80. D. Heitkamp, K. Wagener (W. Ger. 2.711.609, 21 Sep 1978.- C.A. 89 [1978] 218385)
81. KFA Jülich G.m.b.H (French 2.371.959, 23 Jun 1978.- C.A. 90 [1979] 170815)
82. E. Takesute, T. Miyamatsu, Y. Tsutsui (Jap. 77.114.511, 28 Sep 1977.- C.A. 88 [1978] 10275)
83. H. Ito, Y. Yamazaki, Y. Kantake (Jap. 77.114.587, 26 Sep 1977.- C.A. 88 [1978] 124500)
84. H. Ito, Y. Kantake, I. Sasaki (Jap. 77.114.586, 26 Sep 1977.- C.A. 88 [1978] 124499)
85. A. Yamanouchi (Jap. 78.05.090, 18 Jan 1978.- C.A. 89 [1978] 93027)
86. H. Tani, M. Kojima (Jap 78.05.091, 18 Jan 1978.- C.A. 89 [1978] 93026)
87. H. Nakayama, H. Taniguchi, H. Tani (Jap. 78.23.889, 4 Mar 1978.- C.A. 89 [1978] 114559)
88. H. Nakayama, H. Taniguchi, H. Tani (Jap. 78.23.890, 4 Mar 1978.- C.A. 89 [1978] 114560)
89. H. Nakayama, H. Taniguchi, H. Tani (Jap. 78.23.891, 4 Mar 1978.- C.A. 89 [1978] 93062)
90. H. Takayama, H. Taniguchi, H. Tani (Jap. 78.28.690, 17 Mar 1978.- C.A. 89 [1978] 114566)
91. Y. Murakami, H. Shirane (Jap. 78.63.289, 6 Jun 1978.- C.A. 90 [1979]´ 26854)
92. H. Sato, T. Masuzawa (Jap. 78.104.586, 11 Sep 1978.- C.A. 89 [1978] 201049)
93. H. Egawa (Jap. 78.126.088, 2 Nov 1978.- C.A. 90 [1979] 88278)

6. Iceberg Utilization

The second part of the Proceedings of the First International Conference on Iceberg Utilization, held in Ames, Iowa, 2 to 6 October 1977, was published in a special volume of the Desalination journal. In an interview Prince Mohammed Al-Faisal of Saudi Arabia answered queries on his proposal to tow icebergs from the South Pole to enhance water supplies. He said the project is more economically lucrative than their oil industry. Oil is a diminishing resource, but with ample water supplies additional food may be produced and may be utilized in industry [1]. Recent estimates in the United States have indicated that water from the polar ice can be made available for use at 20 to 50% of the cost of desalination of seawater. Using ice involves locating and transporting huge icebergs from the Antarctic and mooring them offshore near water-deficient areas. Though possible with existing technology, the handling of fragile, melting ice masses on the order of 100 million tons presents substantial problems. Other important considerations are the environmental effects and legal responsibilities involved [2].

In a review on water supply and weather modifications from transferred iceberg, it was concluded that the production of fresh water from icebergs has only beneficial aspects: improvement of environmental living conditions of populations who, presently, greatly suffer from lack of water and no environmental deterioration caused by industrial installations. Only Antarctic tabular icebergs are suitable. Towing of such masses, 100 million m^3 as a first step, is nowadays possible with existing tugs. Protection against excessive melting during transfer, which is estimated to last 6 to 8 months, is given theoretically by a shell maintaining a pool of cold water under and around the foot of the iceberg. Such a shell is approximated as a bottom blanket and a skirt around the sides. Technically, such an operation is said to be feasible. Economically, the cost of iceberg produced water is competitive with the cost of water delivered by desalination [3].

The use of icebergs as a refigerant and a water source was reviewed and the events and opinions, emerging from the first iceberg meeting at Rensselaerville, New York, in 1974, were stated [4].

A study was made of possible impact on the Arabian Peninsula of implementing the transportation to coastal ports of Antarctic icebergs, to a scale growing over 20 years to 3000 metric tons annually, as well as the reticulation of water harvested from this source. Issues discussed include data collection, preparation of the load and its navigation at sea and in confined channels. One version of staging, embayment and harvesting was adopted for the study. The operation of onshore' impoundment and pipelines was analyzed in detail, especially with respect to conjuctive seasonal supplies from all sources and their use patterns. The following conclusions are drawn: The system will result in beneficiation of Western Region urban development and adjacent agricultural districts. The pipelining to Central and Eastern Region major cities of iceberg water involves energy consumption and capital investment which greatly overshadow those of the maritive iceberg system. Possibly the lead times in planning and construction will be critical for onshore facilities [5].

Icebergs provide unique reservoirs with significant net accrual of condensation water over evaporation loss. Melting of icebergs in a warm environment offers opportunities to extract the heat from the environment with little energy cost [6]. The harnessing of icebergs which appears to be best able to satisfy all requirements is to surround each iceberg in a horizontal plane below the center of mass with a cable network or sling. Environmental factors such as currents and winds do not appear to influence greatly the energy or time required to transfer icebergs, if sufficient thrust is used to make the transfer to the Northern Hemisphere in about a year or less time [7]. Icebergs will need protection to prevent excessive melting loss. The rapid ablation of the ice that is in contact with the seawater is accounted for by the great heat-transfer rate from the convection of the moving seawater. If a thin layer of poorly conducting material can be kept next to the ice, the heat flow can be limited to the desired protective rate [8].

It appears that icebergs could be used to improve the operational effectiveness of ocean thermal energy conversion plants. Iceberg distribution patterns suggest that areas in the North Atlantic near the confluence of the Labrador current and the Gulf Stream offer promise as sites for such plants. Energy produced in these plants could be used for industrial manufacturing and fresh water produced at these plants could be transported to areas adjacent to the sites [9].

Other work includes a study of the transport of a large tabular iceberg from the Antarctic to the Arabian peninsula [10], selecting and acquiring Antarctic icebergs for export [11], towage of an iceberg [12], terradynamics as applied to ice [13], determination of iceberg underwater shape with impulse radar [14], capability of SAR systems for iceberg detection and characterization [15], geophysical aspects of a large iceberg tow [16], legal and environmental concerns in using icebergs [17], role of liquid oxygen explosive in iceberg utilization and development [18] and iceberg utilization and alternative systems related to fresh water transport, especially recycling and supplementary desalination [19].

LITERATURE TO 6.

1. Anonymous (Challenge 21, No 3 [1978] 25/31)
2. R.L. Civiak (NTIS Rept. PB- 285664/9GA [1978] 12 p)
3. P.E. Victor (Desalination 29 [1979] 7/15)
4. J.M. Day (Desalinatio n 29 [1979] 25/40)
5. D.W. Coillet (Desalination 29 [1979] 191/196)
6. J.L. Hult (Desalination 29 [1979] 41)
7. J.L. Hult, M.C. Ostrander (Desalination 29 [1979] 43/44)
8. J.L. Hult (Desalination 29 [1979] 97)
9. D.J. DeMarle (Desalination 29 [1979] 153/163)
10. D.W. Coillet, R.M. Dunlap (Desalination 29 [1979] 47/78)
11. J.L. Hult (Desalination 29 [1979] 45)
12. L. Montfort, C. Oudendijk (Desalination 29 [1979] 79/95)
13. A. Pope (Desalination 29 [1979] 17/23)
14. J.R. Rossiter, K.A. Gustajtis (Desalination 29 [1979] 99/107)
15. R. Rawson, R. Larson, R. Shuchman, R. Worsfold (Desalination 29 [1979] 109/133)
16. D.C. David (Desalination 29 [1979] 135/152)
17. J.L. Hult (Desalination 29 [1979] 165)
18. G.L. Tate (Desalination 29 [1979] 167/172)
19. C.D. Looyen (Desalination 29 [1979] 173/189)

Patent Index
(Page/Reference number)

AUSTRALIA

495.414	Foley	171/18

BELGIUM

844.769	V.O.E.ST.	88/20
846.604	Rhone-Poulenc	165/72
856.653	Norcem	90/75
861.555	Mousset	45/9
865.608	Lipinski	82/35
866.503	Hauffe	157/47
867.637	Rhone-Poulenc	139/21

BRAZIL

76.01.598	Okazaki	107/173
76.06.954	Pagani	56/13

CANADA

1.033.564	Gray	78/27
1.047.413	Schell	158/117

CZECHOSLOVAKIA

168.962	Kulhavy	46/75
175.008	Skrabak	157/46

FRANCE

2.325.405	U.O.P. Inc.	157/45
337.693	Hoiss	58/9
340.119	Mattern	51/16
341.344	Smirnov	171/20
348.157	Crane Co.	93/23
2.348.488	De Chazelles	78/10
348.725	Takada	58/14
348.726	Takada	58/15
350.123	Marchenko	59/23
351.681	Whiting Fermont	45/11
2.353.322	Quentin	105/91
355.537	Saari	56/15
358.473	Crambes	82/32
361.306	Risch	56/18
361.452	Graber	165/73
2.363.351	Okita	143/180
365.361	Bayer A.G.	143/206
367.520	Nippon Zeon Co.	159/143
368.289	Cochrane	159/144
368.290	Jensen	159/145
2.369.216	Ciba-Geigy A.G.	79/45

FRANCE

2.370.697	Grillo Werke A.G.	79/49
371.959	K.F.A. Juelich	179/81
372.644	Schultheiss	139/20
372.771	Billon	70/42
376.079	Hitachi Ltd.	74/34
2.377.834	M.A.N. A.G.	63/14
379.482	Cie. Dessalement	63/17
380.051	Dow Chemical Co.	159/146
380.053	Monsanto Co.	159/153
381.549	Balaradze	115/46
2.383.129	Mary	61/16

GERMANY EAST

127.541	Kux	74/47
129.043	Jonas	158/94
130.862	Bartsch	141/108
131.528	Bartsch	141/109
131.529	Borgwardt	141/110
134.447	Groebe	144/225
134.448	Groebe	144/224

GERMANY WEST

2.556.210	G.K.S.S.	157/79
607.997	Martinola	140/62
622.461	Eiselein	157/73
624.515	V.O.E.ST.	88/19
626.902	Freilaender	70/53
2.640.977	Schreiner	58/18
643.422	Ii	79/30
649.649	Stummer	107/179
650.482	Hock	70/46
651.438	Desai	79/31
2.653.649	Kuenstle	178/23
654.296.	Perry	167/145
657.775	Krueger	79/32
700.567	Schaefer	36/34
701.820	Meier	139/18
2.705.813	Tenge	107/165
705.814	Tenge	107/166
706.172	Tenge	107/167
706.193	Tenge	107/168
707.715	Fritzen	70/38
2.711.609	Heitkamp	179/80
716.117	Hofstede	56/19
718.882	Hauffe	157/74
719.907	Drude	157/39
720.867	Putral	39/165
2.722.975	Drude	157/40

GERMANY WEST		
2.730.528	Peter	167/146
731.159	Bulang	58/10
731.711	Crambes	82/32
735.443	Munari	143/192
737.263	Merz	45/12
2.741.669	Elfert	144/229
743.673	Wrasidlo	142/171
747.390	Jensen	142/172
747.408	Cochrane	143/181
749.208	Bernat	116/90
2.750.783	Mach	157/36
750.897	Stemme	159/151
754.627	Connelly	158/95
755.504	Bolto	93/32
804.434	Goodman	74/50
2.806.222	Clark	159/129
811.826	Foucras	160/189
812.761	Keefer	157/75
814.326	McLain	159/132
816.088	Kiser	159/133
2.818.127	Lefeuvre	168/151
822.265	Grandine	143/191
822.266	Grandine	143/191
822.784	Kawaguchi	143/182
822.824	Gunjima	104/58
2.825.698	Sakaguchi	157/56
827.012	Mishiro	159/152
829.630	Hashino	143/198
829.893	Bairinji	157/37
834.716	Aoyagi	165/81
2.837.489	Siwecki	157/38
837.845	Latty	144/228
838.665	Yamashita	143/185
2.847.864	Ishimori	178/46

ISRAEL		
43.135	Hoechst A.G.	157/42
45.126	Doriel	70/34
45.370	Oxy Metal Industr.	158/89
46.824	U.O.P. Inc.	142/169
47.088	Ciba-Geigy U.K.	78/26
47.646	Rhone-Poulenc	158/90
48.013	Alfred Rudin Ltd.	142/170
48.121	U.O.P. Inc.	157/44
48.373	Jenaer Glaswerk	144/257
48.473	Yeda R & D Co.	115/56
49.261	U.O.P. Inc.	144/236
49.675	Perry	115/33
49.839	Zan-Al Ltd.	115/48
50.135	Nat. Res. Council	144/227
50.384	U.O.P. Inc.	144/237

JAPAN		
77.38.985	Hitomi	165/70
43.186	Sawa	78/9
47.742	Nakaya	58/20

JAPAN		
77.59.345	Gunjima	104/58
72.398	Murayama	105/81
73.189	G.K.S.S.	157/79
74.314	Mutsukushi	79/50
78.677	Ikeguchi	140/63
77.78.678	Tsukamoto	120/22
78.680	Saeki	115/34
85.886	Uematsu	39/142
88.595	Nagasawa	103/26
89.577	Kuroda	116/97
77.91.788	Imai	104/68
91.789	Sano	104/59
94.877	Tsukamoto	120/23
96.983	Okazaki	107/169
96.984	Okazaki	107/170
.96.985	Okazaki	107/170
77.104.471	Tsukamoto	120/24
104.472	Tsukamoto	121/25
114.511	Takesute	179/82
114.586	Ito	179/84
114.587	Ito	179/83
77.116.790	Ukihashi	103/27
119.470	Inst. Termophysics	171/20
119.481	Maeda	157/50
120.546	Tahata	74/29
120.978	Tako	168/147
77.120.983	Seko	103/28
122.278	Sumie	168/148
122.279	Sumie	168/148
123.382	Chonan	93/19
123.545	Itoi	157/51
77.123.983	Iwasa	159/136
124.478	Fukushima	158/96
124.479	Fukushima	158/97
124.481	Matsuzaki	115/35
124.482	Kuroda	107/176
77.126.681	Watanabe	159/137
127.481	U.O.P. Inc.	143/183
128.880	Shioyama	56/20
128.888	Tsukamoto	159/138
129.674	Hanada	70/41
77.131.974	Nishimoto	58/16
131.978	Sakaguchi	165/54
134.876	Moriyasu	93/20
134.877	Arai	93/21
134.882	Sasaki	168/153
77.136.885	Sawa	74/27
137.466	Shimizu	158/104
138.031	Katsura	82/42
138.057	Iwata	53/10
138.067	Okazaki	107/171
77.140.477	Uwano	48/2
141.483	Takada	58/12
143.988	Kiyota	104/64
143.999	Imamoto	178/58
144.380	Ono	166/75
144.388	Gunjima	103/29
77.146.782	Moriyasu	93/22

JAPAN

JAPAN

77.149.287	Yamaguchi	103/30	78.19.423	Hamada	159/139
150.388	Ichinose	140/61	19.974	Ueno	45/14
150.792	Ukihashi	103/31	19.976	Komatsu	93/17
151.677	Matsuzaki	115/36	21.081	Izumi	178/33
151.678	Matsuzaki	115/37	21.091	Maeda	79/39
77.151.683	Kiyota	104/65	78.22.067	Mihara	106/144
151.684	Kiyota	104/65	22.164	Jo	166/77
152.879	Tokizane	143/193	22.165	Hagari	115/40
156.176	Miyata	140/64	22.580	Sata	103/34
156.177	Miyata	140/65	22.875	Hara	143/194
77.156.180	Takatono	140/66	78.22.876	Niitsu	157/53
156.181	Takatono	140/67	22.887	Takada	104/73
156.182	Takatono	140/68	23.873	Yamada	143/184
156.183	Takatono	140/69	23.878	Tsuchiya	107/180
156.778	Inukai	158/103	23.879	Oka	116/91
77.156.779	Yoshimura	104/62	78.23.880	Oka	116/92
78.00.658	Yoshikawa	48/7	23.889	Nakayama	179/87
00.948	Takadono	158/98	23.890	Nakayama	179/88
01.685	Shioda	45/13	23.891	Nakayama	179/89
02.385	Yanagii	70/43	25.214	Kowaka	88/34
02.393	Negishi	79/37	78.25.265	Kanematsu	166/105
78.03.755	Sasaki	93/16	26.284	Sata	105/92
03.968	Midorikawa	70/31	26.285	Sata	103/35
04.776	Takeuchi	93/24	26.777	Murakami	143/209
04.786	Kiyota	103/32	27.702	Ohkochi	52/23
05.079	Inoue	79/38	78.28.038	Ayabe	178/38
78.05.090	Yamanouchi	179/85	28.082	Tsukamoto	166/106
05.091	Tani	179/86	28.083	Nakamura	140/55
05.988	Teraoka	47/76	28.155	Inoue	178/59
05.989	Teraoka	47/77	28.255	Tahata	46/69
05.990	Teraoka	47/78	78.28.574	Kagechika	115/41
78.06.106	Ogawa	158/100	28.579	Kawamura	74/49
06.240	Ozaki	90/93	28.589	Seida	105/82
06.272	Izumi	74/32	28.690	Nakayama	179/90
06.273	Arayama	70/44	28.863	Yamanaka	82/36
07.783	Fukui	143/208	29.281	Saari	56/16
78.08.690	Machi	104/71	78.29.290	Kiyota	103/36
08.691	Machi	104/72	29.291	Takahashi	103/37
08.692	Machi	104/63	30.482	Oya	157/64
09.278	Stone & Webster	157/52	30.661	Ono	142/173
10.031	Kitabayashi	70/32	30.987	Okazaki	170/172
78.10.369	Matsumura	45/23	78.31.575	Nakahara	70/49
10.383	Tashiro	166/76	32.876	Takeuchi	70/39
11.884	Harayama	143/200	32.877	Takeuchi	70/40
13.334	Teraoka	47/79	32.885	Tamura	141/106
13.419	Sawada	60/3	34.675	Shiraishi	116/94
78.13.427	Sato	115/38	78.34.676	Shiraishi	116/95
14.672	Sawa	116/98	34.690	Takahashi	105/83
14.675	Sawa	74/28	37.789	Udagawa	104/69
14.684	Imai	104/60	39.272	Kawabata	166/107
15.104	Sakamoto	141/104	39.858	Yonekura	82/37
78.15.195	Kudo	38/100	78.39.982	Sumita	144/233
15.282	Kimoto	103/33	41.058	Oka	116/93
15.834	Takada	58/13	41.096	Hiramatsu	89/56
16.374	Yamaguchi	115/39	42.185	Kamiyama	143/186
17.581	Arai	140/70	43.678	Kashiwayanagi	70/45
78.18.482	Eguchi	144/235	78.44.421	Rokushi	49/8
19.171	Nomiyama	117/99	44.486	Takenishi	143/195

185

JAPAN

78.46.489	Murayama	103/38	78.106.388	Koyama	140/57
50.061	Iida	70/36	106.389	Koyama	140/58
55.497	Arita	178/60	106.395	Takesute	105/76
56.172	Oku	105/95	106.678	Asami	104/70
58.150	Ueshima	79/59	108.075	Inoue	82/33
78.58.493	Sata	105/96	78.108.076	Inoue	82/34
58.940	Shiga	82/38	108.888	Sata	105/97
58.974	Tsukamoto	166/108	110.666	Otsuka	141/107
59.782	Mizutani	104/74	110.976	Kuramoto	104/47
60.371	Ono	116/96	110.977	Sumita	143/201
78.60.380	Ueno	140/71	78.110.978	Kitano	143/187
60.388	Kiyota	104/64	110.979	Kitano	143/188
60.879	Sawa	74/33	110.981	Takahashi	79/41
61.579	Murayama	103/39	112.279	Nishimoto	51/19
63.289	Murakami	179/91	112.289	Uehara	107/182
78.65.281	Yamaguchi	105/101	78.113.276	Motegi	168/149
65.358	Sode	166/78	113.784	Kawamura	79/42
66.897	Ono	166/109	114.780	Shimizu	93/26
67.682	Chonan	93/25	116.272	Satone	51/17
68.679	Hamada	46/57	116.273	Satone	52/22
78.69.449	Maeda	93/33	78.116.274	Satone	60/5
69.841	Matsushita	117/109	116.286	Yomo	106/114
70.085	Morita	115/42	116.287	Seko	104/48
70.090	Seida	103/40	119.782	Okochi	63/15
70.984	Asami	105/84	119.783	Yamazaki	63/16
78.71.681	Matsumoto	63/10	78.119.786	Otsuka	140/59
71.692	Seida	105/85	120.688	Nomura	165/71
73.483	Ueno	107/177	123.386	Sumita	143/202
73.484	Hamada	104/41	123.387	Kitano	141/112
73.491	Suzuki	178/39	123.390	Sata	105/98
78.75.139	Takayasu	82/40	78.123.995	Hiiro	38/96
75.166	Sawa	74/30	124.138	Kamata	82/41
79.781	Harada	104/42	124.634	Ekawa	178/32
81.440	Yamamoto	82/39	125.274	Koyama	140/74
82.669	Yamanouchi	159/140	125.283	Seko	105/89
78.82.670	Kuzumoto	159/141	78.125.980	Hara	143/196
82.684	Hane	104/43	126.088	Egawa	179/93
87.982	Konomi	141/105	126.319	Tanaka	166/80
88.671	Kuroda	107/181	127.374	Koyama	141/127
88.672	Kuroda	107/181	129.186	Weiss	93/35
78.88.673	Kuroda	117/108	78.130.285	Ishii	143/190
88.678	Kihara	105/75	131.292	Seko	104/49
92.867	Nakatsuka	166/79	131.293	Imai	104/50
94.289	Seko	104/44	131.980	Yabushita	140/60
95.174	Satone	56/21	132.478	Hayano	144/238
78.95.183	Suzuki	142/174	78.132.480	Sakamoto	143/211
95.185	Abiko	79/40	132.481	Kuwabara	115/43
96.974	Hori	142/175	134.088	Kojima	104/51
97.988	Takahashi	104/45	136.034	Takahashi	168/152
99.087	Tanaka	105/93	136.099	Kitano	143/189
78.100.115	Murai	90/94	78.137.083	Gunjima	115/44
100.120	Kurusu	88/21	137.876	Sato	168/150
100.121	Kurusu	88/22	137.879	Uehara	115/45
100.125	Tamura	90/95	137.888	Sata	104/52
103.982	Iwahori	158/99	138.982	Komatsu	79/51
78.103.989	Tamaru	104/46	78.140.289	Suzuki	105/77
104.586	Sato	179/92	141.187	Sata	104/53
106.387	Koyama	140/56	141.188	Hamada	104/54

J A P A N

78.142.986	Yanagi	141/113
144.473	Torii	107/178
144.883	Uemura	141/114
144.974	Midorikawa	142/176
146.270	Sakamoto	171/21
78.146.800	Taketani	143/197
149.181	Ishii	143/213
149.182	Fukutsuka	117/100
149.188	Seko	104/55
149.873	Ogoshi	160/186
78.149.874	Oshiumi	144/234
79.01.283	Ukihashi	105/105
01.372	Ito	143/207
02.276	Takahashi	144/251
02.279	Kinjo	143/214
03.884	Sasaki	178/61
79.04.283	Takemura	143/215
04.290	Kiyota	104/66
04.291	Kiyota	104/66
05.888	Tanaka	105/94
05.889	Imai	104/61
79.06.869	Seida	105/86
06.870	Seida	105/87
06.885	Kiyota	104/67
06.886	Kiyota	104/67
06.887	Seko	105/90
79.08.180	Kuwahara	117/101
08.671	Onishi	144/216
11.017	Inagaki	88/23
11.081	Horiguchi	144/218
11.882	Tsukamoto	160/187
79.11.971	Komaki	144/217
14.376	Yoshino	158/111
14.389	Eguchi	105/106
15.030	Jo	166/83
15.476	Mishiro	166/82
79.15.478	Kamada	144/219
15.479	Oshiumi	144/220
16.381	Takada	141/121
20.978	Sotoma	74/51
20.981	Sata	105/99
79.21.478	Sata	105/100
21.970	Ochiumi	144/221
21.971	Yamaguchi	117/112
21.987	Mineshima	178/40
23.080	Kamada	140/76
79.24.274	Pottharst	78/11
24.284	Motani	105/104
25.278	Hori	144/222
26.283	Yamazaki	144/223
26.286	Yamamoto	105/107
79.26.287	Suzuki	105/78
26.976	Takahashi	105/87
27.025	Sueoka	160/190
32.181	Takahashi	158/112
32.189	Takahashi	105/88
79.34.416	Jo	166/84
42.419	Jo	166/85

J A P A N

79.43.872	Sawa	74/35
51.982	Mori	158/113
56.248	Shimokawabe	115/47
56.979	Tamura	45/22

N E T H E R L A N D S

76.02.976	Crane Co.	93/23
06.892	Saari	56/14
10.852	Krupp G.m.b.H.	158/93
77.03.192	Oe.St.Atomenergie	78/15
08.679	Allied Chemical	105/102
09.134	Coca Cola Co.	117/111

P O L A N D

73.061	Motyka	178/57
85.468	Jodko	78/14
93.008	Selecki	45/15
96.878	Piwowarczyk	70/35
97.917	Blazejewska	140/75

S O U T H   A F R I C A

77.00.499	Hebden	157/41
04.960	Pienaar	141/125
04.961	Pienaar	141/126
04.962	Sanderson	142/177
06.307	Hock	70/46

S P A I N

446.360	Espinosa Cina	52/24
455.286	Rois Noriega	157/49

S W I T Z E R L A N D

601.113	Friese	70/30

U N I T E D   K I N G D O M

1.486.917	Hoechst A.G.	141/102
487.109	Jenaer Glaswerk	144/256
487.569	Terraqua Products	45/10
490.837	Standard Messo	58/8
491.494	Chemed Corp.	79/28
1.495.670	Baxter Travenol	139/16
495.887	Rhone-Poulenc	144/232
497.953	Tsai	70/33
501.644	France	36/19
502.422	Brown	178/56
1.503.741	Sasakura Engrng.	51/20
505.384	Asahi Chemical	93/18
504.464	Hoechst A.G.	157/43
504.733	Teijin K.K.	142/168
505.816	Smart	78/16
1.505.909	Uniroyal Inc.	79/29
506.545	Kansai Kagaku	59/24

U N I T E D   K I N G D O M

1.506.665	Nippon Oil K.K.	139/17
506.785	Asahi Kasei Kogyo	159/131
508.156	Daicel K.K.	158/102
508.203	Smith	46/56
514.480	Wood	51/18
1.516.203	Daicel K.K.	143/203
516.683	Auscoteng Pty.	45/20
516.792	Hitachi K.K.	158/109
517.220	U. O. P. Inc.	158/110
517.510	Speed	58/19
1.518.486	Brown Boveri	61/11
519.512	Grace & Co.	79/48
519.991	Wafilin B.V.	158/106
519.992	Wafilin B.V.	158/106
521.045	Wafilin B.V.	158/107
1.521.287	Wafilin B.V.	158/108
521.362	Hitachi K.K.	157/65
525.177	Tsuda	166/74
526.053	Pottharst	58/11
526.183	Bayer A.G.	159/142
1.526.301	Ciba Geigy A.G.	79/44
526.836	Thoeni	122/26
526.843	Sumitomo Chemical	143/204
527.001	Bayer A.G.	143/205
533.083	Hitachi Ltd.	61/15
1.536.681	Parr	36/19
538.239	Agnew Clough Ltd.	178/24

U. S. A.

4.039.440	Cadotte	143/178
046.639	Carson	70/48
052.267	McFee	45/16
053.368	Courvoisier	70/29
054.493	Roller	74/26
4.055.473	Hay	70/28
056.418	Aspin	158/127
057.483	Giuffrida	115/32
4.057.511	Bohnsack	78/17
059.147	Thorne	46/48
061.574	Clark	159/128
062.721	Saupe	156/32
062.796	Gardner	79/33
4.072.607	Schiller	79/46
076.626	Meier	139/19
077.883	Bray	156/33
078.975	Spears	70/50
4.078.976	Spears	70/51
079.006	Mitchell	79/47
080.271	Brown	70/54
080.296	Clark	159/129
081.331	Weiss	45/17
4.082.658	Fritsche	159/130
083.780	Call	158/101
083.781	Conger	156/34
083.782	Kunin	79/58
083.904	Sano	141/128
4.085.045	Sone	79/34

U. S. A.

4.086.146	Block	79/35
086.181	Suen	79/36
087.357	Barrett	93/34
088.451	Sadan	178/25
088.550	Malkin	116/59
4.088.563	Marquardt	93/27
089.750	Kirschman	45/18
089.796	Harris	79/43
092.233	Clemens	157/54
4.092.244	Suen	82/43
092.834	Lloyd	171/19
094.747	Pfenninger	61/14
096.039	Carnine	93/11
096.064	Du Fresne	23/88
4.098.691	Filby	79/61
098.720	Hwa	82/44
100.064	Gale	158/105
100.238	Shinomura	143/179
104.165	Braswell	79/60
4.105.505	Saari	56/17
105.534	Beatty	117/103
105.548	Baker	159/147
105.581	Sexsmith	82/45
106.560	Lauro	49/23
4.107.000	Currin	70/47
107.049	Sano	143/212
108.764	Kaneko	159/148
110.170	Kirschman	45/19
110.174	Carson	61/12
4.110.175	Ahlgren	117/110
110.219	Maples	157/55
110.265	Hodgdon	105/79
110.402	Bauman	179/63
112.702	Smirnov	171/20
4.115.225	Parsi	116/60
115.606	Maxson	79/69
116.856	Lee	178/44
116.857	Lee	178/55
116.858	Lee	178/45
4.118.283	Diggs	70/37
119.581	Rembaum	105/80
121.977	Carson	61/13
124.458	Moeglich	105/108
125.462	Latty	143/199
4.125.463	Chenoweth	157/72
126.589	Hamada	104/57
127.467	Smith	74/31
131.428	Diggens	36/18
131.513	Green	45/21
4.131.645	Keblys	178/37
132.587	Lankenau	51/21
135.985	La Rocca	70/52
136.025	Zwack	140/72
136.237	Takahashi	104/56
4.140.637	Walter	159/149
141.825	Conger	157/35
144.185	Block	107/183
4.145.295	Kutowy	141/111

	U. S. A.			U. S. S. R.	
4.147.605	Schenker	117/102	594.056	Sterman	79/70
148.693	Williamson	56/22	597.643	Simonov	79/68
153.545	Zwack	140/73	614.135	Perepechkin	159/135
157.960	Chang	159/150	633.875	Lyubman	105/103
			636.007	Voronkova	141/103
	U. S. S. R.		647.594	Gordievskii	38/101
			653.215	Fridman	178/31
523.699	Karelin	159/134	655.647	Mazurkevich	178/34
528.020	Brien	158/91	655.653	Pisaruk	115/49
528.101	Volgin	158/92	659.526	Marenich	178/30
537.717	Troshenkin	74/48	664.973	Leikin	144/226
548.293	Tseitlin	157/48			

Abbreviations

A. C. S.	American Chemical Society
A. I.	Atom Index
AIChE	American Institute of Chemical Engineers
AMDEL	Australian Mineral Development Laboratories
ASCE	American Institute of Civil Engineers
ASME	American Institute of Mechanical Engineers
AWWA	American Water Works Association
BBC	Brown Boveri Concern
C. A.	Chemical Abstracts
D. A.	Dissertation Abstracts
CEBEDEAU	Centre Belge de l'Etude et de Documentation des Eaux
E. I.	Engineering Index
EPRI	Electric Power Research Institute, Palo Alto
ERA	Energy Research Abstracts
ERDA	Energy Research and Development Administration
G. R. A.	Government Report Announcements
I. A. E. A.	International Atomic Energy Agency
I. E. C.	Industrial and Engineering Chemistry
K. F. A.	Kernforschungsanlage Juelich
M. A. N.	Maschinenfabrik Augsburg-Nuernberg
N. C. R. D.	National Council Research and Development, Israel
N. T. I. S.	National Technical Information Service, U.S.A.
NWSIA	National Water Supply and Improvement Association, U.S.A.
O. A.	Oceanographic Abstracts
O. R. N. L.	Oak Ridge National Laboratory
O. S. W.	Office of Saline Water
O. W. R. T.	Office of Water Research and Technology
TAPPI	Transactions of the Association of Pulp and Paper Industry
UCLA	University of California, Los Angeles
UCRL	University of California, Richmond Laboratory
VDI	Verein Deutscher Ingenieure
V. G. B.	Vereinigung Grosskraftwerksbetreiber, W. Germany
VOEST	Vereinigte Oesterreichische Eisen und Stahlwerke

J. J. Bikerman

Foams

1973. 79 figures. IX, 337 pages
(Applied Physics and Engineering, Volume 10)
ISBN 3-540-06108-8

Contents: General. Foam Films. – Formation and Structure. – Measurement of Foaminess. – Results of Foaminess Measurements. – Three-phase Foams. – Foam Drainage. – Mechanical Properties of Foams. – Optical Properties of Foams. – Electric Properties of Foams. – Theories of Foam Stability. – Foams in Nature and Industry. – Applications. Separation by Foam. – Other Applications.

"...The great merits of Bikerman's work are the compilation of an extensive experimental material, the exceptionally concise and clear style, the demonstration of the subject on hand, of simple numerical examples, and the richness in experimental and methodological detail. Therefore, notwithstanding a certain one-sidedness, the book is a valuable contribution to the literature on surface chemistry."

Acta Chimica

"... this book can be read on two levels, either selectively by a student new to the subject or extensively by an established researcher interested in a thorough and authoritative review."

J. of the American Chemical Soc.

Springer-Verlag
Berlin
Heidelberg
New York

O
Water Desalting

Supplement Volume 1
By Anthony A. Delyannis, Eurydike A. Delyannis,
Athens, Greece
Preface by E. D. Howe
1979. 11 figures, XIV, 360 pages
(Gmelin Handbook of Inorganic Chemistry)
ISBN 3-540-93398-0

The benefication of seawater and of other saline
waters to produce fresh water is of increasing importance. Normal fresh water supplies from precipitation cannot meet the needs of the increasing world
population, particularly in the semi-arid regions of
the world. Thus desalination is being used to aug-
ment normal water supplies in some areas and to
provide all of the fresh water in other places, as in
the arid parts of the Arabian Peninsula. Another
increasing application of desalination is the treatment of polluted waters not only to reduce pollution
but also to recover good water from these unaccept-
able sources. In an extrem case of this nature there
is being marketed in the USA a computer controlled
system for isolated residences, in which a given
supply of good water is continuously recycled
through the system, with makeup water kept to a
minimum. The reconditioning equipment includes
a desalination component as well as the usual sani-
tary devices for preventing the carry over of odors
and disease organisms.
One factor which has changed appreciably since
publication of the Main Volume of this work is the
price and availability of fossil fuel energy. The
embargo of petroleum by the OPEC countries
occured after the Main Volume had been written
and energy costs since that time have increased con-
tinually. This serves to increase the importance of
energy economy in desalination plants, since the
cost of energy in existing plants amounts to some
50 percent of the total water cost. Furthermore,
since plastics are derived from petroleum, high
prices for this material are reflected in increased
costs of equipment for membrane processes used for
desalination.
The above considerations have given rise to the very
large number of publications which have appeared
during the past several years. The authors have
assembled nearly 4000 such references and organi-
zed them in the same order as in the Main Volume.
Each reference has been carefully reviewed and
placed in its logical position relative to the Main
Volume sections. Two sections have been added
particularly in the fields of reverse osmosis and electrodialysis. Also a completely new section on the
possible use of icebergs from polar regions to furnish
fresh water has been added.

Springer-Verlag
Berlin Heidelberg New York

From the Contents
Introduction and General
The water problem. History of desalination. Desalting processes. Raw material seawater
Distillation Processes
Common aspects of distillation processes. Single-
effect distillation. Multiple-effect distillation. Verti-
cal-tube evaporator process development. Vapor
compression. Flash evaporation. Multi-stage flash
(MSF) distillation process optimization and development. Other multistage distillation processes.
Comparison of the distillation processes. Direct
contact heat transfer. Large size desalting plants.
Dual-purpose plants. Waste heat as energy source.
Nuclear energy as heat source. Nuclear powered
agro-industrial complexes. Geothermal energy as
a heat source. Solar energy as heat source. Scale
formation and its prevention. Materials of construction; Corrosion. Disposal of effluents from desalination plants.
Ionic Processes
Ion exchange techniques. Fixed bed ion exchange
processes. Continuous ion exchange processes.
Nonchemical methods of ion exchange resin regeneration. Ion selective membranes. Inorganic ion
exchange membranes. Electrodialysis. Variants of
electrodialysis. Specific electrodialysis applications.
Electrochemical and physicochemical methods of
desalination. Reverse osmosis (Hyperfiltration).
Reverse osmosis membranes. Cellulose acetate
membranes. Modified cellulose acetate membranes. Polymer film membranes. Ultrathin and composite membranes. Other types of membranes.
Reverse osmosis process development. Tubular
reverse osmosis assembly. Spiral-wound membrane module. Hollow fiber module. Various applications of reverse osmosis and ultrafiltration.
Piezodialysis.
Freezing Processes. Hydrate Processes. Liquid-
liquid Extraction.
Nucleation and growth of ice crystals. Freezing processes. Hydrate processes. Liquid-liquid extraction.
Economic Considerations
Feasibility of desalting. The effect of energy. The
impact of energy cost. Recovery of byproducts.
Iceberg Utilization
Table of Conversion Factors
Literature Closing Date: End 1977
In some cases more recent data have been con-
sidered.

Previously Published

O
Water Desalting
Main Volume
By Anthony A. Delyannis, Eurydike A. Delyannis,
Athens, Greece
Preface by H. Fischbeck
1974. 62 figures. XVI, 339 pages
(Gmelin Handbook of Inorganic Chemistry)
ISBN 3-540-93280-1